1,000,000 Books

are available to read at

www.ForgottenBooks.com

Read online
Download PDF
Purchase in print

ISBN 978-1-5276-3301-8
PIBN 10876754

This book is a reproduction of an important historical work. Forgotten Books uses state-of-the-art technology to digitally reconstruct the work, preserving the original format whilst repairing imperfections present in the aged copy. In rare cases, an imperfection in the original, such as a blemish or missing page, may be replicated in our edition. We do, however, repair the vast majority of imperfections successfully; any imperfections that remain are intentionally left to preserve the state of such historical works.

Forgotten Books is a registered trademark of FB &c Ltd.
Copyright © 2018 FB &c Ltd.
FB &c Ltd, Dalton House, 60 Windsor Avenue, London, SW19 2RR.
Company number 08720141. Registered in England and Wales.

For support please visit www.forgottenbooks.com

1 MONTH OF FREE READING

at

www.ForgottenBooks.com

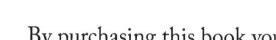

By purchasing this book you are eligible for one month membership to ForgottenBooks.com, giving you unlimited access to our entire collection of over 1,000,000 titles via our web site and mobile apps.

To claim your free month visit:

www.forgottenbooks.com/free876754

* Offer is valid for 45 days from date of purchase. Terms and conditions apply.

English
Français
Deutsche
Italiano
Español
Português

www.forgottenbooks.com

Mythology Photography **Fiction**
Fishing Christianity **Art** Cooking
Essays Buddhism Freemasonry
Medicine **Biology** Music **Ancient
Egypt** Evolution Carpentry Physics
Dance Geology **Mathematics** Fitness
Shakespeare **Folklore** Yoga Marketing
Confidence Immortality Biographies
Poetry **Psychology** Witchcraft
Electronics Chemistry History **Law**
Accounting **Philosophy** Anthropology
Alchemy Drama Quantum Mechanics
Atheism Sexual Health **Ancient History**
Entrepreneurship Languages Sport
Paleontology Needlework Islam
Metaphysics Investment Archaeology
Parenting Statistics Criminology
Motivational

＃ HOOKER'S

JOURNAL OF BOTANY

AND

KEW GARDEN MISCELLANY.

EDITED BY

Sir WILLIAM JACKSON HOOKER, K.H., D.C.L. Oxon.,
LL.D., F.R., A., and L.S., Vice-President of the Linnean Society, and
Director of the Royal Botanic Gardens of Kew.

VOL. III.

LONDON:
REEVE AND BENHAM,
HENRIETTA STREET, COVENT GARDEN.

1851.

PRINTED BY REEVE AND NICHOLS,
HEATHCOCK COURT, STRAND.

HOOKER'S
JOURNAL OF BOTANY
AND
KEW GARDEN MISCELLANY.

Report on the "BROWN SCALE," *or* COCCUS, *so injurious to the* COFFEE-PLANTS *in Ceylon; in a letter from the late* GEORGE GARDNER, *Esq., Director of the Botanic Garden at Peradenia, addressed to the Colonial Secretary, Colombo.* (Communicated by the Right Hon. Earl Grey, Chief Secretary for the Colonies. With a Plate, Vol. II. TAB. XII.)

(*Continued from* vol. ii. p. 360.)

On nearly all those estates where I have been able to trace the rise and progress of this epidemic, the *Coccus* has been first observed in moist hollow places, sheltered from the wind, and from thence has spread itself in all directions, even over the dryest and most exposed localities. This is not, however, universal, because on a few estates, such as those of Pen-y-lan and Dahanyke, it first appeared in dry exposed places. That it prefers moist sheltered situations is certain, as on the Lapallagalla and Muruta estates, which about a year ago had become nearly free from the pest, it still clung in a very obvious manner to those trees that were situated by the side of little streams of water in hollows and ravines.

The extent of the injury caused by the disease was found to vary considerably in the different districts visited, but was always more or less, in proportion to the length of time since it had been first noticed.

In the Muruta district some important information was obtained with regard to the course which the epidemic runs. Thus, on the Lapallagalla estate, where it was first observed in 1843, it did not reach its maximum till 1846, when nearly the whole estate was covered with the pest, and its accompanying black smut, or fungus. During the two first years of its existence, little loss of crop was sustained; but in 1845 and 1846 the loss was upwards of two-thirds of what would have been produced had the trees been healthy. In 1847 both the "bug" and the smut began gradually to disappear, and before the end of the year, it is said that scarcely a vestige of it remained. The seeds of it, however, seem to have been left behind in no little abundance, for the trees are again this season over-run with the pest, and almost everywhere, but more particularly in hollows and ravines by the side of streams, the smut is again blackening the trees.

On the estate called Muruta, about a mile to the eastward of Lapallagalla, the epidemic has pursued a similar course; but, from being a season in arrear, it presents this year a somewhat different appearance. This estate first began to show symptoms of it in the year 1844. During that and the following year it did not cause much damage; but in 1846 it had extended over nearly the whole of the plantation, and about two-thirds of the produce was lost on those portions that were worst affected. In 1847 it disappeared in a great measure, but the trees had received such a shock, that those growing on the least fertile parts of the estate yielded almost no crop, while those on better soil, and therefore more able to recover themselves, produced an average one. This season the trees at a distance appear to have become quite free from the evil, having nearly regained their usual green colour, and the crop promises to be a fair one for the district; but when the young branches, the underside of the leaves, and the footstalks of the fruit are closely examined, the Scale in its different stages is still visible there; and, as the insects multiply in excessive numbers, the probability is that the estate next year will be exactly in the same state that Lapallagalla is this season; indeed, in the hollows, many patches of black are even now to be seen.

Of three or four other estates that were passed through in this quarter, on my way to Dolisbagie, the next district, no precise information could be obtained, none of the superintendents having been longer than a few months on them. All, however, were found to be greatly diseased, large patches, of acres in extent, being very visible in the whole of them. One of them, called Diabetmia, was suffering from another evil—the rat, and it was painful to witness the ravages these animals were committing. A large patch of coffee-bushes, stretching along a flat at the bottom of a bare rocky hill, had so many of the branches cut off and lying on the ground, that it seemed as if some one had gone through pruning them with a knife; and many of the branches thus lopped were fine healthy shoots, covered with fruit. In the adjoining woods I found the Niloo—an *Acanthaceous* shrub, on the pith of the young shoots of which the rats are said to feed—to be all dead; the same was the case last year at Rambodde, where the rats were committing similar ravages. As the Niloo is a plant that forms a great proportion of the underwood of our mountain-forests, and flowers only once in every five or six years, and then dies down, the rat will most probably be a regular periodical enemy to coffee-plantations, at least to those bordering on forests.

The first estate visited in Dolisbagie was Pen-y-lan. There the nuisance did not commence till about the setting in of the monsoon, in the early part of 1846. It first appeared on the top of a ridge exposed to the wind, and on the opposite side of the estate to that on which the Muruta estates are situated, the distance between them being about six miles. Here the extension of the evil was so rapid, and so virulent in its nature, that in the first year of its having been noticed, 1846, it was estimated that a fourth of the whole crop was lost. After the rains of 1847 it began to disappear, some trees becoming perfectly free; but in low sheltered situations it still continued nearly as bad as ever. The trees suffered so much from the previous year's attack, that scarcely any crop was obtained: thus, from one patch, consisting of about seven acres, only 100 bushels of green Coffee were picked. A tolerably fair crop was promised at the beginning of the present season, but at the period of my visit the Scale had become nearly as prevalent as ever, and will no doubt destroy a large portion of it, before it has time to ripen.

Barnagalla and Raxana, two other estates that were passed through in this district, were both found to be suffering to a considerable extent.

In the Ambegamoa district, the first estate visited was Wevely-Talawa. From the superintendent I could get no information, as he had only been in charge for a short time; but from a gentleman who accompanied me, I learned that in 1846 the pest was ravaging it to a frightful extent, and that a heavy loss of crop ensued. At present the trees *seem* to be tolerably free of the "Scale," but on examination it was found still to exist, and a few black patches were here and there observable through the estate.

On Weraloo-gastane, about six miles distant in a northerly direction, the insect was first seen about two years ago; and its ravages have not yet been extensive. At present about one-fourth of the estate is affected, and beginning to turn black; but Mr. Anstruther, to whom it belongs, says that he is checking its progress by sprinkling finely pounded saltpetre, mixed with lime-dust, over the trees.

On the Dahanyke estate, about four miles further north, towards Kotmalee, the *Coccus* first obtained a footing in May 1845, on a single tree near the Bungalow, which occupies an elevated, exposed situation. In 1846 it spread over about six acres, round the spot where it first appeared. In 1847, it disappeared to a considerable extent, but this season it has increased again, and is now fast extending over the estate.

My first object on reaching Kotmalee was to visit the Harrangalla estate, as it was there that I first saw the effects of the injury in June 1846. It had then only been observed about two months before, but had already diffused itself over a patch of about an acre, in a hollow part of the estate, sheltered from the monsoon. Since then it has gone on gradually progressing, and now prevails over more than two-thirds of the whole estate, while the effect produced on the trees is much worse than I have seen elsewhere. It was truly painful to find that trees which, in a healthy state, would produce from two to three pounds of coffee, were either entirely destitute of berries, or only producing a few shrivelled ones, that will hardly repay the expense of collecting. I could not ascertain what the loss of crop had been during the last two years.

On the adjoining estate of Oonoo-galla, which is a much younger one, the *Coccus* was first seen about the end of 1846, also in a hollow part of the estate. This season it is disseminated over about thirty acres.

Katooboole and Kadien-lena, two very large estates in this district, are suffering at present to an alarming extent, and the damage will be very great.

In the Dimboola district the estates are all young, none of them having yet borne a crop. In all of them the *Coccus* has been observed, but only in a trifling degree.

In the Puselana district the estates have been in bearing for two or three years, and all of them are more or less infested with the *Coccus*, though not nearly to the same extent as the Kotmalee ones. The only one of them which I examined particularly, is called the Rothschild estate, containing about 400 acres; and it, perhaps, is suffering more severely than any other in the district. Here I found the history and effects of the plague to be much the same as elsewhere. It was first discerned on a few bushes in a hollow, sheltered part of the estate, where the soil is rich and rather swampy, about the beginning of 1846, and since then it has been gradually spreading till the present time, when, in its different stages, it infests a large portion of the estate, though perhaps not more thirty acres are in the worst state, the loss of crop on which is estimated to be about two-thirds.

On the Hunisgiria range, which is situated to the north-east of Kandy, the Scale did not make its appearance till about two years ago. Upwards of a dozen of estates were visited, and all of them I found more or less injured. The Hunisgiria estate, which is the nearest to Kandy and on the western side of the range, is that on which it first appeared, and the one, also, that has suffered most. As elsewhere, it did not cause either much apprehension or damage during the first year; but this season several large patches, of acres in extent, have turned quite black. The only other badly infected estate in this district is that called Dottlegalla, several miles further north, and between which and the Hunisgiria estate there are two others but slightly touched. Here, however, no information regarding its origin or progress could be obtained, as a new superintendent had just taken charge of it. Between this estate and that of Cabragalla, at the north end of the range, as well as those on its eastern side, none of the estates have yet suffered, though on examination they were all found to possess the seeds of the damage.

I have been particular in my inquiries into the effect which atmospheric changes exert on the habits and effects of the Coffee *Coccus*, but no very satisfactory information has resulted. The general impression seems to be that it flourishes most luxuriantly in wet weather. What effect a continuance of drought would produce on the insect, it is impossible to say, as during the last two years there has scarcely been any dry weather in the Central Province; and this continued moist state of both atmosphere and soil, I am inclined to believe, has had much to do with the prevalence of the epidemic. As the present season promises to be a dry one, it may perhaps partially check its progress to some extent.

Whatever may have been the origin of the *Coccus*, it is certain that, having once appeared, the rapidity with which it multiplies, and the immense number of eggs that each Scale produces, will sufficiently account for the speed with which it extends. As the females do not possess wings, it is quite impossible that they can spread from tree to tree by flight. By means of their legs, however, they possess excellent powers of locomotion; and as at the opening of each Scale, to give egress to the young that have been hatched within it, hundreds of them must necessarily fall to the ground, many will, of course, take possession of adjoining trees. This seems to be the way in which the pest radiates from a centre to a circumference in those localities where it first appears on estates.

As regards its transmission from one place to another, there are many means by which it may be effected. Thus, any one passing an infected tree will be sure to carry away hundreds of the minute young, unknown to himself, as they are all but imperceptible to the naked eye; and I should say that Coolies going from one estate to another, have done more to diffuse the plague than anything else. The same end has, no doubt, also been accomplished by birds and large insects. In some of the estates that I have visited, the disease has first been noticed near the Coolie lines; and where it has broken out in single patches, in the middle of estates, far from roads, birds have most likely conveyed the seeds. That the spread of the pest from west to east has been effected by transmission of the young females from place to place, is, I think, very probable.

Numerous are the remedies that have been employed to check the progress of the blight, but none of them have had the desired effect. Thus, upon the Lapallagalla estate, applications of chloride of lime, lime-water, urine, and manuring with guano, have been tried, and found useless. Cutting down the trees and close-pruning them were also had recourse to in vain, the young shoots immediately becoming as bad as ever.

At Pen-y-lan, powdered lime thrown over the trees, rich manuring with horse-dung and rotten coffee-pulp, application of urine, and tobacco fumigation, were tried without effect.

At Weraloo-gastane, Mr. Anstruther believes that he is checking the mischief on his estate, by sprinkling the trees with a mixture of equal parts of finely-pounded saltpetre and quicklime. Some trees which had been thus treated were pointed out to me; but as I had not seen them before the use of the remedy, it was impossible to form an opinion regarding its efficacy, especially as on examination abundance of the live Scale was still visible. Since my return to Peradenia, Mr. Anstruther has addressed the following letter to me on the subject:—

Ambegamoa, 23rd June, 1848.

My dear Sir,—It occurs to me that perhaps I did not explain to you sufficiently how saltpetre can be most effectually used in destroying the Coffee bug. It should be pounded as fine as possible, and dusted over the branches affected, while the leaves are wet: it then adheres, and almost immediately melts. I have witnessed many instances of its success since I had the pleasure of seeing you. My superintendents have tried other modes of application, and, they think, with good effect,—such as applying it to the roots simply, or mixed with manure. Yours, very truly,

P. ANSTRUTHER.

With regard to this remedy, I must, however, state, that on my return to Peradenia, about ten days ago, I selected two trees, growing in a small patch of infested Mocha Coffee, for the purpose of testing its effects. One of them was treated according to Mr. Anstruther's plan, the other carefully brushed over with a solution of saltpetre in water, of the strength of about an ounce to the pint. These trees have been daily watched, and in ten days they are neither better nor worse than those by which they are surrounded.

Powdering the trees over with sulphur and wood-ashes has also been recommended, but I have not found it efficacious.

At Puselava the Messrs. Worms ascertained that the application of cocoa-nut oil had the effect of exterminating the *Coccus*; but, as it destroyed the trees also, it was abandoned. It is well known that all oily substances applied to insects kill them by preventing respiration; and, in the same manner, if applied to trees, especially to the underside of their leaves, their *stomata*, or breathing-pores, are shut up, and death is the consequence.

Fumigation with sulphur has been recommended, but not yet, so far as I know, put to the test of experience in Ceylon; and some experiments which I am at present instituting with tar-water are not sufficiently advanced to be detailed.

It is, however, very doubtful if any remedy will ever be discovered sufficiently cheap, and, at the same time, easy of application on the large scale which Coffee estates require. The hot-houses of England have long been infested by more than one species of *Coccus*; and powders, washes, and fumigations of all kinds have been again and again prescribed for their destruction, but seemingly without effect, as I find that Loudon, in his 'Encyclopædia of Gardening' (5th edition, p. 431), concludes an article on this subject in the following words:—"Brushing off these creatures is the only effectual remedy, and, if set about at once and persevered in, will save the trouble of many prescribed washes and powders, which are mere palliatives." This measure, of course, is out of the question on a Coffee estate, unless, indeed, while the pest is still in its infancy; but the productive powers of the insect being so great, and the means of spreading the infection so numerous, it is to be feared that such a plan would soon be found to be impracticable.

From all I have seen of the nuisance, I am inclined to believe that it is not under human control; and that, if ever it disappears from the island, or at least becomes so much lessened in its effects as to be productive of but little injury to estates, it will be by running itself out, which blights of a somewhat similar nature have been known to do in other countries. Whether this will prove to be the case with the *Coccus* of the Coffee is unfortunately the less to be expected, as the experience of the last five years goes far to prove its permanency. I have the honour to be, Your most obedient servant,

GEORGE GARDNER, F.L.S.,
Superintendent of Botanical Garden.

References to the Plate, TAB. XII.

1. A branch of the Coffee-tree, infested by the *Coccus*, with the back of a leaf covered in the middle with the male insect, and a few matured females on the margin:—*nat. size.*

2. A healthy cluster of young Coffee-berries :—*nat. size*.
3. A cluster of berries after being over-run with the *Coccus* :—*nat. size*.
4. A *Coccus* immediately after being hatched, when there is no difference between the male and the female :—*highly magnified*.
5. The male in the pupa state :—*highly magnified*.
6. The perfect male :—*highly magnified*.
7. A female shortly after she has attached herself to the stem of the plant :—*highly magnified*.
8 & 9. The female about the period of impregnation :—*highly magnified*.
10. The female, or "scale," arrived at maturity :—*nat. size*.
11. The same :—*highly magnified*.
12. A few eggs :—*highly magnified*.
13. A morsel of the black fungus :—*highly magnified*.
14. A branch of the same, bearing seeds :—*still more highly magnified*.

Short notice of the African Plant DIAMBA, *commonly called Congo Tobacco; by* R. O. CLARKE, ESQ., *Surgeon and Colonial Apothecary to the Colony of* SIERRA LEONE : *communicated by the Author.*

[The properties of *Bang*, or Indian Hemp, by some called *Cannabis Indica*, but with more justice considered by others to be identical with the *Cannabis sativa*, are well known and fully detailed in the 'Materia Medica' of Dr. Pereira (vol. ii. p. 1242). Mr. R. O. Clarke, Surgeon to the colony at Sierra Leone, has discovered that the same narcotic has been long in use in the interior of that colony, and has communicated the following particulars, together with samples of the dried plant, and extracts from it, which are deposited in the Museum of the Royal Gardens of Kew. Where the intoxicating uses of this drug were first detected it would be difficult to say. Pereira observes, that the Asiatics and Egyptians employ hemp for the purposes of intoxication. The *majoon* used at Calcutta, the *rupouchari* employed at Cairo, and the *dawamesc* of the Arabs, are preparations of this kind. From Egypt or Arabia the knowledge probably extended to tropical Western Africa, and perhaps thence also in the other direction into India.—ED.]

The *Diamba* plant (*Cannabis sativa*) is considered to be indigenous in moist situations in the interior of tropical Western Africa, near the Congo or Zaire river.

A story is told of its discovery by a huntsman, who observed a number of antelopes, who had browsed upon the *Diamba*, to be stupified; and having informed his neighbours of the extraordinary circumstance, they repaired in a body to the spot. The approach of the people, or firing of their muskets, had, however, no effect in rousing the animals to a sense of their danger, and accordingly they were all quickly despatched.

It is well known to the Portuguese on this *coast*. Its seed was brought to Sierra Leone by Congoes captured by one of our cruisers, and is now distributed over the colony. It is chiefly cultivated and prepared by these people or their descendants, but it is also grown by the Akoos, Eboes, and many of the other liberated African tribes, and likewise by the Maroons, Settlers, and Creoles.

The average height of this annual bushy shrub here varies from six to seven feet, but in fertile soils it attains the height of twelve or thirteen feet; and some of the larger plants occupy a space of twenty feet in circumference!

It is sown in April or May, and shoots up in three or four days, but its growth is, from time to time, retarded by nipping off the points of the top and lateral branches. It flowers in August, ripens in October or November, when the flowers are plucked off. In December the leaves are removed, when it withers, and is rooted up, the branches being tied up and used as brooms. A second crop, sown in September, and watered during the dry season, ripens in January or February.

The flowers, exposed to the sun or fire, slowly dried, and mixed with the seed, are the parts of the plant preferred, and in this state the drug is termed *Maconie*. The leaflets are similarly prepared, but only employed when the former cannot be procured, as this preparation, called *Makiah*, is apt to cause violent headache.

It is smoked from a large wooden pipe or reed, called *Condo*, or from a small calabash, but common clay-pipes are also used: it is extensively consumed by many of the liberated Africans and Creoles, who frequently meet at each other's houses, to enjoy the luxury and soothing influence of *Diamba*. Upon these occasions the pipe is handed about from mouth to mouth, and soon produces the desired intoxicating effect. The smoke, twice or thrice drawn into the mouth, is there detained, and a large portion is swallowed, as it slowly passes off by the nostrils: most agreeable sensations soon follow, and excitement displays itself in hearty bursts of laughter, loud exclamations, droll exhilarating conversation; but, as the debauch proceeds, its full effects are developed. Temporary frenzy seizes the smokers, and they issue from their haunts, singing and shouting, as they reel and stagger to their homes. Intense and maddening headache, accompanied with stupor, is often the result of these orgies, and the latter consequence generally lasts for twelve hours.

One pipe charged with this powerful drug, is enough to produce in four persons the most delightful exaltation without injury, and it is much esteemed by the natives as a remedy for cough, pains in the chest and stomach.

Diamba is vended under the names of *Maconie** and *Makiah*: the former is made up into very small packets, which are sold at one half-penny each, the latter into larger ones for the same sum. A small plant in full flower and seed will yield to its owner the value of ten shillings'-worth of *Maconie*.

The Origin of the Existing Vegetable Creation. By PROFESSOR J. F. SCHOUW. *Transactions of the Meeting of the Scandinavian Naturalists at Copenhagen, in* 1847, Append. K. p. 119. (Translated from the Danish, by N. WALLICH, M.D., F.R.S., V.P.L.S.)

(*Continued from* vol. ii. p. 377.)

If we compare in a similar manner the flora of Lapponia, or, which is the same thing, the Scandinavian mountain flora, with that of the rest of Scandinavia, we shall obtain the following results, taking Hartman's flora for our guide:—.

	Scandinavia.	Lapponia.
Flowerless	·03	·05
Trimerous	·26	·31
Pentamerous apetalous	·08	·09
,, petaliferous	·63	·55

According to geologists, the Scandinavian mountains are older than the Alps; and yet we find that the flora of Lapland, which is the

---

* One cannot but be struck with the similarity of this and the next following word *Makiah*, and that of μηκων, of the Greek, applied to a powerful drug with analogous properties; and, as already observed, a preparation of this plant, used at Calcutta, is called *Majoon*.—ED.

same as that of the Scandinavian mountains, approaches rather towards the antediluvian vegetation, inasmuch as the numerical proportion of flowerless plants is larger, of apetalous pentamerous somewhat larger, of petaliferous pentamerous considerably less. Comparing together the floras of Lapland, or the Scandinavian mountains, and the Alps, with reference to the numerical extent of their large groupes, we shall perceive a more marked discrepancy among them, than between the floras of the Alps and Germany, Scandinavia and Lapland; and yet if we keep in view the habitual characters of the floras, their families, genera, and even species, the analogy of the flora of the Alps with that of the Scandinavian mountains, becomes far more manifest, than the analogy which exists between them and the corresponding lower countries, and which, according to climate, might be expected to prevail. This will become evident by combining together the preceding schedules :—

	Antediluvian.		Germany.	Alps.	Scand.	Lapp.
	Before Chalk.	After.				
Flowerless	·81	·02	·02	·02	·03	·05
Trimerous	·06	·13	·21	·16	·26	·31
Pentamerous.						
apetalous . . .	·12	·45	·08	·04	·08	·09
petaliferous . . .	·01	·40	·69	·78	·63	·55

Another peculiarity in one part of the alpine flora consists in the remarkable uncertainty of proportion among its species, and the vacillation of forms, which renders it extremely difficult, if not impossible, to determine the species correctly. It follows, in consequence, that certain forms are looked upon as reducible to a few species only, while other authors divide them into many. I need mention only *Draba*, *Arabis*, *Hieracium*, *Gentiana*, and *Salix*. This unsteadiness as regards form is the more remarkable, since Alpine plants propagate themselves more by means of buds than seeds; and it is known that the former preserve the specific features of plants more rigidly than the latter. If I am correct in my supposition, that plants have originated, not from one, but from many parent individuals, it appears probable, that species may have originated, from types of nearly-allied forms becoming gradually fixed by propagation from buds or seeds, and by their expelling other forms. But if this view is adopted, it is clear that the older flora must possess more numerous and fixed forms than the recent. By the influence of man new forms (varieties) arise, which seem to indicate a sort of return to the primitive state of things. But this result loses much of its value, when it is considered, that the Scandinavian mountain-flora, which is the more ancient, shows much instability in regard to forms, pointing towards the great variety of localities, as the leading cause.

Although I think it reasonable to conclude, that the flora of the Alps is of more recent date, than that of middle Europe, or of the Scandinavian mountains, yet I do not by any means consider this as finally proved. To attain any degree of certainty, we require a great many data, founded on geological structure, which are yet wanting; and in order to form conclusions from the internal character of the floras, we must compare many of these, on purpose to obtain the value of certain numerical proportions between the principal groupes and the characters of the floras. It is only by the earnest co-operation of the botanist, geologist, and zoologist, that we can expect to arrive at any conclusive results. I could wish, especially, that botanists would be induced to study thoroughly the phyto-geographical divisions, their peculiar features, and the physical conditions, under which they present themselves. What has been advanced above, must be considered as propositions only, awaiting further researches. In the meanwhile, I will conclude with the following remarks. It is well known that New Holland and South Africa are remarkable by the great diversity, as well as peculiarity, of their vegetable productions; while, on the contrary, the flora of extra-tropical South America is devoid of either, and approaches not a little to the flora of Europe and North America. Now, this multiplicity of forms cannot have originated in migration, for which neither of those countries is adapted; New Holland being entirely surrounded by the ocean, and South Africa equally so in three directions, while in the fourth it is bounded by mountains and barren deserts. Neither can we suppose that the multiplicity has arisen from circumstances of climate, which, from the greater influence of the sea in the southern hemisphere, should be less diversified. May we not, therefore, explain this singularity in the three continents of that hemisphere upon *historical* grounds? In New Holland, as well as in South Africa, there is a certain instability of species; and families belonging to the most perfect, exhibit a great development and predominancy, such as *Mimoseæ* and *Myrtaceæ*. Finally: most salt-water plants (*halophytes*) belong to the less-developed pentamerous groupe, that is, to the apetalous, which predominated in the former creation beyond what is the case now. May this not indicate, that these plants belong to the more ancient forms, which, inhabiting the sea-shores, were with less difficulty preserved during grand natural convulsions?

If it should be objected to inquiries like the preceding, that they lead to no certain results, it may be answered, that the progress of our knowledge of fossils during the last fifty years having been gigantic, we may be confident that our researches, too, will be productive of good. We all know that limits are set to human knowledge; but it is only by means of trials that we can ascertain where those limits present themselves. The naturalist is not to be deterred by the silent Sphinx of nature: he should endeavour to compel her to speak out.

Decades of Fungi; *by the* Rev. M. J. Berkeley, M.A., F.L.S.
Decade XXXI.
(*Continued from* vol. ii. p. 112.)

The present Decade, with the exception of the last species, consists of the novelties collected by Mr. Spruce in his visit to the Pyrenees, and the province of Parà in Brazil. The species are few in number, but, especially the Brazilian, exhibit some interesting forms. All the Brazilian fungi in the collection are enumerated, but it has not been thought necessary to do so with those from the Pyrenees.

* *Agaricus Campanella*, Batsch.
Hab. Tanaü.

I have seen abundant specimens of this from Xalapa, and from various parts of North America, some having the gills yellow as in European specimens, and others, especially those from Ohio, agreeing with Mr. Spruce's specimen in having the gills cinereous. It appears, however, from Mr. Lea's notes, that even in the Ohio individual the gills were at first tawny. All agree in the peculiar nature of the down at the base of the stems, which, in the Xalapa specimens, send out many decumbent shoots. It is possible that more than one species may exist, and that perhaps belonging to the genus *Marasmius*, being the analogue of *A. Campanella*, but the dried specimens before me, though numerous, are not sufficient to establish the point.

301. *Marasmius inoderma*, n. s.; cæspitosus; pileo subirregulari excentrico fibrilloso-sericeo; stipite nigrescente; lamellis latiusculis breviter adnatis, interstitiis lævibus.

HAB. Parà.

Pileus ¾–1 inch broad, nearly even, depressed, pallid, irregular, somewhat excentric, clothed with silky anastomosing or reticulate fibres. Stem ½–1 inch high, not ¼ a line thick, nearly smooth, at first pale, at length brown, slightly incrassated at the base. Gills moderately broad, slightly attached, distinct; the interstices smooth.

Distinguished by the nature of the silky covering of the pileus, which is different from anything that I have seen in other species. Externally it resembles *Marasmius Vaillantii* and *M. alliiodorus*, Mont.

* *Schizophyllum commune*, Fries.

HAB. Caripi.

302. *Schizophyllum umbrinum*, n. s.; umbrinus, contextu concolore cartilaginis; pileo flabelliformi profunde inciso lobato; lobis furcatis. (TAB. I. fig. 1.)

HAB. Caripi.

Umber-brown. Pileus about ½ an inch long and broad, flabelliform, attached by the elongated vertex, from which proceed a few byssoid radiating threads, which form an orbicular spot on the matrix, somewhat strigose behind, and paler, clothed in front with adpressed spongy down, which breaks up into little fascicles or areolæ, deeply inciso-lobate, the lobes flabelliform, incised and furcate. Gills at first clothed, like the pileus, with umber down.

I have not seen the fructifying surface of the hymenium. It is distinguished by its umber hue, to which there is no approach in the other species, by its regularly lobed and furcate pileus, and its dark cartilaginous substance.

Tab. I. fig. 1: *a, S. umbrinum, nat. size*; *b,* upper, and *c,* dennr side, *slightly magnified.*

* *Polyporus sanguineus*, Fr.

HAB. Caripi; Parà.

* *Trametes hydnoides*, Fries.

HAB. Caripi.

303. *Stereum Galeottii*, n. s.; umbonato-sessile, parvum, convexum, rigidum; pileo cervino velutino-tomentoso crebrissime badio-zonato; zonis hic illic glabris nitentibus; hymenio cinereo-alutaceo. Galeotti, no. 6853.

Hab. Caripi, Spruce; Vera Cruz, Galeotti; Xalapa, Mr. Harries.

Pileus 1¼ inch broad, 1 inch long, subflabelliform, umbonato-sessile, mostly convex above, slightly undulated, thin but rigid, fawn-coloured, clothed with velvety down; repeatedly zoned; zones mostly very close and narrow, frequently forming bay-brown fasciæ, smooth and shining, alternating with paler. Hymenium tan-coloured, with a cinereous tinge.

Undoubtedly nearly allied to *Stereum lobatum*, Kze., but a much smaller and neater species, remarkable for its closely-zoned pileus.

Galeotti's specimens, it should be observed, are far less beautiful than those of Mr. Harries and Mr. Spruce.

* *Hypolyssus Montagnei*, Berk. in Hook. Journ. of Bot. vol. i. p. 139. tab. 6, fig. 1.

Hab. On twigs. Caripi.

* *Dictyonema membranaceum*, Ag. Syst. Alg. p. 85.

Hab. Caripi.

* *Hypochnus nigrocinctus*, Ehrenb. Hor. Phys. Ber. p. 85.

Hab. Caripi.

* *H. albo-cinctus*, Montagne, Ann. des Sc. Nat. 2 Ser. vol. viii. p. 361.

Hab. Caripi.

This is not precisely the plant of Dr. Montagne, having the border rather more byssoid. Another form occurs in the same collection more nearly approaching his plant, but thinner. All the three are probably the barren thallus of some lichen.

* *Peziza cinerea*, Batsch.

Hab. Tanaü.

Agreeing precisely in fructification, as in outward form and colour, with European specimens.

304. *P. herpotricha*, n. s.; capsulis convexis pallido-rufis demum obscurioribus glabris immarginatis, fibris ramosis repentibus insidentibus. (Tab. I. fig. 2.)

Hab. On living leaves, mostly on the upper surface. Caripi.

Forming orbicular patches on the leaves, resembling those of an *Asteroma*. Cups convex, immarginate, pale red-brown, seated either singly or many together on branched, rugged, dark-brown, radiating fibres, which give out numerous ovate processes from their sides, and separate easily from the surface of the leaf. The main threads are connected by numerous fine branched hyaline filaments. Asci short, slender. Sporidia oblong, very slightly curved or sigmoid, uniseptate, with frequently a nucleus in either cell.

This has very much the aspect of a Lichen, but there are no gonidia. It has somewhat the same habit as *P. leucorhodina*, Mont., but has no close affinity with any described species, as far as I have been able to discover.

Tab. I. fig. 2: *a, P. herpotricha, nat. size*; *b*, mycelium; *c*, asci and sporidia, *both highly magnified*.

* *Phacidium dentatum*, Kze.

HAB. Caripi.

The asci, both in the Brazilian and European specimen, contain a quantity of thread-shaped sporidia, as in some other species of the genus. In the Brazilian individuals they are rather more slender. Corda's figure appears to be taken from something different, as indeed the general appearance of the plant seems to indicate. *Phacidium Delta*, Kze., which grows on the leaves of laurels in Madeira, has the same structure, but the sporidia are much thicker.

* *Hypoxylon obovatum*, Mont. *Sphæria obovata*, Berk. in Ann. of Nat. Hist. vol. iii. p. 397.

HAB. On bark. Caripi.

* *H. Leprieurii*, Mont., Ann. des Sc. 2 Ser. vol. i. p. 352.

HAB. On bark. Caripi.

This might at first be taken for a distinct species, but a section shows that it is merely a state of that described by Dr. Montagne. The outer surface is covered partially with a delicate velvety coat. Some of the bark is carried up by the disc, forming, together with a portion of the stroma, an acute margin. Ultimately, however, the bark with the stroma attached to it falls off, leaving a hollow disc and an obtuse margin, both of which are very evident in a vertical section.

* *Thamnomyces Chamissonis*, Ehrenb. Hor. Phys. Ber. p. 79. tab. 17. fig. 1.

The capitula of this fungus contain several oblong perithecia, exactly as in *Hypoxylon Leprieurii*, &c., to which it is evidently closely allied. The sporidia are oblong, with one side rather curved. I do not find the stem usually hollow. It consists of three parts—the outer laccate coat, a dense black stratum, like charcoal, which is brownish externally, and a loose pith, which occasionally vanishes, or is turned aside in the process of making a section, the outer portion of which, next to the charcoal-like stratum, is reddish.

* *Micropeltis applanata*, Mont. Cuba, p. 326.

HAB. Caripi.

The specimens are young, and have a blue cast with a white orifice. The asci are not yet formed.

305. *Depazea Mappa*, n. s.; maculis e plagis variecoloribus subconcentricis fusco-limitatis formatis; peritheciis olivaceis; ascis subelongatis, sporidiis oblongis sublanceolatis hyalinis. (TAB. I. fig. 3.)

HAB. Caripi. On leaves of some unknown plant.

Forming patches an inch or more in diameter, consisting of more or less concentric irregular areæ, those in the centre mostly whitish, the others presenting one or more shades of more or less bright red-brown; each area being strictly defined by a brown, wavy, somewhat diffused line. The whole presents the appearance of a coloured map. Perithecia punctiform, olive, with a white speck in the centre. Asci at first short, at length slightly elongated, lanceolate; sporidia oblong, inclining to lanceolate, with one side more strongly curved, hyaline. Occasionally two or more perithecia are connected by a greenish substance.

A very pretty species, marked by several distinct characters. A single specimen exists in the collection of a variety or allied species, growing upon some larger leaves, evidently congeneric with those on which *D. Mappa* is developed. The spots are much larger, consisting of a larger number of more strictly concentric rings, which are of a more uniform wood-colour. Unfortunately, the spots are without perithecia.

Tab. I. fig. 3: *a*, *Depazea Mappa*, nat. *size*; *b*, young and mature asci, with the sporidia *magnified* 300 *diameters*.

* *Sphæronema epicecidium*, Berk. in Hook. Kew Gard. Misc. vol. i. p. 291. tab. x. B.

HAB. On galls. Caripi. But not on the same gall on which Sir E. Home gathered the specimens described as above from Parà.

306. *Agaricus* (Pleurotus) *Sprucei*, n. s.; imbricatus; pileo pelliculoso convexo tomentoso-scabriusculo; stipite strigoso brevissimo vel obsoleto: lamellis latiusculis.

HAB. On stumps. Col de Louvic, South of France.

Pileus 3–4 inches broad, imbricated, convex, clothed with a subgelatinous pellicle, which is minutely scabrous, especially behind, with short *yellowish down*; flesh moderately thick; margin involute. Stem externally short or obsolete; when present, clothed with yellow strigæ. Gills at first narrow, but at length more than $\frac{1}{4}$ of an inch broad.

This species, of which I can find no description, resembles *A. salicinus*, but is remarkable for a subgelatinous coat, like that of *A. palmatus*, with the addition of a short scabrous down, especially behind. The whole plant seems to have a more or less yellow tint.

307. *Exidia straminea*, n. s.; minuta, cupulæformis, erumpens, straminea, basi rugosa stipitiformi, extus subtiliter pubescente, intus subplicata. (TAB. I. fig. 4.)

HAB. On smooth bark. Wood near Pau, South of France.

Erumpent, straw-coloured, $\frac{1}{4}-\frac{1}{3}$ an inch broad, cup-shaped, at length flexuous, externally slightly wrinkled, clothed with extremely minute pubescence, scarcely velvety, elongated below into a short sulcate stem; internally even, except at the base, which is somewhat wrinkled.

The substance consists of delicate anastomosing flocci, which expand above into 2-3 septate clavate moniliform threads, which form the hymenium. Margin involute, denticulate.

This pretty species has the habit of a *Peziza*, and is altogether distinct from every species with which I am acquainted. I have not seen perfect spores, but the specimen seems to have arrived at maturity.

Tab. I. fig. 4: a portion of the threads of the hymenium, *magnified* 300 *diameters*.

308. *Sphæria* (Circinatæ) *parmularia*, n. s.; parva, pustulæformis; peritheciis circinantibus lateraliter arcte compressis, ostiolis vix distinctis umbonem efformantibus; ascis linearibus, sporidiis lato-oblongis uniseptatis fuscis.

HAB. On the smooth living bark of young trees (apparently on birch); Transoubat, South of France. On oak; King's Cliff.

Scarcely a line broad, forming little pustules with a black umbo. Perithecia circinating, closely packed, so as to present, when cut through, a triangle with one curved and two straight sides. Ostiola in general indistinct. Asci linear, containing eight brown, broadly oblong, uniseptate sporidia, like those of many *Diplodiæ*.

This curious species has been known to me for some years, but has never been published, in consequence of the doubt attached to its proper position, in consequence of its growing on living bark, contrary to the habit of other cortical *Sphæriæ*. It has, however, been submitted to several experienced Lichenologists, as Mr. Babington, Mr. Borrer, Professor Fries, and Dr. Montagne, without eliciting any satisfactory opinion, and I therefore think myself justified in considering it as a *Sphæria* rather than a *Verrucaria*, especially as there is no crust. In some states it resembles very closely the genus *Parmentaria*, Fée. Externally it is not unlike *Sph. turgida*. The perithecia vary in number from 5 to 10, but are always laterally compressed, so that a section reminds one of the disposition of the carpels of an orange.

309. *Physarum iridescens*, n. s.; confertum, sessile vel spurie stipitatum; peridiis subglobosis columbino-chalybeis tenerrimis; floccis albis, sporis atris.

HAB. On *Jungermanniæ*. Labassère, South of France.

Crowded, either entirely sessile or spuriously stipitate; peridia globose or contracted at the base, from their crowded mode of growth, very delicate, reflecting prismatic colours like iridescent copper; flocci irregular, branched, varying in thickness, white; spores deep violet-black to the naked eye, but inclining to lilac under the compound microscope, globose when moist, elliptic, but pointed at either end when dry.

Agreeing exactly in habit with *P. bryophilum*, but differing in its white flocci and darker spores, whose diameter is twice as great as in British specimens of that species. After a careful search I can find no intimation of the Pyrenean species.

310. *Agaricus* (Psathyra) *calvescens*, n. s.; pileo submembranaceo ex ovato conico-subcampanulato obtuso primitus piloso-tomentoso demum calvescente; stipite floccoso sursum glabro fistuloso; lamellis latis ascendentibus adnexis distantibus cinereis. Hook. fil., no. 117, cum ic.

HAB. On mossy earth, in tufts. Darjeeling, 7,500 feet. September.

Odour like that of *Ag. campestris*. Soft, brittle. Pileus $1\frac{1}{4}$ inch across, at first ovate, white, with a pale reddish-yellow tinge, clothed with pilose fasciculate deflexed down, thin, conical, subcampanulate, gradually becoming smooth, even, cinereous, with the exception of the yellow apex. Stem 2 inches high, 2 lines thick, floccoso-squamose below, the tufts of flocci pointing upwards, smooth above the point of attachment of the edge of the young pileus, white, *slightly incrassated downwards, fistulose*. Gills broad, ventricose, ascending, adnexed, rather distant, cinereous, with a pale border. Spores dark, elliptic.

Nearly allied to *Ag. pennatus*, but larger, with more distant gills, &c. It is also more densely clothed with down, the free flocci of which make pilose fascicles.

(*To be continued.*)

BOTANICAL INFORMATION.

Physical Geography of SIKKIM-HIMALAYA. *Extract of a letter from* BARON HUMBOLDT *to* Sir W. J. Hooker, *together with copy of a letter on the* PHYSICAL CHARACTER *of* SIKKIM-HIMALAYA, *addressed to* Baron Humboldt, by Jos. D. HOOKER, R.N., M.D., F.R.S., &c.

So honourable is what follows, both to the writer of the subjoined letter and to the illustrious philosopher from whom the injunction to print it has proceeded, that, much as I fear I shall expose myself to the charge of vanity in giving publicity to language so highly complimentary to my son, I hope and trust I am actuated by a higher principle, *that* of justice, which forbids my withholding it. " Laudari a laudato " has always been accounted a most justifiable object of ambition; and where is there, in the world of science, a man equally *laudatus* as Humboldt?

Extract from Baron Humboldt's Letter.[*]

Six days ago I received an admirable letter from your son, containing a perfect treasure of important observations, relating to the mountain-masses of Himalaya, their geology, meteorology, and botanical geography. What a noble traveller is Joseph Hooker! What an extent of acquired knowledge does he bring to bear on the observations he makes, and how marked with sagacity and moderation are the views that he puts forward! I can neither part with such a remarkable letter, nor keep to myself the *résumé* which it contains of his researches in Thibet and on its confines; therefore I desire that it should be published, and correctly published, in England; and therefore do I now deposit it in

[*] This is the close of a very long letter on other subjects,—much, relating to *Victoria regia*, which will appear in another part.

your hands. I could not consent to bury such a prize. You have already published one of your son's letters to me, which, though important, is far less valuable than the present. Only let me have the honour of being mentioned as the individual to whom it was addressed; for I feel no little pride in being known to enjoy the friendship and correspondence of your son. When he returns to us in spring, he will find his own fame widely diffused and solidly based. At all events, do me the favour to inspect the printing of this document, and let it be done from a copy made under your own eye.

Our illustrious geographer, M. C. Ritter, was so kind as to transcribe it for me; but we are still doubtful about many of the names of places, though we have carefully referred to several maps.

It has given me great pleasure to receive a confirmation, in this letter, of many guesses which I had ventured on the subject of the *soi-disant* table-land of Thibet;—also, on the question whether the Himalaya presents a continuous crest, clad with perpetual snow, or whether the loftiest peaks are not rather situated out of the line of a medial axis;—also, whether my notions upon the limits of eternal snow, on the two slopes, and the causes of their apparent irregularity, would ever be confirmed on the spot by an impartial and well-informed eye-witness. This is now the second time that your son and Mr. Hodgson have given their testimony to the accuracy of the opinions which I advanced in my 'Central Asia,' a work which has never been translated into English, but which is that in which I think I have brought forward more novel information than in any of my other publications.

I have felt deeply anxious about your health, and my satisfaction is proportionably lively at hearing that your strength is considerably restored. I have lived, ever since 1848, in the midst of political excitement and popular insurrection, and I can truly aver that my chief comfort has arisen from my literary labours, often prolonged far into the night. I have just published my third and exclusively astronomical volume of 'Kosmos.' Health and powers are still mercifully granted to me; whilst the dearest friend I have on earth, M. Arago, is threatened with the loss of his sight.

And now I do entreat you, my dear and kind old friend, to excuse the length of this almost illegible letter, and to receive the assurance of my affectionate and respectful regard,

BARON ALEXANDER HUMBOLDT.

Potsdam, Dec. 11th, 1850.

Letter from Dr. Hooker *to* Baron Humboldt, *dated—*
Khossya Mountains, Sept. 23rd, 1850.

My dear and venerable Friend,—My correspondence has been thrown into such confusion during the last twelve months, that I am almost afraid to think how long a time may have elapsed since you have heard from me; much longer I know it is than it should have been, under any circumstances; especially as I have never yet thanked you for the distinguished place you are said to have accorded to my name in the 'Aspects of Nature.' My copy of this work has not yet reached me, but I do not want that to persuade me how grateful I should feel. I have very much to thank you for, and have had for years.

I acquainted you with my having applied in vain to the Surveyor-General for information about the elevation of the peaks, Dwhalghiri, Gosain-than, &c., &c. He has not replied to me; and Dr. Campbell, who addressed him since on the subject, has met with no better success. I believe that he has not calculated any but Chumalari and Kinchinjunga. There are four which he considers as rivals: all are within a very few feet of the same height, and from what I hear that he has said, I conclude they are—

1. Kinchinjunga, 28,148 feet accurately measured.
2. A mountain, in about 86° 30′ E. longitude.
3. Gosain-than.
4. Dwhalghiri.

Jewahir is considerably lower than any of these.

The Deputy Surveyor-General (Capt. Thuillier) is an intimate friend of mine, but he can give me no information, his Principal having neither forwarded results nor data to the office at Calcutta. These particulars are probably just what I told you before; I know no more. Mountain No. 2, I have seen repeatedly: it divides the Arun river from the Kosi rivers, and is visible from Darjeeling and elsewhere.

You are aware of my visits to two Thibetan Passes, in Eastern Nepaul, west of Kinchinjunga, in November and December 1848, and of my protracted residence on the Sikkim frontier in June, and my visit to five other Passes in September and October 1849. During the latter excursion Dr. Campbell joined me, and it terminated in our imprisonment by the Sikkim Rajah's prime minister and counsellors. Our lives were threatened, and Dr. Campbell had a very narrow escape, being treated with great barbarity.

I entered Thibet, and spent some days north of lat. 28°, discovering the source of the Arun river, which flows from west of the Ramtchoo Lakes, behind Kinchinjunga, to its exit from the mountains of Thibet in Nepaul. From what I saw, and the voluminous details I have collected, I am now convinced that the snowed Himalaya do not form an individual mountain chain; but that they consist of groups of snowy peaks, widely separated from one another, and are parts, or meridional spurs, of a much more lofty range of mountains between the great masses of snow and the Yarou-tsampu river. The southern water-shed is behind all the snows, often half a degree and more; and the Thibetan frontier runs along the line of water-shed behind, as often as along the ridges of snow. Thibet, between the Yarou and the snow, is, as you have well described it, a lofty, rugged, and quite impracticable country; though none of the peaks, north of the water-shed on that line, reach the elevation of the snowy ones.

The snowy peaks, called Himalaya, occupy *nœuds* of great elevation, and whose surface is less rugged. The rivers, too, throughout Thibet, run in broad open valleys with ragged, scarped flanks. Except along the courses of the rivers, the country is uninhabited, and impracticable; and *détours* of any extent are preferred to crossing the chains in Thibet. Of these great *nœuds* I recognize nine, well marked, and separated by an unsnowed tract of comparatively low elevation. They are—

1. A group in long. 93°, which I am now measuring, dividing the Soubansiri from the Monass.
2. A group I am also observing, between the Monass and Patchiou.
3. Chumalari group, between Patchiou and Matchoo, which rises north of Chumalari.
4. Donkiah (24,000), which I have visited, between the Matchoo and Lachen (or Teesta).
5. Kinchinjunga, between the Teesta and Arun.
6. The Peak in 86° 30', between the Arun and Kosi.
7. Gosain-than, between the Kosi and Gundule.
8. Dhwalghiri, between the Gundule and Gogra.
9. Jewahir and Mansarovar?

The unsnowed space between 5 and 6 is very great. East of 1 the snowy Himalaya cease, and the lower ranges sink greatly. I am now nearly due south of 1 (S. 10° W.), and see the horizon to 70°

east, without a snowy peak or elevation. The snow commences again east of the Dihong, which I do not doubt is the Burrampooter and Yarou. All the rivers, mentioned above, as between the groupes of snows, have their sources in a well-defined range, south of the Yarou. That range now forms the boundary of Nepaul, west of Gosain-than, and did, east of it, before the war between Nepaul and Thibet. East of Gosain-than there is no natural boundary, except where the spurs of the mountains, east and west of the exit of the Arun, constitute one. The Thibet boundary of Sikkim is Kinchinjunga, to the west; and the range between the Lachong and Teesta in Thibet, which is a westerly spur from Donkiah, and this last runs south for forty miles in a snowy meridional range. Chumalari does so, likewise; and the Chinese have drawn the line south from the eastern flank of the Donkiah for as many miles, and then north again, along the western flank of the Chumalari. This V-shaped portion includes the greater part of the Matchoo's course, and the town of Pari; it is interposed between Sikkim and Bhotan. The Patchiou valley is as low as 6,000 feet, in the latitude of Chumalari, and looking from Thibet, S.E., I could descry no snow between the Chumalari and the group east of the Patchiou. This last groupe I also see from these mountains (distant about 210 miles), and from both stations I make its elevation to be probably 24,000 feet.

When in Thibet, in lat. 28° 10′ N., and long. 88° E., I took angles with a theodolite from a mountain 18,500 feet above the sea, on the southern flank of the Arun valley. The ranges to the north of me were nowhere below that elevation, but did not rise above 22,000 feet. Eastward, I saw THREE meridional ranges, *i. e.*, that of Chumalari (S.E.), and two others, each of which came from a lofty mountain-land, and were connected with a beautiful snowy range, about eighty miles to the N.E., which I believe separated the U and Tsang provinces of Thibet. North-west of me was the course of the Arun, and beyond that a very lofty range, rising to 25,000 feet, which took a southerly direction, forming the snowy Himalaya group between the Arun and Kosi, *i. e.*, that, west of Kinchin. This line of water-shed is a physical feature, well recognized by the Thibetans: between the Arun and Chumalari it is called Dingcham province, and sometimes Damtsen. The inhabitants are *very* black; they rear the shawl-wool goats, and yaks, and are a very turbulent race of savages, detested as cattle-robbers by the Sikkimites, Nepaulese, and Thibetans. To us they were uncouth, rude, and boisterous, requiring great bullying : I owe them a grudge for breaking my azimuth compass, though more accidentally, I believe, than through wilful mischief. They opposed my crossing the Arun.

Kiong-lah is the name of the range forming the great axis, or watershed between the Yarou and the Arun. Turner crossed Dingcham, and reached his highest ground about Ramtchien. That is the centre of the *nœud* : from thence six mountain-chains or spurs radiate :— 1, that west of Bhotan, on which is Chumalari; 2, that east of Sikkim, with Donkiah on it; 3, that bounding the Arun on the south, with Kinchin-junga; 4, ...; 5, the axis running north-west between the Painom and Arun, and then south-west; 6, the Odoo mountains? running south-east between the Patchiou and Monass.

The rivers are—1, S. the Matchoo; 2, S.W. the Teesta; 3, W. the Arun; 4, N. the Painom; 5, N.E. . . . .; 6, S.E. the Patchiou. The mean elevation of this tract is 16–17,000 feet; nothing will ripen. Digarchi is in about the lat. and long. that Turner assigns to it. Crops only come to maturity under the shelter of, and in the radiation from, the black rocks of the flanks of the Painom valley and its tributaries. The Wallnut grows, but does not perfect its fruit, nor Peaches; and Willow is the only tree from 8–12 feet high. The result of all the oral information I collected about its native and cultivated vegetation, &c., would lead me to assume the elevation of this region to be 14,000 feet, and the mean temperature of October (*fide* Turner) gives the same, calculated at the rate of 1° Fahr. = 400 feet, which, from a multitude of observations of air and sunk thermometer (3 feet), contemporaneously with others at Darjeeling and Calcutta, I have found to be the decrement for altitudes above 1,000 feet north of the snows. In 'Asie Centrale,' you perhaps assume Turner's October temperature of *Digarchi* for that of *Lhassa*. Of Lhassa I have many good accounts. Grapes do not ripen, but are imported. Peaches, Willow, and Wallnut do well, both wood and fruit, but no other tree. Dromedaries abound. It is a poor place, after all that has been said; the census is small (I forget what). The accounts given by those persons who have and have not seen such a town as Purneah, in Bengal, are sufficiently different and instructive.

The Yarou is navigable for skin-boats, in the rainy season, from Giantchi to Digarchi, and from Digarchi to the meridian of Lhassa; but the course is too tortuous : it flows between lofty mountain ranges, so far east as the great lake (I forget its proper name, but not Yarbongh, as it is called on the maps,—it is certainly very large, but the island is small, towards the south-west angle); the Yarou trends south, and the country is warm, producing the Mulberry, and some say, Silk and Rice. Every Thibetan describes the people on its southern banks, near the bend, as atrocious savages, dwelling in mountainous woods bordering Assam; they are the Abors and Bor-Abors, of course: there the Yarou (they say) enters the mountains, and flows southward to the Burrampooter, the stream becoming too rapid for navigation, and the inhabitants of its banks are too savage. East of Bhotan, by Towang, there is no snow; *en route* from Assam to Thibet, none anywhere; and *Juniper* grows to the mountain-tops. The rivers, from a little south of the Yarou, flow to Assam. Nor is the boundary of East Bhotan well defined. Every information about the Yarou was the same, and my informants were very numerous; the best, a well-educated monk, who was brought up at Mendoling, a goompa two days southward of the Yarou, on a river flowing to Towang and Assam, five days off. Menchona is the mart of that quarter, and the only great one east of Pari. North of the Yarou is the salt country, lofty, rugged, barren, and inhospitable to the last degree. The Yak cannot proceed beyond a few days' march northward of Digarchi; and sheep are almost the only beast of burden, except *man*.

Of the physical features of Eastern Thibet, its manners, customs, and agriculture, religion, &c., I have collected many details, but I forbear to weary you with them. Your general account is admirable. Plains, as you say, are but local features, and very limited ones; the country is one of stupendous, rugged mountain-chains, and not of plains or table-land. The flat-floored valleys, and scarped flanked mountains, which are of comparatively uniform elevation above the sea, though of widely variable height above the valleys, give it a very different look to the snowy Himalaya. The absence of snow and wood completes the delusion; and, as you truly observe, the traveller who suddenly encounters the features of an open country, the access to which begins abruptly, is very apt to be deceived, and the theodolite alone rectifies the error. The line of perpetual snow is about 15,000 feet, where the masses of the mountains first rise so high; it gradually rises with the increasing height of the land, to 19,000 feet, and 19,500 feet is the lowest mean level in lat. 28° 30′, long. 88° E., and 15,000 in the same long., and lat. 27° 30'. I have seen well-defined patches quite exposed, lower down on the south flanks of Kinchin. Glaciers abound, but not towards the southern limits of the perpetual snow, where the mountains are too rugged.

It is a remarkable fact that every river-valley I have ascended opens out, and is broad and flat-floored at about 10,000 feet: it becomes imperceptibly more and more so, in ascending; and as it conducts you beyond the limits of clouds, rain, and snow, it gradually assumes a dryer, barren, and more Thibetan character.

The average rain-fall at Darjeeling is 120 inches, and it is progressively less in the interior, till the mountains reach 15,000 feet, and then it suddenly diminishes in quantity.

The first snow falls in the Sikkim-Himalaya at 15,000 feet, in September, and does not melt always. In Sikkim the snow-line descends in October, in Thibet not till December; in the drier northern parts, it again melts up to 17,000 feet. In Thibet, as far north as Kiong-lah, it also often falls in the end of August and September every year, and melts again to 19,000 feet. As the rains are not over till October, the sporadic falls are numerous, and quite disguise the perpetual snow.

You have well appreciated the several complicated phenomena of precipitation, evaporation, solar and terrestrial radiation. These all act differently, both as to amount and duration, and affect different times of the year, north and south of the snowy belt. Still, these effects are progressively greater, in going from the snows of the Himalaya to those of Thibet. Comparing Sikkim with the northwestern Himalaya, I should say that the snow-line is lower in Sikkim, because far more falls, and the sky is more cloudy. With regard to the mountain axis in Asia, I do not doubt it is the Himalaya, *i.e.*, the mountain mass between the Yarou and plains of Bengal. The line of mean greatest elevation is probably 18,000 feet: it is north of the snowy Himalaya, and zigzag. Proportionally very little of the snowy Himalaya rises above 18,000 feet; and the 300 feet of that which is exposed to view being always snowed, and always projected to the eye in a straight line, the delusive effect is that of a mountain-chain, on this side the main axis and water-shed.

I am here joined by my old friend and schoolfellow Dr. Thomson, of the Thibet mission; a man of great enthusiasm, and of the highest scientific attainments. He had, independently, adopted the same view of the snowy Himalaya as I did in Sikkim, and all we can see from these (the Khossya) mountains, appears to confirm it. Dr. Thomson has visited the Karakoraur Pass, and finds it to be as laid down in your map to 'Asie Centrale.' He regards the Karakoraur and Muztagh as the real continuation of the axis, which runs north-west from Mansarowar. The latter is, indeed, another *nœud*, from which three chains of mountains radiate, one dividing the Indus from the Sutlej, another north of the former river, and another south of the latter. In fact, the course of the Indus and Sutlej much resembles that of the Soubansiri and Yarou, the latter being the Dihong, and finally the Burrampooter. With regard to atmospheric phenomena, I found the south-east wind to prevail up to 18,000 feet; whereas at 24,000 the westerly current is perennial. The south-east is all but perennial in Sikkim, but is considerably checked by the Himalaya; still it keeps Thibet cloudy and showery from May till October.

The south-east monsoon hardly reaches Sikkim, the currents being generally between south and south-west. These, passing between the Garrows and Rajmahal mountains, cause the Sikkim-Himalaya to be much wetter than the Bhotan mountains, which are sheltered by the Khossya. In the latter, the rain-fall is excessive, 350 inches last year at Churra, and 120 this last JULY. I have several times carefully measured 11 inches in 12 hours! There is no error in my instruments or those established at Churra. The atmospheric pressure, temperature, and wet-bulb, I register as often as is possible, without interfering with my other pursuits; and I find that a thermometer buried at three feet, gives very valuable results.

The atmospheric tide is uniform as to time at all elevations, but the amount gradually diminishes, and progressively with the elevation,—at 17,000 feet it appears to be but a few thousandths of an inch. I almost invariably read off all the instruments seven times in the twenty-four hours, and I have horary observations from all heights, up to 16,000 feet; I take the same times as at the Calcutta Observatory.

For geodetical observations I have but little leisure, and confine myself to a careful survey of my route, and occasionally the elevations of remarkable distant mountains. The survey, and numberless barometric observations over all parts of Sikkim, may afford elements for a computation of the mean mass of that section of the Himalaya.

Vegetation I have found up to 18,500 feet in Thibet. Herbaceous plants are sufficiently abundant at 18,000, in certain spots; and of shrubby ones, I have several species of *Lonicera* and *Rhododendron* from upwards of 17,000 feet, with *Gnaphalia* and *Ephedra*. A nettle (*Urtica*) attains this elevation, as do *Zannichellia* and *Ranunculus* in water. *Compositæ* are, however, much the most alpine natural order; many genera, *Gnaphalium, Sdussurea, Artemisia*, and *Erigeron*, ascending to 17,000 and 17,500 feet, together with *Astragalus* and *Valeriana* (*Nardostachys*). A coleopterous insect, allied to *Meloe* ? and *Acarus*, are found inhabiting 18,500 feet, along with the Kiang, or wild ass, and various ruminants; there, too, the fox, hare, and two species of smaller rodents burrow, all of which have parasites. At 17,000 feet bees live and feed; *Diptera* are common; *Ephemera* is found, and *Papilionidæ* of two genera (*Polyommatus* and *Argyuris*). The absence of the large *Carabidæ*, which occur in the European Alps, is a striking anomaly. At 16,000 feet I have observed the house-fly and *Lumbricus*. The musk-deer I never saw at great elevations, not even in midsummer above 12,000 feet; there is, however, another species in Thibet. A monkey inhabits the Pine-woods at 11,000 feet, but is more common at 9,000. The limits of the *Pines* are highly curious and well marked, differing both from Bhotan and the northwestern Himalaya. There is no *Deodar*, nor *P. excelsa* or *Gerardiana*, nor *Cupressus torulosa*; whereas the *C. funebris* ? of China is commonly planted, and there is a wild Larch (*Pinus longifolia*) 1–4,000 feet; Yew (*Taxus*), 7–10,000 feet. *Abies Brunonis*, Larch, and *Abies Khutrow* (alias *Smithiana*), all grow between 8–11,000 feet. An erect Tree-Juniper and fruticose *Rhododendron*, eight feet high; Birch, Willow, and *Pyrus*, ascend to near 15,000 feet in favourable situations, but are all stunted. *Loniceræ* are, however, the commonest shrubs at 14,000 feet, of several species, with *Rosa* and *Berberis*.

My geological notes want all to be worked up. Gneiss is the prevalent formation between 7 and 17,000 feet; above that, encrinitic granites, and in Thibet, fossiliferous limestone and tertiary rocks. Below 7,000 feet come the mica and clay-slates, sandstone, and coal. There is no lime in Sikkim proper, except as a deposit. Hot springs, of temperature 110°, I have found at 17,000 feet, with sulphurous salts. The most curious and novel geological feature consists in the magnificent moraines, which abound in every valley, descending from the snow. They commence at 10–11,000 feet, and may thence be traced continuously up to the glaciers themselves, whose lower edges, often twenty miles higher up the valley, are always at about 17,000 feet. The proofs hence appear clear of a gradual *sinking* of the Himalaya: I could obtain no signs of a rising, but abundant evidence of the contrary. The glaciers have certainly receded 7,000 feet.

And now it is high time to draw this long letter to a close; I fear it is but a poor attempt to inform you of what I am doing. Dr. Thomson and myself intend to continue our explorations here till November, and then to visit Cachar, Chittagong, and Arracan, from whence we embark for Calcutta, and England in the early spring of next year. Our conjoined collections will be enormous. His consists of 4,000 species, from the Plains of India, Thibet, the north-western Himalaya, Kashmir, and Lahore; and mine of nearly as many, from the Sikkim-Himalaya and Bengal. We have since, unitedly, gathered here upwards of 3,000, and we expect perhaps to add another 1,000 in Chittagong and Arracan. The vegetation is wonderful; we have twenty *Oaks* alone from the Khossya mountains, and an equal number from the Himalaya mountains.

Mr. Hodgson, of Darjeeling, always desires his grateful acknowledgments to you.

I rejoice to hear of your continued good health. The news of his Prussian Majesty's accident reached me, and I thought it would afflict you very much. It is a matter of rejoicing that a sovereign who has the true interests of science so much at heart should have recovered.

Believe me, with feelings of grateful and affectionate regard,
Your faithful and obedient friend,
JOSEPH D. HOOKER.

Plants for Sale.

Mr. Samuel Mossman has lately returned from New Zealand and Australia, with a considerable number of dried plants, which he has placed in the hands of Mr. S. Stevens for sale. The New Zealand collection is from the Northern Island, and contains more than 100 species. The Australian collection consists of South Australia, New South Wales, and Van Diemen's Land plants, amounting to about 400 species.

We learn with much satisfaction, that Mons. E. Bourgeau, the botanical traveller and collector for the Société Française d'Exploration, has arrived in Paris with a fine collection of plants from the mid-eastern parts of Spain.* The address of Mons. Bourgeau is at "*No.* 11, *Rue des Blanc-manteaux, Paris.*" As a judicious collector, and as a highly liberal distributor of his specimens, Mons. Bourgeau is already very favourably known in the botanical world; so that his present collection may be expected to prove well worthy of consideration. The number of species is not stated; but in his former journeys the sets sent to England have run from 400 to 600 species.

Extensive Herbarium.

A vendre, un herbier que l'on évalue à 24,000 espèces, disposées selon la Méthode Naturelle, et se trouvant dans un état de conservation parfaite. De ces plantes, une partie a été ramassée pendant un long séjour en Basse Allemagne et dans les provinces occidentales de l'Empire Russe; une autre et majeure est le fruit de voyages longues et pénibles dans la Russie Asiatique, surtout dans la chaîne de l'Altaï jusqu'aux frontières de la Chine; une troisième est acquise par l'échange et l'achat de botanistes voyageurs en différens pays. Par conséquent, on y trouve presque toutes les plantes de l'Allemagne, la plupart de celles de l'Europe septentrionale et méridionale, des Etats Unis tant orientaux qu'occidentaux, puis une partie considérable de celles du Cap de Bonne Espérance et d'autres parties du monde; auxquelles s'ajoutent de nombreux échantillons pris dans les jardins botaniques du Nord de l'Allemagne. Les exemplaires sont aussi nombreux, aussi bien choisis et bien desséchés, que variés selon les localités. En outre, ils sont dénominés avec tout le soin possible, en quoi ceci leur donne une valeur particulière, qu'ils renferment tous les matériaux qui ont servi aux travaux scientifiques du possesseur, dont les résultats s'impriment depuis longtemps et continuent de paraître. Pour d'autres détails, comme pour apprendre les conditions, sous lesquelles on pourra acquérir cette collection, on est prié de s'addresser par de lettres affranchies à M. L. C. Treviranus, Professeur de Botanique à l'Université de Bonn, en Prusse Rhénane.

[The Editor of this Journal can testify to the great value of this herbarium, in regard to the beauty of the specimens and to its perhaps unrivalled richness in northern European and northern Asiatic plants, no less than to its authentic importance in regard to specific identity.]

* This collection is now ready for delivery; and there will accompany the sets to the accustomed subscribers 110 species of plants from Algiers and the lesser Atlas: of which a large proportion are *bulbous*, authentically named by M. Durieu.

Contributions to the Botany of WESTERN INDIA;
by N. A. DALZELL, Esq., M.A.

(*Continued from* vol. ii. p. 844.)

Nat. Ord. AURANTIACEÆ.

PIPTOSTYLIS. Genus novum, cum icone.

TAB. II.

GEN. CHAR. *Calyx* 4–5-fidus. *Corollæ petala* 4–5. *Stamina* 8–10, libera: *filamenta* alterna breviora, basi complanato-dilatata, apice subulata; *antheræ* ovales, biloculares, longitudinaliter dehiscentes. *Ovarium* rotundatum, toro stipitiformi impositum, *triloculare*; *ovula* in loculis gemina, collateralia, ex apice anguli centralis pendula. *Stylus brevis, crassus, cum ovario articulatus, caducus. Stigma* stylo haud crassius. *Bacca* abortu bi-uni-locularis, loculi abortu 1-spermi. —Frutex 6–7 ped. altus, inermis; foliis imparipinnatis; foliolis alternis, *subcoriaceis*, glabris, crenulatis; paniculis terminalibus, corymbiformibus, elongatis.

Piptostylis *Indica*, Nob. (TAB. II.)

Calyx parvus, segmentis rotundatis, ciliatis. *Alabastrus* sphæricus. *Corollæ petala* ovali-oblonga, 1½ lin. longa. *Stamina* cum petalis inserta, breviora, petalis opposita. *Foliola* 2–4-juga, alterna, ovata, obtuse acuminata, apice emarginata, breve petiolulata, glaberrima, nitida, pellucido-punctata, 2–4 poll. longa, 1½–2 poll. lata. *Paniculæ* corymbiformes, folium æquantes, paniculæ rami pedicellique velutini.—Crescit in Canara; fl. Martio. Fructum maturum non vidi.

This genus is intermediate between *Sclerostylis* and *Bergera*, agreeing with the former in the structure of its flowers and ovary, but having the whole habit and appearance of *Bergera*. From the former it differs in the character of its inflorescence, in having a caducous style; from the latter, in its three-celled ovary, its short thick style situated in a depression of the ovary and *wholly* caducous; while the style of *Bergera* has the base persistent, the upper part only falling off. The ovules of *Bergera* are said to be solitary, and to be attached by their *middle* to the *middle* of the axis. Had I been able completely to verify these statements, they would have afforded important distinctions;

but they seem to me to rest partly on imperfect observations. I was naturally led to examine carefully the structure of *Bergera Königii* in connection with this plant, and I found the ovules *suspended* from very near the apex, *to the top of the cell*. I examined fresh specimens, both wild and cultivated, and was not a little surprised to find in the *latter*, that the cells of the ovary *have often two collateral ovules*. Even if I had not observed this fact, I might have justly inferred that the ovules, when solitary, were so only by abortion, from the structure of the dissepiment and the position of the solitary ovule. The dissepiment is thin, membranous, and diaphanous; down the centre run two stout parallel veins, which give origin to the ovules. Where one ovule only is present, it is found entirely *on one side* and attached to one vein only; and from this want of symmetry we are justified in concluding that twin ovules is the normal condition in each cell, and, under favourable circumstances, this is sometimes actually the case. The obvious affinity, therefore, of *Bergera* to the plant under consideration, deducible from similar habit and appearance, is increased by this discovery. What appears to be another species of this new genus, is a shrub described by Junghuhn, in his 'Reisen durch Java,' as growing in the cemeteries of the Chinese in that island, and which he calls "genus novum *Sclerostyli* affine."

TAB. II. fig. 1, flower-bud; fig. 2, flower; fig. 3, stamens; fig. 4, pistil; fig. 5, vertical section of ovary; fig. 6, transverse section of ovary.

Nat. Ord. SANTALACEÆ.

SPHÆROCARYA.

S. *leprosa*; foliis oblongis coriaceis glabris basi subrotundatis apice acutis (cum petiolo 6–9 lin.) 7 poll. longis 2 poll. latis, floribus in tuberculo axillari squamoso subsessilibus glomeratis, calycis laciniis brevissimis semiorbicularibus lacerato-ciliatis, corollæ petalis linearibus acutis apice recurvis sesquilineam longis, filamentis petalis oppositis iisque adnatis, fructu juniore pyriformi *leproso*, adulto sphærico.—Crescit in Canara; floret tempore frigido.

This, which is a large tree, differs from the only published species in points of considerable importance, among which the existence of a distinct permanent calyx and well-developed petals, or, at. least, parts occupying those relative positions, are not among the least.

The calyx is five-lobed, and alternate with the lobes are five petals, many times longer than the calyx-lobes, and caducous. The filaments are long, *adnate to the petals* throughout their whole length (as in *Embelia* and others of the *Myrsineæ*), the anthers appearing sessile about one-third below the apex of the petal. The disc, which is cushion-shaped and slightly five-lobed on the summit, has no scales of any kind; the stigma is simple. Above the base of the very short pedicel are two half-sheathing short bracts, the one a little above the other, and of the same texture as the lobes of the calyx; while the base of the pedicel is surrounded by 3–4 scurfy-looking scales. The podosperm arises from the base of the cavity, and is *straight*, bearing three pendulous ovules at its apex, two of which are always abortive. The seed is spherical, with a copious albumen, the embryo not being in the axis and at the apex, as in the published species, but lying obliquely, the radicle being centrifugal, as in *Rubiaceæ*, and forming an angle of 45° with the vertical axis of the seed.

I may take this opportunity of mentioning that I met with the *Scleropyrum Wallichianum* in Canara, and that it is not a tree, as Rheede supposed, but a weak straggling shrub, often having unbranched shoots twenty feet high, and apparently of one year's growth. As the structure of the wood of this shrub is very peculiar, and a knowledge of it may tend to throw light on its natural affinities, which I believe are not supposed to be incontrovertibly established, I shall describe it. A transverse section of a stem, an inch and a half in diameter, has an ample pith half an inch in diameter, of a yellow colour, and more dense than that structure generally is. The medullary rays are exceedingly numerous, one-sixth to one-tenth of a line in breadth, and alternate with rays of woody tissue of an equal breadth, or even narrower. In the woody tissue are seen numerous porous vessels or ducts, discernible with the naked eye: this latter circumstance, together with the cellular tissue of the medullary rays equalling or slightly exceeding in bulk the whole woody tissue, produces a wood as light as cane. Except the walls of the duct, no spiral vessels or vessels of any other character are visible;—this structure is very like that of some species of *Loranthus*. The structure of the seed and the position of the embryo in *Scleropyrum* are the same as in the plant above described. N.B. The drupe of *Sphærocarya leprosa*, when mature, is spherical, dry, and three-fourths of an inch in diameter.

Nat. Ord. SAPOTACEÆ.

BASSIA.

B. *elliptica*; foliis ellipticis vel elliptico-obovatis breve obtuseque acuminatis petiolatis coriaceis utrinque glaberrimis, pedicellis axillaribus geminis v. ternis petiolo 3–4-plo longioribus, fructiferis erectis.

Folia (cum petiolo 6–9-lin.) 3–4½ poll. longa, 1½–2½ lata. *Calyx* 6-partitus, pilis fulvis adpressis dense villosus, lobis biseriatis, exterioribus late rotundato-ovatis transverse rugosis, interioribus ovato-lanceolatis exterioribus paulo longioribus. *Filamenta* uniserialia, brevissima, villosa, corollæ summo tubo inserta, lobis corollinis opposita, per paria connata, lobis alterna, solitaria. *Antheræ* longe cuspidatæ, apice erosò-denticulatæ, limbi lobis paulo breviores. *Corollæ tubus* 2 lin. longus, limbi lobis lineari-oblongis acutis paulo brevior. *Fructus* oblongus, glaber.

HAB. Canara; fl. et fr. Februario.

Nat. Ord. RHAMNEÆ.

VENTILAGO.

V. *Bombaiensis*; ramulis petiolis floribusque fulvo-tomentosis, foliis lanceolatis basi acutis inæqualibus apice acuminatis crenatis (crenaturis calloso-mucronulatis) utrinque glaberrimis nitidis, *floribus in foliorum axillis* fasciculatis (15–20) breve pedicellatis, pedicellis petiolo brevioribus.

Folia coriacea, (cum petiolo 3-lin.) 3–3½ poll. longa, 1 poll. lata; venæ costales paucæ (utrinque 3), alternæ, earumque *axillis glandulis rufo-villosis* præditis. *Pedicelli* 2 lin. longi. *Calycis laciniæ* media facie valde cristatæ.—Crescit in montibus Syhadree, prope Chorla Ghaut; fl. Februario.

This species is distinguished at first sight from *V. Madraspatana* by very different inflorescence, the flowers being here collected in the axils of each leaf, and not on long bare terminal branchlets, as in *V. Madraspatana*. The leaves of both species are much alike in size and form, but in the new species they are narrower, more deeply crenated, and less finely acuminated. On a nearer inspection, the differences are more marked, the costal veins in *V. Madraspatana* are double in number, and form a greater angle with the central rib, and are not furnished with glands in their axils, like the new species. The fruit in the new species, I am informed, is much larger, but I have not met with it.

Nat. Ord. RUBIACEÆ.

SAPROSMA. (Saprosme?)

S. Indica; ramis teretibus dichotomis glabris, foliis oppositis sessilibus obovato-ellipticis basim versus cuneatim attenuatis apice acutis glabris integris marginibus recurvis, stipulis utrinque solitariis intra-petiolaribus caulem vaginantibus, floribus terminalibus fasciculatis paucis brevissime pedicellatis, pedicellis basi squamis rigidis glabris suffultis.

Frutex 3–4-pedalis. *Folia* opposita (altero valde minore), 2–4½ poll. longa, 9–22 lin. lata; venarum costalium axillis glandulis cavis præditis. *Calyx* 1 lin. longus, cupuliformis, 4-dentatus, dentibus longiusculis triangulari-subulatis patentibus, sinubus inter dentes latis rotundatis. *Pedicelli* semilineam longi. *Ovarium* biloculare; loculis uniovulatis, ovula e basi dissepimenti adscendentia. *Bacca* ovalis, dentibus calycinis coronata, lævis, cærulea, fœtidissima, abortu monosperma, semine globoso, raro disperma, et tum seminibus dorso convexis facie planis; testa reticulato-venosa, crustacea. *Embryo* in albuminis cornei parte inferiore positus, rectus; cotyledonibus foliaceis parvis ovalibus; radicula infera, longa, albumine dimidio brevior, basim versus incrassata, fere conica, discoideo-truncata.— Crescit in montibus Syhadree, prope Chorla Ghaut, lat. 15° 30'.

Nat. Ord. MALPIGHIACEÆ.

ASPIDOPTERIS.

A. Canarensis; foliis late lanceolatis acuminatis coriaceis glaberrimis, *floribus axillaribus simpliciter racemosis, racemis solitariis pollicaribus*, pedicellis paulo supra basim bibracteolatis, ovarii valleculis inter alas pilis rufis villosis, samaris ala orbiculari emarginata venoso-reticulata præditis, dorso ala minore semiovata cristatis.

Rami glabri. *Folia* 2½–3 poll. longa, 14–15 lin. lata, basi rotundata, adulta glaberrima, epunctulata, juniora utrinque pilis anguste fusiformibus medio affixis rufis conspersa. *Rachis* rufo-villosus, petiolum bilinearem vix superans; pedicelli longi, **graciles**, glabri. *Calycis* *laciniæ* ovatæ, obtusæ, dorso rufo-villosæ, marginibus glabræ, semilineam longæ; petala oblonga, obtusa, calyce 4-plo longiora. *Fructus alæ* orbiculares, scariosæ, diametro sesquipollicares.—Crescit in Canara; fl. et fr. Februario.

Nat. Ord. EUPHORBIACEÆ.

GLOCHIDION.

G. *tomentosum*; ramis flexuosis, foliis breve (2 lin.) petiolatis rotundatis ovatis vel ellipticis acutis supra sparse subtus densius tomentosis 2–4 poll. longis 1½–2 poll. latis, floribus axillaribus fasciculatis, masculis paucis.

Frutex 3–4-pedalis, strictus, basi nudus, apicem versus parum ramosus. *Perigonii masculi* foliola 6, carnosa, æquilonga, recurva, 3 exteriora rotundata, duplo latiora, dorso pubescentia, interiora ovata, glabra; pedicelli pilosi, 4–5 lin. longi; antheræ ut in genere. *Perigonii fœminei* foliola 4, biserialia, irregularia, apice dentata v. lacerata. *Ovarium* sphæricum, sulcatum. *Styli* 5 in unum coaliti. *Stigmata* 5, simplicia; *ovarium* 5-loculare, loculi 2 ovulati. *Pedicelli* 2 lin. longi.—Crescit in Canara; fl. Feb., fr. Aprili.

Nat. Ord. ACANTHACEÆ.

NOMAPHILA.

N. *pinnatifida*; tota glanduloso-pubescens, foliis petiolatis profunde pinnatifidis, floribus in foliorum axillis oppositis solitariis sessilibus et internodiis supremis abbreviatis capitato-congestis, calycis laciniis linearibus obtusis.

Caulis obtuse tetragonus, ad nodos tumidus. *Folia* lineari-lanceolata, 3 poll. longa, 1½ poll. lata, pinnatisecta, segmentis 6–8-jugis lineari-oblongis obtusis serrulatis. *Bracteæ florales* oblongæ, foliaceæ, ciliatæ, glandulosæ, calycem æquantes. *Calyx* 2–2¼ lin. longus. *Corollæ* (5 lin.) palatum bullatum. *Stylus* puberulus, stigma laterale. —*Herba* tenera, 1–2-pedalis.—Crescit in ripis fluminum Concani australioris; fl. Januario et Martio.

Nat. Ord. SCROPHULARINEÆ.

TORENIA.

T. *bicolor*; caule repente radicante, foliis petiolatis triangularibus basi vix cordatis crenato-serratis, pagina supeiiore parce pilosula, calyce lineari incurvo basim versus attenuato æqualiter 5-costato, labio superiore 3-denticulato, inferiore bidentato, *corolla* calyce duplo longiore, filamentorum anticorum dentibus filiformibus fere lineam longis, capsula calyce inclusa lineari.

Flores axillares, bini vel terni, pedunculati; pedunculi floriferi 6–7 lin., fructiferi 9–10 lin. longi. *Corollæ* uncialis tubus curvatus, basi valde attenuatus, ultra calycem inflatus. *Labium* superius rotundatum, integrum, cum tubo *intense violaceum*, inferius 3-fidum, laciniis rotundatis, *albis*. *Petiolus* 4–6 lin. longus, lamina 8–10 lin.—Crescit prope Vingorla; fl. Julio.

A really beautiful species, and well worthy of cultivation. The corolla is shaped something like *Æginetia Indica*, and is more fleshy than in *T. Asiatica* and *T. cordifolia*, both of which are indigenous in this part of India. I failed in getting ripe seeds of this plant for Kew Garden last year, but hope to be more fortunate this season.

(*To be continued.*)

DECADES OF FUNGI; *by the* Rev. M. J. BERKELEY, M.A., F.L.S.

Decades XXXII., XXXIII.

(*Continued from p.* 21.)

Sikkim-Himalayan Fungi collected by Dr. Hooker.

311. *Agaricus* (Psathyrella) *discolor*, n. s.; pileo hemisphærico primitus vitellino dein expanso versus marginem purpurascente; stipite flexuoso fistuloso versicolore; lamellis latis griseo-purpureis secedentibus. Hook. fil., No. 137, cum ic.

HAB. On the ground and on dead timber. Darjeeling, 7,500 feet. October. Abundant.

Inodorous, tufted, tender, brittle, delicate. Pileus 1–1½ inch broad, at first hemispherical, yellow, then expanded, gradually changing, with the exception of the centre, to greyish-purple, dry, smooth, membranaceous. Stem flexuous, 3–4 inches high, 1–1¼ line thick, shaded with pink and yellow, fistulose. Gills rather broad, ventricose, rounded behind, separating from the stem, greyish-purple.

Allied to *A. atomatus* and *A. vinosus*, Cord., but remarkable for its brighter colours. The gills are always visible below the pileus, in consequence of their projecting beyond the edge.

We have given figures at Tab. III. of *A. triplicatus*, *A. Broomeianus*, and *A. verrucarius*, described in the volume for 1850.

* *Coprinus comatus*, Fr. Epic., p. 242. Hook. fil., No. 51, cum ic.

HAB. On grassy earth. Darjeeling. Abundant in one spot only. May to October.

The figure, at first sight, indicates a new species, distinguished by its annulato-vaginate stem; but on examining the specimens, it is clear that the ring is frequently free, exactly as in *A. comatus*. The stem has a cottony web in the cavity, precisely as in that species, nor can I point out a single distinctive character. I have this species from Bombay, collected by the late Mr. J. D. Campbell.

312. *C. Hookeri*, n. s.; pileo ex ovato campanulato glabro striato; stipite subæquali fistuloso candido exannulato; lamellis confertissimis latis umbrino-purpureis affixis. Hook. fil., No. 101, cum ic.

HAB. In grassy places. Jillapahar, 7,500 feet. July, August. Rare.

Inodorous, brittle, very tender, cæspitose. Pileus $2\frac{1}{4}$–3 inches across, and about as much high; at first broadly ovate, then campanulate, smooth, dry, almost shining, striate, livid brown, becoming purple-brown in age; flesh thin, of the same colour as the pileus. Stem 3–5 inches high, $\frac{1}{4}$ an inch or more thick, hollow, but with the walls thick, pure white, not having the slightest trace of a ring. Gills extremely close, broad, purple-brown, with an umber tinge, affixed, subcrenate. Spores oblong, elliptic, with one side much less curved.

This fine species has much the habit of *C. atramentarius* and *C. deliquescens*, from the former of which it differs in the smooth, almost shining pileus, and the total want of a ring; from the latter in its different form when young, and broad gills. The spores are more elongated than is usual in the genus.

313. *C. vellereus*, n. s.; pileo ovato subcarnoso striato fulvo vellere flocculoso-squamoso secedente prædito; stipite e basi subbulboso attenuato cavo; lamellis lanceolatis primitus albis latiusculis liberis. Hook. fil., No. 115, cum ic.

HAB. On dead wood and earth, as on the mossy edges of Mr. Hodgson's verandah. Darjeeling. August, September. Common.

Inodorous, brittle, tender. Pileus 1 inch or more across, 1¼–1½ inch high, ovate, obtuse, tawny, slightly striate, clothed with a flocculent, squamulose, white fleece, which peels off from the margin towards the apex, rather fleshy; flesh red. Stem 3 inches high, ¼ thick, somewhat bulbous below, attenuated upwards, hollow, white. Gills lanceolate, crowded, rather broad, free.

This species has exactly the habit of *C. domesticus*, but the flesh is thicker and the pileus less striate, its nearest affinity being with *C. extinctorius*, from which it differs in its ovate pileus and rooting stem.

314. *Paxillus chrysites*, n. s.; totus aureus, pileo infundibuliformi molli glabro subvirgato, margine striato; stipite fistuloso; annulo demum caduco; lamellis decurrentibus venosis. Hook. fil., No. 47, cum ic.

HAB. On dead wood. Darjeeling, 7,500 feet. June. Abundant.

Subcæspitose. Pileus infundibuliform, 2 inches or more across, smooth, with a few virgate filaments, very soft and moist, fleshy; margin striate. Stem 2 inches high, ½ an inch thick, smooth, nearly equal, fistulose. Ring rather broad, deflexed, at length falling. Gills moderately broad, venose, decurrent. Spores ochraceo-fulvous, rather small. The whole plant is of a uniform golden yellow, which changes in drying to umber.

A very curious and interesting species, differing from everything with which I am acquainted.

315. *P. sulphureus*, n. s.; sulphureus; pileo cum lamellis obconico plano flocculoso-squamoso carnoso; stipite deorsum attenuato solido; lamellis latis ventricosis adnato-decurrentibus. Hook. fil., No. 119, cum ic.

HAB. On dead wood and on the ground. Darjeeling, 7,500 feet. September. Common.

Inodorous, sulphur-coloured. Pileus 1½–2 inches across, plane, obconical when taken with the gills, sometimes slightly depressed, clothed with little flocculent scales, fleshy. Stem about 2 inches high, ¼–½ thick, attenuated below, strongly dilated above, and confluent with the pileus. Gills very broad, distant, rather ventricose, thick, waved, adnate, decurrent. Spores elongated, oblong, yellow-olive.

A very interesting species, remarkable for its narrow elongated spores, like those of a *Boletus*.

316. *P. pinguis*, Hook. fil.; pileo depresso demum infundibuliformi

luride sulphureo flocculento subcarnoso; stipite brevi subæquali concolore; lamellis sulphureis adnato-decurrentibus. Hook. fil., No. 60, cum ic.

HAB. On earth and mossy banks. Darjeeling, 7,500 feet. June.

Inodorous, soft, brittle. Pileus 2 inches broad, at first convex and depressed, then infundibuliform, subcarnose, of a lurid sulphur-yellow, dry, flocculent. Stem about 1 inch high, ½ of an inch thick, solid, nearly equal, obtuse, of the same colour as the pileus. Gills moderately broad, sulphur-coloured, adnato-decurrent, interstices and sides venose. Spores elongated, oblong.

Nearly allied to *P. sulphureus*, but more clumsy in form, more dingy in colour, with a highly depressed pileus and far narrower gills.

* *Russula furcata*, Fr. Ep. p. 352. Hook. fil., No. 36, cum ic.

HAB. On clay banks. May. Sinchul, 8,500 feet.

Substance firm. Stem and gills snow-white; the former obtuse, below solid or irregularly hollow.

317. *R. grossa*, n. s.; pileo cyathiformi viscoso maculato-squamoso, margine rugoso involuto; stipite crasso obeso subæquali; lamellis decurrentibus antice latioribus integris.

HAB. Darjeeling.

Pileus 4 inches across, cup-shaped, viscid, spotted with innate darker scales; margin narrowly involute, rather rugged. Stem about 3 inches high, ¾ of an inch thick, nearly equal, obtuse. Gills decurrent, broader in front, entire.

Well characterized by its viscid, spotted pileus and coarse habit. Dr. Hooker compares it with *Ag. verrucarius*, Berk. The gills are yellowish when dry, but I cannot ascertain the colour of the spores.

318. *R. cinnabarina*, Hook. fil.; pileo carnoso rigido e convexo depresso sicco opaco, margine substriato; stipite demum subcavo duro rubro flavoque; lamellis candidis antice latiusculis postice rotundatis. Hook. fil., No. 65, cum ic.

HAB. On clay banks. Darjeeling, 7,500 feet. June. Very rare.

Inodorous. Substance firm, between fleshy and corky. Pileus 3 inches broad, at first convex, then strongly depressed, dry, opake, deep vermilion, sometimes shaded with yellow, by no means polished, fleshy but striate at the extreme margin. Flesh tinged with red beneath the cuticle. Stem 2 inches or more high, obtuse, nearly equal, firm, at first solid, at length presenting several cavities, vermilion or yellow. Gills snow-white, broader in front, but not remarkably so, attenuated behind and rounded; interstices reticulated, obscurely if at all forked.

A splendid species, differing from *R. rubra* in its opake pileus, striated margin, and gills not so decidedly broader in front or projecting beyond the margin.

* *R. lepida*, Fr. Ep. p. 355. Hook. fil., No. 5 (pro parte).

HAB. On clay banks. Darjeeling, 8,000 feet. April to July. Scarce.

Firm and almost coriaceous. Pileus dry. Margin quite free from striæ. Gills pure white.

Dr. Hooker, in his notes, compares it with *Agaricus muscarius*. As no remark was made as to the circumstance whether it is mild or acrid, it is impossible to say decidedly whether this is *R. rubra* or *R. lepida*, but I am inclined to think that it is the latter species.

* *R. emetica*, Fr. Ep. p. 357.

HAB. On clay banks. Darjeeling.

Pileus viscid. Texture of stem and pileus loose and spongy. Found with *Russula rubra*.

Another species, which I cannot determine, occurred on banks at Sinchul.

319. *Marasmius iridescens*, n. s.; pileo campanulato subumbilicato sicco polychroo; stipite sursum attenuato cavo; lamellis angustissimis concoloribus. Hook. fil., No. 41, cum ic.

HAB. On mossy banks. Sinchul, 8,000 feet. May. Extremely rare.

Inodorous, singularly beautiful. Pileus 1¼ inch broad, campanulate, slightly umbilicate, dry, smooth, slightly fleshy, not striated, variegated with yellow, grey, and pink. Stem 2¼ inches high, 2 lines or more thick, incrassated below, smooth, hollow, composed of fibres coloured like the pileus. Gills very narrow, rather thick and obtuse, shaded with red and blue, yellow towards to the extremity, free, reddish-brown when dry.

This exquisite fungus, which appears rather to be a *Marasmius* than an Agaric, combines somewhat the characters of Fries's two first sections, without agreeing absolutely with either. The edge of the gills is slightly obtuse, and the whole aspect of the hymenium, when dry, is very much like that of *M. peronatus*. I know of nothing very closely resembling it. Its colours are as various as those of *A. Harmoge*.

* *M. erythropus*, Fr. Ep. p. 378.

HAB. On the ground. Darjeeling.

There are no notes or drawings of this species, but the specimens agree with British individuals. They are much tufted.

320. *M. consocius*, n. s.; gregarius, insititius, totus albus; pileo convexo scabro, margine glabrescente; stipite subæquali furfuraceo; lamellis angustis venosis adnexis, interstitiis reticulatis. Hook. fil., No. 4, cum ic.

HAB. On dead twigs. Darjeeling, 8,000 feet.

White, inodorous. Pileus gregarious, but not fasciculate, forcing up the cuticle of the twigs on which it grows, 1 inch across, convex, coriaceous, speckled in the centre with minute scabrous points; margin smooth. Stem 1 inch or more high, 1-2 lines thick, clothed with furfuraceous scales, especially at the base. Gills narrow, more or less venose, with reticulated interstices, attenuated behind, adnexed.

Resembling in habit *M. ramealis*, but very much larger, and differing in the nature of its gills, pileus, and stem.

321. *M. caperatus*, Berk.; albus; pileo membranaceo plicato; stipite brevissimo furfuraceo fuscescente; lamellis latis adnatis; interstitiis venoso-plicatis. Hook. fil., No. 15, cum ic.

HAB. On twigs of live and dead bushes. Tonglo, 10,000 feet. Abundant. The characteristic fungus of that region in the month of May.

Snow-white, very delicate, inodorous. Pileus 1-1½ inch or more broad, convex, sometimes galeate, membranaceous, plicate and corrugated, smooth. Stem very short, furfuraceous, at length brownish, sometimes obsolete. Gills distant, broad, adnate; interstices strongly plicate and venose.

A species very much resembling this, but having a longer stem, was abundant on wood from the tropics, in a stove at the Royal Botanic Gardens, Kew, two years since. The pileus is soon reflected, resembling in this respect some of the small *Pleuroti*. It is a very beautiful fungus.

* *Lentinus Lecomtei*, Fr. Ep. p. 388. Hook. fil., No. 13, cum ic.

HAB. On wood. Tonglo, 6-8,000 feet. Abundant. May, June.

322. *L. Hookerianus*, n. s.; pileo infundibuliformi, tenui, strigososeteoso cinnamomeo-gilvo; stipite irregulari subrudi strigoso-velutino obscuriore; lamellis angustissimis decurrentibus pallide luteis. Hook. fil., No. 16, cum ic.

HAB. On dead wood. Darjeeling, 6–9,000 feet. Abundant. May to July.

Pileus infundibuliform, 2–3 inches across, marked with two or three impressed zones when dry, clothed with cinnamon-coloured strigose bristles, which at length become tawny. Stem 1¼–2 inches high, ½ an inch thick, rather irregular, swollen at the spongy base, subcinereous when fresh, of a dull tawny when dry, strigose-velvety below, more finely velvety above, especially at the termination of the gills, which become smooth abruptly. Gills pale yellow, very narrow, entire, decurrent, not by any means thick or obtuse.

This species differs from *L. Lecomtei* and *L. strigosus* in its narrow gills, in this point agreeing with *L. velutinus* and *L. setiger*, Lév., but the coat is coarser than in the former, and not formed of two kinds of setæ as in the latter. The gills are also more delicate than in either. It is not so graceful a fungus as *L. velutinus*, and, unlike *L. Lecomtei*, has no tendency to be excentric.

323. *L. coadunatus*, Hook. fil.; cæspitosus; pileo infundibuliformi sicco tenui albo opaco glabro; stipitibus basi coadunatis furfuraceis æqualibus cavis; lamellis angustis tenuibus utrinque attenuatis decurrentibus. Hook. fil., No. 68, cum ic.

HAB. On dead wood. Darjeeling, 7,500 feet. June, July. Frequent.

Inodorous, rather delicate, soft when young, leathery when old, densely tufted. Pileus 2¼–4 inches across, infundibuliform, smooth, thin, opake-white, especially when dry. Stems united at the base, equal, 1–2¼ inches high, ¼ of an inch thick, tinged with rufous, hollow. Gills narrow, pale, attenuated at either end, thin, strongly decurrent.

This species resembles *L. leucochrous*, and *L. cladopus*, Lév., agreeing with the former in its opake appearance, and with the latter in its branched stem; but it is a larger species than either, and, though closely connected, distinct, differing from the former in its stem, from the latter in its milky-white pileus. Fortunately, I have authentic specimens of both species. As Léveillé's two *Lentini* have been seen only when dry, it is impossible to say whether they have a hollow stem, as in the present case.

324. *L. hepaticus*, n. s.; pileo convexo-explanato udo glabro hepatico; stipite brevi æquali solido obscuro intus substantiaque pilei pallide rubro; lamellis confertis integris postice rotundatis albis demum hic illic incarnatis. Hook. fil., No. 53, cum ic.

Hab. On trunks of trees. Darjeeling, 7,500 feet. June, July.

Inodorous. Pileus 3 inches across, convex, at length more or less expanded, smooth, moist, but not viscid, of a reddish liver-brown, sometimes slightly zoned. Substance thick, rather tough and leathery, pale red. Stem 1-1¼ inch high, ¼-½ of an inch thick, hard, solid, nearly equal, blunt, of a browner tint than the pileus. Gills much crowded, moderately broad, commencing beyond the margin, entire, rounded behind, white or shaded with pink.

The pileus is sometimes excentric, with an ascending stem, sometimes quite regular. Its general habit indicates the division *Pleurotus* of the Agarics, but its tough leathery substance *Lentinus*. Its colours remind one of *Lentinus resinaceus*, Trog, and the regular individuals resemble at first sight *Lactarius subdulcis*.

325. *L. subdulcis*, n. s.; albus; pileo imbricato multiplici-lobato estriato glabro, margine incurvo; stipite spurio vel obsoleto; lamellis latiusculis subdistantibus. Hook. fil., No. 25, cum ic.

Hab. On dead wood. Darjeeling, 7-8,000 feet. May. I have what I believe to be the same species, but in a bad state, from Ceylon.

White, imbricated, subflabelliform, variously lobed. Pileus 3 inches broad, smooth, not striated, somewhat depressed behind; margin incurved; substance thin, but tough. Stem obsolete, or consisting of a short, thick, irregular, horizontal projection from the vertex. Gills rather distant, broadish, slightly rounded behind; edge entire, occasionally the gills are somewhat decurrent.

Fries makes the difference between *Panus* and *Lentinus* to consist in the respectively unequal and entire, or equal and toothed gills. This character is, however, by no means constant; *Lentinus Djamor*, to which this species is nearly allied, several of Léveillé's species, not to mention the whole section *Scleroma*, have gills as entire as in any *Panus*. The present is a very pretty species, distinguished from *L. Djamor* by its highly lobed pileus and involute margin. It has a faint, rather sweet odour, from which I have taken its name.

326. *Panus monticola*, n. s.; solitarius; pileo depresso lobato tenui sicco glabro pallido; stipite erecto elongato centrali solido velutino; lamellis angustis integris descendentibus.

Hab. On the ground, probably attached to wood. Tonglo.

Solitary. Pileus 2-3 inches broad, pallid, thin, smooth, lobes depressed or subinfundibuliform. Stem 2 inches or more high, very variable in thickness, sometimes very slender, solid, minutely velvety. Gills narrow, pallid, descending, but not ending abruptly or altogether.

A pretty species, resembling some states of *Panus torulosus*, but solitary, more graceful in habit, and entirely destitute of the violet or lilac tinge which is more or less conspicuous in that species. It is difficult to draw up very distinctive characters, but there is every reason to believe that the two species are distinct.

* *P. conchatus*, Fr. Ep. p. 398.

HAB. Darjeeling.

* *Schizophyllum commune*, Fr. Ep. p. 403.

HAB. On dead wood. Darjeeling, 7,000 feet.

327. *Xerotus cantharelloides*, n. s.; pileo subinfundibuliformi fuligineo subtiliter fasciculato-tomentoso; stipite gracili obscuro glabriusculo; lamellis ochraceis dichotomis angustissimis obtusiusculis decurrentibus.

HAB. On dead wood. Jillapahar. June.

Pileus 1½ inch broad, subinfundibuliform, fuliginous, clothed with minute fasciculate down; margin involute. Stem 2½ inches high, 2 lines thick, darker than the pileus, nearly smooth, rooting slightly at the base, and attached to particles of decayed wood by a few flocci. Gills ochraceous, extremely narrow, dichotomous, rather obtuse, decurrent, grooved towards the base. Spores white, ovate.

This curious species bears some resemblance to *Cantharellus aurantiacus*. The gills are grooved as in *Trogia*, in consequence of the lines of division being carried backward along the edge of the gills from the point of bifurcation.

* *Lenzites repanda*, Fr. Ep. p. 404. Hook. fil., No. 14, cum ic.

HAB. Hot valleys, on trunks of dead trees, abundant, perennial, 2,000–5,000 feet.

The margin in fresh specimens, as figured by Dr. Hooker, is of a deep cinereous tint.

* *L. betulina*, Fr. Ep. p. 405. Hook. fil., No. 29, cum ic.

HAB. Dead timber. Darjeeling, 3,000 feet. Perennial.

328. *L. rugulosa*, n. s.; pileo sessili subflabelliformi postice crassiusculo antice acutissimo radiato-rugoso subzonato albido-fuscescente; poris mediis, dissepimentis tenuibus subalutaceis demum lamelliformibus.

HAB. On trunks of trees. Darjeeling.

Pilei 3 inches long, 2¼ broad, sessile, laterally connate, subflabelliform, sometimes lobed, thick behind, very acute in front, slightly zoned, rough with radiating wrinkles, smooth, somewhat shining, whitish, changing in parts to brownish. Hymenium pale tan-coloured; pores behind entire, in front much elongated, their dissepiments thin, lamelliform.

A very pretty species, with somewhat the habit of *Lenzites applanata*, but altogether more delicate, and, indeed, less faithfully exhibiting the type of the genus.

329. *Boletus Emodensis*, n. s.; pileo primitus ovato-globoso, volva universali demum deorsum circumscissa et apicem stipitis vaginante obtecto; seniori expanso hemisphærico dense squamoso tomentoso ruberrimo, margine appendiculato; stipite elongato flexuoso æquali e mycelio spongioso enato; poris flavis amplis liberis; carne leviter cærulescente. Hook. fil., No. 100, cum ic.

HAB. On the ground. Darjeeling, 7,500 feet. July and August. Common.

Mycelium thick, spongy, reddish-brown. Pileus at first ovato-globose, covered entirely, as well as the stem, with the universal volva. This bursts regularly below as the stem elongates, and forms at its apex a close sheath. The pileus soon becomes hemispherical, and at length expanded and several inches broad, retaining the volva in large coriaceous shreds at its edge. Its surface is dry, of a vivid strawberry-red, and densely clothed with downy scales. Stem 6 inches or more high, 1 inch or more thick, nearly equal, flexuous, rather rough and tomentose, red, like the pileus, but paler, yellow above, within reddish-brown at the base, above white or pinkish, slightly changing to blue when cut. Inodorous. Pores rounded behind, ample, yellow within and without. Spores fusiform, ochraceous-yellow.

One of the most splendid of the Himalayan fungi, and the pride of its genus. It is perhaps most nearly allied to *B. chrysenteron*, but its very peculiar volva, not forming a ring round the stem, but perfectly free, its spongy mycelium, and brilliant scaly pileus separate it from all known species.

Tab. III. *B. Emodensis*, *nat. size*, in different stages of growth.

330. *B. ustalis*, n. s.; pileo convexo fuligineo-atropurpureo tomentoso; stipite subæquali obeso rugoso-reticulato nigrescente; poris adnatis ochraceis. Hook. fil., No. 122, cum ic.

HAB. On rotten trunks of trees. Darjeeling, 7,500 feet. September. Rare.

Pileus 2–3 inches or more across, convex, of a sooty purple-black, dry, closely tomentose; margin at first involute. Stem 3 inches high, 1 inch or more thick, nearly equal, very coarsely reticulated, darker than the pileus; substance white, but dark at the base and towards the margin, as also at its junction with the pileus. Pores ochraceous within and without, adnato-decurrent. Spores fusiform.

A highly curious species, remarkable for its habit, dingy colour, and coarsely reticulate stem. Its affinities are very obscure.

Catalogue of CRYPTOGAMIC PLANTS *collected by* PROFESSOR W. JAMESON *in the vicinity of Quito; by* WILLIAM MITTEN.

In preparing for distribution the very valuable collections of Dr. Jameson considerable delay has unavoidably occurred, for I have been anxious to obtain as correct an idea as possible of all the species described by the late Dr. Taylor; and the great facilities I have had of examining both the specimens in his herbarium before it left England, and afterwards the *Hepaticæ* in Sir W. J. Hooker's herbarium, most of which had been through the hands of Dr. Taylor, made me reluctant to close my labours before I had gone over all the species which he had named and described. And I must here observe, that in the instances in which I have differed from his decisions, I beg to bear testimony to his great acuteness in appreciating minute differences; but the rocks upon which the greater portion of his species will eventually be lost are want of comparison with specimens already in his herbarium, and of reference to the works of other observers. On the last account, however, something must be allowed, as he was, by his residence in the south-west of Ireland, cut off from the opportunity of consulting many of the rarer or more costly works of continental botanists; indeed, Dr. Taylor appears to have trusted far too much to his memory.

Several sets of these collections having been examined by Mr. Wilson before the remainder came into my hands, I applied to him for a list of the species he had found in them, which he immediately supplied me with; and as none of the sets corresponded in all particulars, I have adopted the numbers used in his list, as far as possible, for the sake of uniformity;—and, in all cases wherein I have come to a different conclusion, I have done so with the greatest deference to his experience, so much greater than my own: I also wish it to be distinctly understood, that I am alone responsible for all the errors or oversights that may be found in the descriptions of the species to which Mr. Wilson's authority is attached, the names for which I have taken as they stood in the list with which he obligingly supplied me.

The species in these collections have no special localities assigned to them, with few exceptions; but they may be understood to be from the vicinity of Quito, and the neighbouring volcanic mountain Pichincha, localities which appear to be most prolific in *Cryptogamia*; and I perfectly agree with Dr. Taylor in saying, that Professor Jameson is "justly entitled to the gratitude of Bryologists for the trouble he has taken to send so many interesting and valuable specimens for their admiration."

1. Harrisonia *secunda*, var. *cirrhifolia*.— Hedwigia secunda, *Hook. Musc. Exot.* vol. ii. p. 3. t. 46. Braunia cirrhifolia, *Wils. MSS.*

2. Polytrichum *Antillarum*, Rich. Brid. vol. ii. p. 138 et 747.

   b. Polytrichum *commune*, Linn. var.?

   A few barren stems of this moss have been sent mixed with the preceding, and may, perhaps, belong to *P. commune*.

3. Polytrichum *polysetum*, Hook. et Arnott.—P. Jamesoni, *Tayl. Lond. Journ. Bot.* 1848, p. 188.

   I have not been able to find any description of this species, but it was so named in Hbm. Taylor and in Mr. Wilson's list. It is probably *P. oligodus*, Kunze, and agrees in habit with *P. perichætiale*, Mont., and *P. Simense*, Bruch. et Sch.: but differs from both in its long cylindrical capsules. *P. Jamesoni*, Tayl., is certainly the same species.

4. Barbula *Quitoensis*, Tayl. Lond. Journ. Bot. 1847, p. 332.

   This species comes near to *B. inermis*, Mont., judging from the figure of that species in Bryol. Europ.; the peristome is free two-thirds only, and not to the base, as described by Dr. Taylor.

5. Barbula *denticulata*, Wils. in schedula; dioica, cæspitosa humilis subsimplex, foliis remotis spathulatis acuminatis, costa in mucronem brevem excurrente, margine e medio ad apicem denticulato, textura tenera molli, basi e cellulis elongatis laxis maxime pellucidis superne minoribus, chlorophyllo fere nullo, theca in pedun-

. culo longo rubro torto cylindrica fusca, operculo oblique longe conico capsulam dimidiam æquante, annulo simplici persistente, peristomio fere ad medium tubuloso aurantiaco.

This pretty species comes nearest to *B. cuneifolia*, but differs in its denticulate and more pellucid leaves, the areolation of which is more lax than in any other species of this section of *Barbula* that I have seen.

STREPTOPOGON, *Wils. in litt.* Calyptra mitriformis, superne scabra; *peristomium* simplex, ciliiforme; *cilia* 32, æquidistantia, in ciliola duo postice fissa, lævia, in spiram unam dextrorsum contorta, basi in membranam angustam coadunata; *cellulæ operculi* contortæ.

6. S. *erythrodontus*, Wils.   Tortula erythrodonta, *Tayl. Lond. Journ. Bot.* 1846, p. 50. Barbula erythrodonta, *Wils. l. c.* vol. v. tab. xv.

In its mitriform, scabrous calyptra, this curious moss resembles some species of *Tayloria*, but the peristome is that of *Barbula*, to which genus it is closely allied.

7. Barbula *elongata*, Wils. in schedula; dioica, elongata parce dichotomè ramosa viridis dein rufescens, foliis caulinis siccitate tortis madefactis patulis e basi erecta elongatis dense reticulatis lanceolatis acuminatis apice obtusiusculis, nervo excurrente, margine reflexo, apicem versus inæqualiter repando-serrulato, e cellulis minutis fere lævibus chlorophyllosis areolatis, perichætialibus erectis longioribus e basi convoluta flavida, theca in pedunculo longo rubro cylindrica elongata curvula, operculo subulato, peristomio in membranam breviusculam circiter bis contorto.

Flos masculus terminalis innovatione axillaris; *antheridia* et *paraphyses* numerosa.

This fine species is remarkable for the length of its setæ, and long cylindrical capsules. The peristome is combined for about one-third of its length.

b. *Idem*. Setæ younger.

8. Barbula *aculeata*, Wils. in schedula; dioica? elata parce ramosa sordide virens inferne rufescens, foliis imbricato-patentibus recurvisve lanceolatis longe acuminatis apicem versus serratis, nervo rubro acumine undulato pallido desinente, setis breviusculis sæpe geminatis, capsula cylindrica, peristomio basi in membranam angustam breviusculam unito.

The specimens of this noble species are few and imperfect. It appears to be about four inches high, and the leaves large in proportion; those at the summits of the stems are of a dull green, the lower ones ferruginous, the setæ are often several from the same perichætium, and short for the size of the moss; the peristome is short, and pale, combined for about one-eighth part of its length.

The nearest allies to this species are *B. robusta* and *B. serrulata*, but it is very distinct from any species with which I am acquainted.

9. Barbula *fragilis*, Tayl. Lond. Journ. Bot. 1847, p. 333.
10. Grimmia *ovata*, Web. et Mohr.
11. Blindia *acuta*, Br. et Sch.; a state in which the setæ are more or less curved.
12. Fissidens *repandus*, Wils. in schedula; dioicus, humilis subsimplex, foliis circiter 8-jugis erecto-patentibus marginatis apice acuminatis recurvis, lamina ½ producta dorsali basi angustata, areolis e cellulis minutissimis viridibus et obscuris, theca erecta vel paululum inclinata ovali, operculo erecto conico subulato, calyptra latere fissa.

I give the above short character, although I have great reason to suppose this moss to be the same as *F. Ceylonensis*, Dozy et Molkb. I have specimens corresponding very nearly from the Neelgherries. This species appears to differ most from its allies, by the acuminate upper portion of the lamina and the minute opake cells.

Fissidens *turbinatus*, Tayl. Lond. Journ. Bot. 1848, p. 190, is *F. flabellatus*, Hsch. in Fl. Brasil. t. 2.

13. Dicranum *vaginatum*, Hook. Musc. Exot. t. 141.—D. Jamesoni, Tayl. Lond. Journ. Bot. 1847, p. 332.
14. Dicranum *areodictyon*, C. Muller, Synops. Musc. Frond. p. 394.— Campylopus nitidus, *Wils. in schedula.*

Agreeing with other species of *Campylopus* in all respects, excepting the calyptra, which is destitute of cilia at the base.

15. Dicranum *brachyphyllum*; cæspitosum gracile strictum lutescens inferne fusco-tomentosum, innovationibus julaceis, foliis caulinis siccitate arcte appressis remotis lanceolato-acuminatis summo apice parce denticulatis, basi cellulis alaribus valde conspicuis, comalibus densioribus sed ejusdem formæ, perichætialibus vaginantibus superne serrulatis, thecis solitariis vel pluribus ex eodem perichætio in pedunculos cygneos obovatis gibbosis siccitate curvulis, operculo conico subulato rubro obliquo, calyptra fimbriata, peristomio ut in *D. flexuoso* sed breviore.—Campylopus brachyphyllus, *Wils. in schedula.*

The slender stems, with the closely appressed leaves, give this species a different look from its near ally, *D. Funkii*, C. Muller; Campylopus fragilis, *Bryol. Europ.* The capsule is much curved, and has some resemblance to that of *D. heteromallum* in form.

16. Dicranum *erectum*, Mitten, MSS.; elatum laxe cæspitosum dichotome ramosum apice falcatum lutescens inferne fuscescens, foliis caulinis secundis falcatis e basi lata sensim longissime acuminatis capillaceis concavis lævibus, basi cellulis alaribus conspicuis, perichætialibus basi convolutis, theca in pedunculo flexuoso erecto oblongo lævi, operculo conico subulato, "calyptra basi fimbriata." —Thysanomitrium macrophyllum, *Wils. in schedula.*

This species is allied to *D. uncinatum* and *D. denudatum*, but is a far larger moss; in size it has some resemblance to a slender state of *D. majus*; the calyptra is not present in any of the specimens I have seen, but Mr. Wilson informs me that it is fringed at the base.

17. Pilopogon *gracilis*, Brid. vol. i. p. 519. Didymodon gracilis, *Hook. Musc. Exot.* t. v.

I was at first inclined to consider this distinct from *P. gracilis*, supposing the capsule to be erect as figured in Musc. Exot., but Mr. Wilson observes that the capsules in the original specimens are not exactly as figured.

18. *Idem.*

b. Dicranum *læve*, C. Muller, Synops. Musc. Frond. p. 409. Campylopus lævis, *Tayl. Lond. Journ. Bot.* 1846, p. 47.

19. Leptotrichum *crinale*. Didymodon crinalis, *Tayl. Lond. Journ. Bot.* 1848, p. 280.

20. Trichostomum *cylindricum*, C. Muller, Synops. Musc. Frond. p. 586. Didymodon calyptratus, *Tayl. Lond. Journ. Bot.* 1848, p. 188.

21. Barbula *Australasiæ*, Hook. et Grev.

Accords with New Zealand and Tasmanian specimens; the peristome of this moss is scarcely at all twisted when the operculum is first removed, but soon becomes slightly so; the structure of the teeth is truly that of *Barbula*, and very different to that of *Trichostomum*.

22. Trichostomum *densifolium*, Mitten, MSS.; monoicum, dense cæspitosum albido-tomentosum luteum inferne fuscescens, foliis patenti-recurvis tortilibus squarrosis late lanceolatis, nervo sub apice evanido, margine e basi ad medium reflexo, apicem versus serrula-

tis, basi e cellulis elongatis angustis firmis flavidis superne minutissime rotundatis areolatis minute et densissime papillosis, perichætialibus longissime convolutis, theca in pedunculo flavo torto cylindracea læviter curvula brunnea, operculo stricto capsulam dimidiam adæquante, peristomii dentibus breviusculis aurantiacis, calyptra straminea.

The densely-set cauline and long convolute perichætial leaves give this species a very different appearance from its congeners, and, except in its being far more robust, it has considerable resemblance to *Barbula calycina*. In the form of its leaves it comes near to *T. luteum*, but differs in habit and in its longer setæ, perichætia, and capsules.

23. Ceratodon *purpureus*, Brid.
24. Funaria *calvescens*, Schw.
    b. Funaria *fuscescens*, Mitten, MSS.; cæspitosum subsimplex, foliis oblongis concavis integerrimis, nervo excurrente, theca magna valde incurva pyriformi siccitate læviter striata, in pedunculo tortili annulato, operculo conico-hemisphærico, peristomio interno e dentibus brevibus abruptis.

Closely as this species resembles *F. hygrometrica* in size and general habit, it differs in the brown colour of its capsules, which are but slightly plicate when mature, and the imperfect internal peristome.

Funaria Jamesoni, Tayl., has, I believe, a clavate, plicate capsule, and a normal inner peristome.

25. Bartramia *rufiflora*, Hsch. B. elegantula, *Tayl.* B. minuta, ejusdem; and probably also B. angulata, *Tayl.* Glyphocarpa alpicola, *Tayl.*

This appears to be the same species as Gardner's No. 45, and to be correctly referred to *B. rufiflora*; the synonyms above quoted, I believe, all belong to this species: *B. minuta* and *Glyphocarpa alpicola* are short states, *B. angulata* and *B. elegantula* appear to be states drawn up amongst other mosses.

26. Bartramia *incana*, Tayl. Lond. Journ. Bot. 1848, p. 189.
27. Bartramia *intertexta*, Schimper.
    A better state at No. 105.
28. Bartramia *Jamesoni*, Tayl. Lond. Journ. Bot. 1847, p. 334. B. viridissima, *C. Muller.* Leucodon bartramioides, *Hook. Icones Plant. Rar.* vol. i. t. 71.

I have not been able to ascertain if C. Muller's name is the earliest, which is probably the case.

29. Bartramia *longifolia*, Hook. Musc. Exot. t. 68.
30. Bartramia *flavicans*, Mitten, MSS.; hermaphrodita, laxe cæspitosa ramosa, ramis strictis dichotomis viridibus et flavescentibus, foliis caulinis erecto-patentibus madefactis strictis siccitate appressis, e basi subquadrata laxa superne dilatata erecta, subito in subulam lanceolatam reflexiusculam minute areolatam chlorophyllosam productis, margine minute serrulato, dorso papillis minutis aspero, nervo lato subulam superiorem totam fere occupante, perichætialibus basi latioribus laxius areolatis, theca in pedunculo elongato rubro globosa sulcata microstoma, operculo convexo-conico, peristomio simplici externo.

In size this species resembles *B. pomiformis*, but is allied to *B. ithyphylla*, *B. patens*, and *B. robusta*, differing, however, from all in its shorter leaves; from *B. patens* and *B. ithyphylla* it further differs in its single peristome, and from *B. robusta*, which is a larger species, in the inflorescence as well as peristome: the capsules are orange and the setæ red, which, with the yellowish colour of the whole plant, give it a very pretty appearance.

31. Bartramia *subsessilis*, Tayl. Lond. Journ. Bot. No. lxvi. p. 334.
    Glyphocarpa lævisphæra, *Tayl. l. c.* 1846. p. 56, is assuredly the same species.
32. Bartramia *integrifolia*, Tayl. Lond. Journ. Bot. 1846, p. 55.
33. Bartramia *tomentosa*, Hook. Musc. Exot. t. 19.
34. Meilichoferia *Jamesoni*, Tayl. Lond. Journ. Bot. 1847, p. 331.
35. Meilichoferia *longiseta*, C. Muller, Synops. Musc. Frond. p. 236.
36. Bryum *Wilsoni*, Mitten, MSS.; dioicum, dense cæspitosum, caule longissimo gracillimo ramoso inferne radiculis fuscis intertexto, foliis inferioribus ovato-acuminatis evanidinerviis, superioribus sensim longioribus, comalibus lineari-lanceolatis integerrimis nitidis, nervo crasso percurrente e cellulis laxis angustis areolatis, theca in pedunculo elongato gracillimo erecta pyriformi membranacea lævi primum pallida demum fusca, operculo breviter conico concolori, peristomio externo pallido, interno processubus pertusis, ciliis nullis.

Most nearly allied to *B. pyriforme*, which it closely resembles in the colour of its leaves and capsule: it is also allied to *B. Sphagni*, Brid., judging from the figure of that species in 'Bryologia Europæa.'

37. Bryum *megalocarpum*, Hook.

Not present in any of the sets I have seen, but I saw it in Hb. Taylor from Prof. Jameson.

38. Bryum *Taylori*, C. Muller, Synops. Musc. Frond. p. 264. Brachymenium subrotundum, *Tayl. Lond. Journ. Bot.* 1846. p. 56. Acidodontium subrotundum, *Hook. et Wils. l. c.* vol. v. t. 16. H.

39. Bryum *seminerve*, C. Muller, Synops. Musc. Frond. p. 264. Brachymenium seminerve, *Tayl.* Acidodontium seminerve, *Hook. et Wils. Lond. Journ. Bot.* vol. v. t. 16. I.

This elegant species has perhaps more general resemblance to *B. crudum* and *B. longicollum* than to any others, and agrees with them in having the leaves gradually larger towards the tops of the stems.

40. Bryum *speciosum*, Mitten, MSS. Leptotheca speciosa, *Hook. et Wils.*—*Hook. Icones Plant. Rar.* vol. iv. t. 748. Brachymenium Jamesoni, *Tayl. Lond. Journ. Bot.* 1848, p. 252. Weissia Jamesoni, *Tayl. l. c.* 1847, p. 330.

A beautiful moss, which has much affinity with *B. Peromnion*: it is perfectly distinct from *Leptotheca Gaudichaudii*, Schw., judging from the figure of that species.

41. Bryum *soboliferum*, Tayl. Lond. Journ. Bot. 1846. p. 51.

42. Bryum *crinitum*, Mitten, MSS.; dioicum, dense cæspitosum fasciculate ramosum, foliis caulinis ovato-oblongis concavis nervo in pilum flexuosum albidum excurrente cuspidatis, perichætialibus lanceolatis margine reflexo, e cellulis laxis areolatis, theca in pedunculo gracili ovato-cylindrica, operculo conico brevissimo, peristomio *B. Hornschuchiani*.

Although this species has many points of resemblance to *B. Hornschuchianum*, it appears to be distinct in the following particulars:—the plants are shorter and more densely cæspitose, the leaves are wider and the nerve is produced into a long white hair-like point, the capsule is longer, and the perichætial leaves have the margins reflexed.

43. Bryum *julaceum*, Smith.

44. Mnium *rostratum*, Schw.

45. Zygodon *cyathicarpus*, Montagne, Ann. des Sciences Nat. 1845, p. 106. Wils. in Lond. Journ. Bot. vol. v. p. 1. Gymnostomum linearifolium, *Tayl. l. c.* 1846, p. 42.

46. Zygodon *tenellus*, Mitten, MSS.; dense compactus humilis lutescens inferne ferrugineus, caule tenello innovando ramoso, foliis caulinis lineari-lanceolatis nervo carinato excurrente mucronatis,

comalibus longioribus ligulatis ubique minutissime et densissime papillosis opacis, perichætialibus ovato-acuminatis laxius areolatis, nervo evanescente, theca in pedunculo mediocri gracili recta ovali-pyriformi sub orificio parum constricta.—Operculum et calyptra desunt.

This curious little species has much the habit of *Z. compactus*, but appears to be nearest allied to *Z. tenerrimus*, from which it differs in several particulars.

47. Zygodon *Reinwardtii*, Braun. Z. denticulatus, *Tayl. Lond. Journ. Bot.* 1847, p. 329.
48. Orthotrichum *speciosum*, Nees. O. longifolium, *Tayl. Lond. Journ. Bot.* 1846, p. 45.
49. Orthotrichum *laxifolium*, Wils. in schedula; monoicum, laxe pulvinatum elatum fastigiato-ramosum luteo-virens, foliis caulinis lanceolatis acuminatis basi angustatis plicatis carinatis margine revolutis evanidinerviis e cellulis minutis papillosis areolatis, perichætialibus basi latioribus laxioribus apice angustioribus acutissimis, theca immersa oblonga lævi pallescente siccitate octies plicata, operculo conico apiculato, calyptra scaberrima straminea apice purpurea totam fere capsulam obtegente, peristomii externi dentibus 8 bigemmatis siccitate revolubilibus pallidis, ciliis 8 dentes longitudine et latitudine æquantibus ochraceis, basi e cellularum tribus seriebus compositis.

Closely allied to *O. leptocarpum* and *O. speciosum*, but a more slender plant; the leaves are narrower and longer, as well as more loosely set, and little altered in drying; the cilia are remarkably wide, and at the base composed of several series of cells, which diminish irregularly upwards; in substance they resemble those of *O. speciosum*.

50. Macromitrium *longifolium*, Brid. vol. i. p. 309 et 738. Schlotheimia oblonga, *Tayl. Lond. Journ. Bot.* 1846, p. 47.

(*To be continued.*)

Note on PLATYNEMA; *by* G. N. WALKER ARNOTT, LL.D., *Professor of Botany, Glasgow.*

In 1833 Dr. Wight and I described, in Jameson's Edinb. Phil. Journal, a plant which we called *Platynema laurifolium*, belonging to the Natural Order *Malpighiaceæ*: our only specimen was obtained from the Madras or Missionaries' herbarium, and the accompanying label led us to suppose it had been collected in Ceylon. I have not received the same from Ceylon, from either the late Colonel Walker or from Dr. Wight, since we first described it, and I have, therefore, had occasionally some doubts as to its being indigenous there, if, indeed, there had not been some error as to the specimen having ever been grown in Ceylon at all.

A few days since, having been requested by a friend to look over and put generic names to his set of Lobb's collections of plants, I was both surprised and gratified to find No. 474 to be precisely my plant: it was collected by Mr. Lobb in Malacca. As to the genus, it is placed by M. Adr. de Jussieu, in his memoir on the *Malpighiaceæ*, among those whose position is uncertain; but from analogy he presumes the combined style to be a simple one, the other two being abortive, and that it ought to be placed next *Acridocarpus*. In describing the style as composed of three united, I intended to convey that the same structure occurred here as in the allied *Hiptage*. Jussieu concedes to this last and to *Tristellateia* a perfectly simple style, arising from one only of the carpels, the other two styles being abortive; but although he assigns plausible reasons for such a structure, I confess I am not satisfied that such is the true one; at least, I do not well understand how fertilization can be effected in the three carpels if one only has a perfect style and stigma. Any one who is in possession of Lobb's No. 474, and who will take the trouble of glancing over Jussieu's figures of the genera in the work above mentioned, will at once see that it is a species of *Tristellateia*; nor, perhaps, is it specifically distinct from *T. Australasica, a*, Rich. (I have not, however, access to that work, where it is figured), which has been found in New Guinea, a latitude not differing much from Malacca. With the other species, which are all from Madagascar, I am still entirely unacquainted.

A mistake has sometimes occurred, either in the numbering of Lobb's plants, or in the names assigned to them by M. Planchon, in the 5th and 6th vols. of the 'London Journal of Botany.' Thus in his list, vol. vi. p. 472, Nos. 294 and 295 are both called species of *Solanum*; in the set I have looked over, these numbers are both *Rubiaceæ*, and species of *Argostemma*. No. 331, in the same way, is not an "*Uncaria*," but precisely the same as No. 279, and a species of *Urophyllum*, perhaps *U. glabrum*.

As M. Planchon has not named any of the Asclepiadeous plants, I may soon take another opportunity of communicating to you the result of my examination of those I have seen in this collection.

BOTANICAL INFORMATION.

Cedron.

We are glad to find that our friend Mr. Purdie, of the Botanic Garden, Trinidad, has seen in the French papers the notice that was taken of the *Cedron*, as mentioned in our last number of the 'Kew Garden Miscellany,' p. 378; and he has just sent us some further particulars, which he had published in the 'Port of Spain Gazette,' and which we here give.—

"During my travels in New Granada (South America), I had often heard of the virtues of the *Cedron*, long before I had the pleasure of meeting with the tree. It is rare to find a Peon or Ariero without a seed, although they are expensive. I have, myself, paid a dollar a seed at San Pablo, where the tree is indigenous, even within the precincts of the village. Its use is not confined to the cure of serpent-bites alone, but has the reputation of superseding sulphate of quinine in cases of fever, and that in the country of the Cinchona barks.

"Now, about eight years ago, the Government of New Granada sent a commission of several medical men and students, accompanied by Dr. Cespides (Professor of Botany in the University of Bogota), to ascertain what plant produced the *Cedron*, its locality, and quantity procurable. You now see it in all the apothecaries' shops in the different towns of that Republic; so that now, in the midst of forests of the Peruvian Bark tree, another remedy at least equally specific (and that without any chemical preparation) is in process of superseding it. During my stay in Bogota, I was informed of the locality of the *Cedron* (by Dr. Cespides, a gentleman of considerable knowledge and experience in the plants of New Granada), which I found would be on my route from the province of Antioguia, by way of the Rio Magdalena. Thus, Providence has decreed, that out of the alluvial and pestilential plains of this magnificent river, a remedy should come for the cure of its own maladies. On my reaching the village of Nari, in the great valley of the Magdalena, in August 1846, I found that the surrounding woods contained the celebrated *Cedron*, as also lower down the river at San Pablo. I was glad to find that it was the season of its having ripe fruit. The villagers had already collected each their little hoard of *Cedron*, although they would not show me more than a few seeds, unless I would purchase some. The mode of preparation is simple and easy :—the fruits are gathered, which resemble in appearance a large peach ; the outer rind or covering is thick, fleshy, and excessively bitter, and its large seed is immediately surrounded by a not very compact fibrous substance, which answers the purpose of the stone in stone-fruits ; this is all removed, and the seed taken out, separated in two pieces at the natural fissure (which are called by botanists the cotyledons), and dried in the sun; beyond a limited quantity of these, it was no object to me to obtain; what I particularly wanted was a knowledge of the tree, and ripe vegetating seeds : those dried in the sun will never vegetate. I was told that it would be useless for me to go to the woods, as the trees had already been pillaged in all directions : this, however, did not deter me from trying; and after three days' search, at some distance from the village, I obtained about thirty fruits, each containing one seed and the germ of a plant. A few I preserved entire in spirits, the rest I planted in a box of earth at once, to prevent their perishing, as is the case with most large seeds, if not kept constantly excited. Those I sent to the Royal Botanic Gardens, Kew, grew well, and at the present time plants of *Cedron* would be more easily obtained at the Royal Botanic Gardens, Kew, than in its native country. Those I brought to the Botanic Garden, St. Ann's, are thriving well; some of the trees are now seven feet in height."

Dr. Link.

(From the 'Literary Gazette.')

Science has sustained a great loss in the death of Dr. Link, Professor of Botany in the University of Berlin, and Director of the Royal Botanic Garden of that city ; and those of our readers who, up to the present time, have been in the habit of reading his annual reports on the progress of botany, and his various recent contributions to that science, will marvel to know that he died in the eighty-second year of his age. His literary career extends back for more than half a century, his first botanical essay, consisting of some observations on the plants of the Botanic Garden at Rostock, having been published in 1795. Professor Link was one of those men whom, coming between the present and the past, it must be gratifying to the memory of living botanists to have met. He was contemporary with Linnæus, having been eighteen years old when the great author of the ' Systema Naturæ ' died, and, from his botanical tastes, was probably acquainted with that naturalist's writings long before his decease. Dr. Link was in Great Britain at the meeting of the British Association for the Advancement of Science, at Glasgow, in 1841. Those who were present on that occasion will not forget his vigorous, yet venerable appearance, and the zeal with which he entered into the scientific business of the Association.

We have no particulars of the early life of Heinrich Friedrich Link; but we know that he graduated at Göttingen in 1789, having read on that occasion an inaugural thesis on the Flora of Göttingen, referring more particularly to those found in calcareous districts. Shortly afterwards he was appointed Professor of Botany at Rostock; subsequently he held the same chair at Breslau; but the latter and larger portion of his scientific life was spent at Berlin. He practised at Berlin as a physician among an extensive circle of friends, who had a high opinion of his medical skill. Although the name of Link fills a large space in the literature of botany, his mind was not of the highest order, and his contributions to science are not likely to make a very permanent impression. Still, he was an energetic, active man, with an observant mind, a retentive memory, and with considerable power of systematic arrangement. Hence his works, like those of Linnæus, have been among the most valuable of the contributions to the botany of the century in which he lived. Of these, his ' Elementa Philosophiæ Botanicæ ' may be quoted as the most useful. This work, which was published in 1824, has served as the basis of most of our manuals and introductions to botany since that period. In this work he especially dwelt on the general anatomy of plants, and gave a new arrangement of the tissues. In order to render it more available for study, he published subsequently a series of plates, under the title ' Icones Anatomico-Botanicæ,' consisting of very faithful representations of the microscopical structure of various parts of the plant. The Elements have gone through several editions, and, though superseded, like most other works of the kind, by the publication of

Schleiden's 'Principles of Scientific Botany,' may still be consulted as a faithful mirror of the science as it then existed.

Professor Link devoted considerable time and attention to the description of new species of plants, most of which he published in a continuation of Willdenow's 'Species Plantarum.' In conjunction with Count Hoffmansegg, he commenced a Flora of Portugal, and he also published a memoir on the plants of Greece. Link contributed several valuable papers on physiological botany to the Transactions of the Natural History Society of Berlin; but he has done more service for vegetable physiology in his annual reports than in any other of his writings. They comprise a summary of all that had been published in botany during the year, accompanied with many valuable remarks and sound criticisms of his own. In these reports he had to defend himself and others from the heavy artillery directed against them by Schleiden, who, whilst claiming for himself a large margin for liberty of opinion, is most unscrupulous and pertinaciously offensive towards those who differ from him. In these literary contests, however, Link showed that the experience of above fifty years had not been lost upon him, and he was not unfrequently more than a match for the vigour and logic of his youthful and more precipitate adversary. He has now gone to his grave, and we doubt not that his talented antagonist will be the first to raise a monument to his memory. Whatever regret may be felt at Berlin for his decease, or whoever may succeed him in his Chair and Directorship, the name of Link will always be honourably associated with the progress of botany in the nineteenth century. Every scientific man may strive with advantage to imitate him in his unwearied industry, in his tolerance of views opposed to his own, and in his willingness to undertake any task, however humble, by which he might advance the interests of the science to which he had devoted his happily prolonged life.

LINNEAN SOCIETY.—W. Yarrell, Esq., V.P., in the chair. Mr. Gould exhibited a drawing, the size of life, of a new and very extraordinary bird, belonging to a genus and species entirely new, and which he had called *Balæniceps Rex*. It is an inhabitant of the interior of South Africa, and has the head and bill of a pelican, and the feet and legs of a crane. It stands four feet in height, and its principal food consists of young alligators. The bill is not so large,

but stronger than that of the pelican, and its feet are not webbed.—
A memoir was read from Joseph Woods, Esq., on the various forms
of *Salicornia*, accompanied with additional notes by Mr. Richard
Kippist, curator of the Society. The author commented on the
varieties and sub-varieties of the species of the genus *Salicornia*, more
especially the British species, *S. herbacea*. He proposed to make a
new species, *S. lignosa*; and was disposed to regard the variety of
herbacea, *S. procumbens*, as a distinct species. He questioned whether
S. radicans was specifically distinct from *S. herbacea*. A species found
in the South of France, with tubercled seeds, he proposed to call
S. megastachya. This plant was the type of Moquin Tandon's genus
Arthrocnemum. The seeds of the true *Salicornia*, according to Mr.
Kippist, are covered with curved simple hairs, whilst the embryo is
conduplicate and incumbent, and the seed exalbuminous. In *Arthrocnemum* the testa of the seed is crustaceous: the embryo is only slightly
curved, and lies in the midst of albumen. The paper was illustrated
with drawings of the seeds, and dried specimens of the various
species, with their seeds.—The Earl of Derby presented to the Society
a series of drawings of the living animals at Knowsley Park, by Mr.
Waterhouse Hawkins. These drawings were beautifully executed, and
consisted principally of species of *Ruminantia* and Solidungulate Pachy-
derms. We were sorry to observe that the binder had carelessly
injured several of the plates, and cut off the names. J. W. Bryans,
Esq., was elected a Fellow of the Society.

NOTICES OF BOOKS.

Description of the PALMYRA PALM *of Ceylon; by* WILLIAM FERGUSON. Colombo. 1850.

Under the above title we have received a very interesting brochure, in 4to, of nearly 100 pages, and we heartily wish our several colonies in all parts of the world would present us with similar publications on the various vegetable and other products of the respective countries, and in the same popular form. Travellers in these regions little think how ignorant we "stay-at-home" people are of the properties and uses of some of the most common vegetables they see, and probably pass unheeded. How little was known in England of extracting Toddy from the Palms in India till Mr. Strachan furnished us with the valuable details we have given in our first volume of this Miscellany, and till he presented the beautiful drawing and the model, and the implements illustrative of this operation, to our Kew Museum.

It was with a view of correcting an erroneous notion prevalent in Ceylon that the *Toddy* was extracted from Palm-trees by *incision*, that Mr. Ferguson was led to study the whole history of this noble Palm (*Borassus flabelliformis*), and to give the result of his inquiries to the world by a full description, printed in Ceylon, accompanied by several wood-engravings, also executed in the island, by *a native artist*: thus giving encouragement to art as well as to science in the colony.

The work commences with a full botanical description, in English, and in popular language; then the synonyms, including all the Indian names. The geographical distribution comes next; the value of the *Palmyra* at Jaffna, and in other parts of India; the products of the tree and of every part of the tree, and the uses of the several parts; the mode of extracting Toddy, Jagary, &c.; its cultivation, &c. &c.; concluding with miscellaneous notes and observations on an extended scale. Nothing seems omitted that can tend to complete our knowledge of the *Palmyra Palm*, which seems second only in importance to the Cocoa-nut.

The woodcuts add much to the value of the work. We have—1. A plate devoted to the Palm in its various stages of growth. 2. The flowers and fruit, with a native knife, and style used for writing, and a book of *Palmyra* leaves. 3. The third plate represents a *Banyan*-tree, which so often entwines and seems to unite with the *Palmyra* Palm. This is a very bold and clever sketch of A. Nichol, Esq., A.R.H.A., and the woodcut is by the native artist, Juel de Sÿlva (who executed the woodcuts). 4. Represents the union of the *Banyan* with the *Palmyra*. 5. Shows the "well-sweep" for raising water, made of the trunk of this Palm. 6. A *Palmyra*-leaf, and "Kelingoos," or young Palms for planting: and lastly we have, if not a plate, what is more interesting than that, a specimen of a *Palmyra*-leaf, as used for writing upon with a style, having an inscription on it in English and in Cingalese.

We trust that Mr. Ferguson, who is now in England, will on his return to Ceylon give us further accounts of other useful vegetable productions.

Das KÖNIGLICHE HERBARIUM *zu* MÜNCHEN *geschildert*. (*Sketch of the* ROYAL HERBARIUM *at Munich*.) *By Dr. C. F. Ph. von Martius. Translated from the German in the* Gelehrten Anzeigen, Bd. xxxi. No. 89-93 [*separately printed at Munich* 1850]; *by* N. WALLICH, M.D., F.R.S., V.P.L.S.

The Royal Herbarium at Munich was founded in the year 1813, when H.M. King Maximilian Joseph I. caused the scientific property of the late President of the Imperial Academy of Naturalists and Professors at Erlangen, Dr. J. C. Dan. von Schreber (born 1739 and deceased 1811*), to be purchased for the Royal Academy of Sciences, which thereby became possessed of a rich library, valuable zoological and mineralogical collections, and an herbarium, which, on account of the high reputation of its late owner, deserves to be considered as a worthy foundation for a botanical museum.

The chief part of the Schreberian Herbarium was arranged, though not catalogued, according to the owner's edition of the 'Genera Plantarum' of Linnæus. Inclusive of a number of unadjusted packages of specimens, it consisted of about 12,000 species. Besides the plants collected by Schreber himself, at Leipzig, Upsala (where he had been a pupil of Linnæus), and at Erlangen, both wild and cultivated, the herbarium received valuable additions from celebrated botanists,—such as O. Swartz, Georgi, Pallas, Mühlenberg, Röttler, J. R. Forster, Von Wulfen, Thunberg, Vahl, and others. Schreber had collected zealously and largely Grasses and Cyperaceæ. He acquired possession of Güldenstädt's entire herbarium, in which there are numerous plants gathered by this naturalist in Georgia, Mingrelia, and the Caucasian Steppes (see his Voyage through Russia and the Caucasian Mountains; St. Petersburg, 1787, 4to). Schreber had, besides, purchased the collections of Schmiedel, his predecessor in the chair of botany; which, however, he only very partially incorporated in his own herbarium. Pre-eminent for rarity and interest in the Schmiedelian collection, is a series of Cape of Good Hope and Ceylon specimens, obtained from his friend and correspondent during many years, N. L. Burmann, professor at Amsterdam. With the said collection came also one made in North America by Schöpf, medical

* For a further account of this celebrated man, see p. 118 of the preceding volume of this journal.—ED.

officer to the Margrave-Anspach troops (see his Voyage; Erlangen, 1788, two vols. 8vo), together with this industrious botanist's important MSS. entitled 'Index Plantarum Novæboracensium.' Besides all this, Schreber had constantly been on the look-out, in order to increase his materials by exchange and purchase. He was particularly connected in this respect with the Moravians, and received frequent despatches from them from Southern Russia (Sarepta), Labrador, and the West and East Indies. He purchased Hoppe's Centuriæ, Erhart's Decades, Schrader's Cryptogams from the Harz Mountains, and Funck's from the Fichtel-range; also Schleicher's Swiss plants. The herbarium contains, accordingly, a considerable number of German plants. The southern divisions, especially the Alps of Carniola and Carinthia, had been closely examined by Rainer, Von Wulfen, Hoppe, and Frölich, from whom the collection obtained many important species. A. W. Roth, physician at Vegesack, who was the first to publish a complete Flora Germanica, presented his preceptor with many valuable contributions from northern Germany. Casimir Chr. Schmiedel, whose great services in the analysis of plants in recent times, have been especially acknowledged by Robert Brown, had contributed an herbarium, accumulated during his journeys in the South of France, Switzerland, Nizza, and Naples; and a very interesting series from Monte Baldo, being the original specimens of Seguier's 'Flora Veronensis,' is deserving of particular notice in this place. Schreber received some rare Spanish specimens from Gmelin; from France he purchased Loiseleur's collections made in the neighbourhood of Paris, and in botanical gardens. The flora of Siberia, though to a small extent only, is represented in the Schreberian Museum by contributions from Pallas, Georgi, and Messerschmied. Some species were also added, probably through Gleditsch, collected by Gundelsheimer, the companion of Tournefort, in his voyage to the East.

From tropical Asia Schreber had received some plants from Joh. Gerh. König, who had gone to India so far back as the year 1769, in the service of the Nawaub of Arcot, and had travelled far and wide (he died in 1785). More considerable contributions were made by the Danish missionaries at Tranquebar, John and Röttler,[*] pupils of König;

---

[*] I had the happiness of knowing personally both these most worthy missionaries and excellent botanists, and of corresponding with them. With Dr. Charles John, I became acquainted on my visit to Tranquebar in 1807; he died not long after-

together with those of Burmann from Ceylon, already mentioned, they form one of the most important parts of the Schreberian herbarium.

It is less rich in African plants; and although Vahl had given some from North Africa, Isert from Guinea, and Sparmann and Thunberg from the Cape of Good Hope, yet these contributions are of little extent in comparison with the vegetable treasures received in modern times from those regions.

We have already mentioned that Schreber had received some contributions of the American floras from the Moravian missionaries at Labrador. He likewise got somewhat extensive materials sent by Von Wangenheim and Mühlenberg, from New England, though they were mostly anonymous and recommended to more close examination. Several hundred species were collected by Dr. Crudy at St. Thomas and the Bahama Islands; and, above all, a most valuable selection of original specimens of the plants of the West Indian Flora, gathered and communicated by O. Swartz, whose 'Icones Plantarum Incognitarum' and 'Flora Indiæ Occidentalis' were printed at Erlangen, under the superintendence of Schreber (1791-1806). But, on the other hand, there was scarcely a single representative of the flora of the South American continent; and the same remark applies to the peculiar vegetation of Australia, with the exception of a few genera purchased of Joh. Reinh. Forster.

It is clear from the preceding observations, that neither in a systematic nor phyto-geographical point of view, does the Herbarium Schreberianum satisfy the claims of modern times; especially as it has great deficiencies in some of the principal forms, and wants representatives altogether in some of the large floral territories. Strictly speaking, at the time when these collections were purchased by the State, they had been already distanced by three decenniums of progress and development in Systematic Botany.* It became the object, therefore,

---

wards. Dr. John Röttler I saw in 1812 and 1813 at Madras, on my voyage to and from the Mauritius. Like the late Dr. William Carey, he was heart and soul devoted to the missionary cause, he was a great orientalist, and ardently attached to the study of plants. The venerable man died in his 87th year, on the 27th January, 1836. His important herbarium has been finally presented to King's College, London. In the church at Vepery, Madras, there is an affecting tablet erected to his memory; and several Röttler-scholarships have likewise been founded at the seminary attached to it. He was born at Strasburg, in June 1749.—*N. W.*

* It was only at the solicitation of the bookseller, that Schreber, who had been engaged already since 1775 in publishing his 'Natural History of Mammalia,' could

of the Royal Academy, to procure a more uniform representation of the entire kingdom of Flora. No increase, however, was made by purchase for some years; and we can only mention one addition, presented by Privy Councillor Grimm, through the instrumentality of Von Schlichtegroll, Secretary to the Academy, consisting of his herbarium, collected during the years 1780 to 1800. It contains some plants, which existed in botanical gardens during that period, but which are now out of cultivation altogether, or occur only seldom; and, besides, the plants which Grimm himself had gathered in Germany and the South of France. This herbarium possesses, therefore, much historical interest.

A considerable increase of the Academic herbarium was caused by the travels of Spix and Von Martius in the Brazils (1817–1820). This may be estimated at nearly 8,000 species, of which 800 were collected in Istria, Malta, Gibraltar, Madeira, and Portugal, and the rest in Brazil. Von Martius got at Madrid, from Don Fel. Bauzá, the companion of the unfortunate circumnavigator Malespina, about 400 Bolivian species of plants, collected by Thad. Haenke in Cochabamba and its environs, being the first contribution from Spanish America incorporated with the herbarium. Dr. von Martius's 'Adversaria Botanica,' written on the subject of the plants observed by him during his Brazilian voyage, and which comprise 3,318 numbers, are attached in six quarto volumes to the Royal herbarium. A transcript, according

---

be persuaded to undertake a new edition of Linnæus's 'Genera Plantarum,' after that of Reichhard (Frankfort, 1778). He now took measures to augment his herbarium; but, in so doing, he omitted putting himself in communication with the French botanists, who might have furnished him with important contributions. He had not obtained Ant. Laur. de Jussieu's celebrated work, in which the genera were treated with a greater depth of research than was the case in the Linnæan 'Genera Plantarum,' in time to avail himself thoroughly of it for the edition he had in hand. Jussieu's work had appeared in 1789, and Usteri's reprint in 1795, precisely the same years in which Schreber published the two volumes of the Genera. The pupil of Linnæus, while adopting the method of his master, was compelled to acknowledge the far more profound treatment of the subject by the French botanist: and this caused some bitterness in the otherwise very anxious mind of the learned man, which was increased by untoward critiques from abroad (especially on the subject of his Linnæan *purismus* in the nomenclature of the Aubletian genera), and contributed gradually to lessen his attachment to botany. Schreber accordingly published nothing in that science, after he had concluded his work on the Grasses (' Description of the Grasses,' Leipzig, 1769–1810, 3 vols. fol.) and after concluding his contributions to Schweigger's and Körte's 'Flora Erlangensis,' and to Sturm's 'Flora Germaniæ.' After he had finished the edition of the 'Genera Plantarum' in 1791, he scarcely devoted any time to incorporating new contributions with his herbarium; beyond that period we may, therefore, fairly assume, that the literary and historical importance of the herbarium does not extend.

to the natural families, is in course of progress, as time and opportunity admit thereof.

An important augmentation was caused to the Royal herbarium, by the transfer of the Ludwig-Maximilian University at Landshut to Munich, its botanical collections being incorporated with the former, with proviso of the right of property on the part of the University; and all the collections were placed under the custody of the director of the botanic garden, as their conservator (Regier. Blatt, 21 March, 1827). This University Herbarium was founded by Dr. Schultes, immediately on his assuming the office of Professor of Botany, and particularly increased after he had undertaken, in connection with Römer of Zurich, the edition of the 'Systema Vegetabilium' (1817). It has peculiar value on account of the contributions of Bertero from the West Indies; Balbis, from Italy; Rochel and Kitaibel, from Hungary; Besser, from Caucasus and Southern Russia; and Römer, from various countries. It also contains the purchased collections of Sieber from Martinico, the Mascarenhas, the Cape of Hope, New Holland, and Egypt. For the preservation and increase of the University herbarium, a public grant of 50 florins was annually allowed, which has been laid out, since 1831, for the purchase of the requisite paper, and of collections, as these offered themselves for sale. In this manner have been acquired collections made by Hoppe, of German Alpine plants; by Schimper, of 600 Abyssinian and 500 Nubian plants; by Lunz and Fischer, of Arabian; by Baron Karwinski, of Mexican. The flora of the Pyrenees was increased by the purchase of Endress's collection; of Gallicia and the South of Russia, by that of Szowitz's collection; and additions were likewise made by the purchase of North American and Neapolitan plants. With the intention of facilitating the review and completion of species now combined together, also of lessening the cost of preservation, all the specimens received from Landshut, and those since acquired by purchase, have been added to the general herbarium, with which they now form one uniform total. Nevertheless, they are attached to separate half-sheets, and distinguished by proper tickets as the property of the University; which latter, likewise, possesses a collection of fungi made of wax at Vienna, by Trattinik (1804, 1805) and sold by him.

Since the year 1827 numerous additions have been made to the general herbarium, by presents, exchange, or purchase. Von Schrank presented his whole collection in 1832 (he was born Aug. 21, 1747, and died Dec. 22, 1835). It comprised a collection of Bavarian plants, forming the materials of the 'Flora of Bavaria' of that celebrated naturalist, published at Munich in 1789 (2 vols. 8vo), and preserved in forty-two cases. Further, a collection of Cape plants made by Brehm;* among which, are the original specimens which had served for Schrank's dissertation on the Cape *Irideæ* (Mem. of the Roy. Bot. Soc. vol. ii. p. 165-224); and the *Gnaphalioideæ* (Mem. of the Academy of Sciences at Munich, vol. viii. p. 141-172); also some purchased collections, among them one made by Salzmann about Montpellier. One year later were commenced Dr. Wallich's distributions of Indian specimens of plants, among public and private herbariums, by authority of the East India Company. Of Pöppig were purchased 350 species from the southern Andes chain, and Chili.

In the year 1834 were obtained 1040 species of Zeyher's and Ecklon's Cape plants; also an entire herbarium made by a Bavarian botanist, Fr. Xav. Berger, of Berchtesgaden, during his many years' peregrination in South Germany, and his sojourn, during three years, as military chaplain to the Bavarian troops in Greece, where he died. The purchase of this precious collection was effected by extraordinary royal grants, amounting to 1,200 florins. At subsequent periods were added to the general herbarium 810 species collected by Preisz, in Western New Holland; and nearly 1,300 species gathered in Brazil, by Patricio da Silva Manso, Ackermann, Luschnath, Clausen, and Riedel. From Mexico was obtained a considerable addition, resulting from two voyages of the Baron von Karwinski to that rich country; and, by exchange, several hundred species collected by Andrieux in Mexico; by Allan Cunningham, in New Zealand; by Fred. Zuccarini, jun., in Greece; by Tenore, in Naples; by Szowitz, in the environs of Odessa; and by Salzmann, in Tangier. Dr. Guyon, chief surgeon to the army at Algiers, and corresponding member of the Royal Academy of Sciences, sent from thence several hundred specimens; Von Zwackh sent 192

* This worthy gentleman I hope is still alive and in good health. In October 1843, I had the pleasure to meet him at Uitenhagen, on my tour through the Cape of Good Hope districts with the Judge of Circuit. He kept a capital druggist's shop, and was Deputy Sheriff of the district. In his garden I saw, among many other things, three fine and large species of *Zamia* and a most gigantic *Tamus*. Mr. Brehm possesses a large amount of botanical knowledge concerning the productions of the Colony, and is, besides, a man of very general information. He accompanied my party some days on our route towards Graham's Town.—*N. W.*

rare Pyrenean plants, and Papperitz of Dresden, 150 select plants of Central Germany. The vegetation of the Tyrol is nearly completely represented, by numerous transmissions of specimens to the Royal Academy, on the part of the Imperial Ferdinandeum at Insbruck, amounting to more than 1,000 species. We mention here, especially, the liberal donations of Dr. von Barth, at Calw, corresponding member of the Royal Academy of Sciences, consisting of 160 species, collected by Kotschy and Hohenacker, in North Persia; 154 species sent from Labrador, by the missionaries there, among which are some very beautiful specimens of cryptogamic plants and mosses; 600 species from Georgia and the rest of the Caucasus; and 295 species collected by the missionary Metz, in the East Indian province of Canara. From Hungary were received 250 species from Mr. Kovátz; several species from Java, collected by Dr. Kollmann, chief surgeon in the Royal Netherland service. We omit detailing several other contributions of minor extent, among them Dr. Kummer's, and Sendtner's from different parts of Germany. Of the latter we have a complete series of all the plants he collected during his voyage made under royal support, on the coast of Dalmatia and in Bosnia. Nor has the Royal Botanic Garden been wanting in contributing its share of important additions, especially of such species as are with difficulty procurable from their native countries. There are some thousand specimens, among which those carefully prepared and preserved by Dr. Kummer deserve especial notice.

On the death of the excellent Dr. Jos. Gerh. Zuccarini, prof. and second conservator (born 10th Aug., 1797, dec. 18th Feb., 1848),* the herbarium, which he had formed during five-and-twenty years of intense zeal and activity, was purchased by the Royal University, for the sum of 10,000 florins, payable in sixteen years, by annual instalments. The general herbarium became almost doubled by this grand addition, not only of species collected by the deceased himself during his frequent visits to the South of Germany, to Berlin, and Leyden; but by the numerous contributions to him from many of the leading botanists of the time, with whom he kept up a frequent correspondence.† The flora of Asia is in particular richly represented

---

* For a notice of this distinguished botanist, see the first volume of this Journal, p. 180.—ED.

† *German* plants were furnished by Koch, Hoppe, Funck, Fürnrohr, Treviranus,

in this herbarium. Thus, from the coasts of Asia Minor, are collections made by Fleischer, and communicated through the Itinerary Union of Würtemberg; 285 species from the Taurus, furnished by Kotschy; the Syrian-Arabic plants of Sieber, Schimper, the Leuchtenberg-Archiater Dr. Fischer, during a number of years professor in the school of medicine at Kafr El Ainee, at Kairo, who made frequent excursions into Lower Egypt and several parts of Arabia; the entire harvest of Drs. Roth and Erdl, who accompanied M. von Schubert in his voyage to the Levant, consisting of plants from Lebanon and the deserts of Egypt and Arabia; and, finally, a valuable collection made by Col. Chesney on the Euphrates, and communicated by the Horticultural Society of London. From Georgia and the other Caucasian territories, there are plants furnished by Hohenacker, Brunner, Steven, A. Richter, of Moscow, Dr. Tschernjajew, professor at Charkow, Bunge, Von Ledebour, and the Imp. Acad. at St. Petersburg. The two last-mentioned parties, besides C. A. Meyer, Bunge, Turczaninow and Drége, sent Altaic, Davurian, and other Siberian plants; Joerdens and Von Chamisso, from Kamtschatka; Bunge, Bentham, and the Imp. Russian Academy communicated Chinese plants. Japan plants were furnished by Siebold and Bürger, and by purchase from Göring, which served as materials for the beautiful works on the Flora of Japan, published by Zuccarini and Siebold. Indian plants were brought home by Barón C. von Hügel; others were communicated from the Wallichian collection, by Bentham and Lindley; and by Krauer.

Ernst Meier, Schiede, Hargasser, Leo, Fleischer, Vahl, jun., Bischoff, Albers, Nolte, Lessing, Unger, Sauter, Graf, Schultz, Alex. Braun, Mohl, Noë, Köberlin, Tratzel, Raab, Schnitzlein, Schenk, Kittel, Frölich, Hinterhuber, Bentham, Biasoletto, Lehmann, Berger, Pritzel, Schimper, Kummer, Sendtner, Ohmüller. *Swiss* plants, by Mohl, Raab, Trachsel, Schleicher, Seringe, Thomas, Brunner. *Hungarian* and from the rest of *Austria*, by Sellow, Eubel, Leydolt, Reissek, Lang, Sadler, Vahl, Hoppe. *Dalmatian*, by Peters, Fenzl, Viviani. From the *Bannat*, by Heuffel, Rochel, Besser. *Moldavian*, by Czyhak and Szowitz. *Rumelian*, by Griesebach. To complete the list, and to do justice to the importance of this herbarium, we here add all the other countries and botanists represented by its riches:—*Italy* and *Piedmont*: De Notaris, Balsamo, Savi, jun., Bertoloni, Tenore, Viviani, Visiani, Biasoletto, Petter, Noë, Michahelles, Solcirol, Fleischer, Schiede, Gussone, Tineo, Bracht (a zealous collector in Lombardy, who fell as captain in the 52nd Regiment, in the battle of Novara). *Spain* and *Portugal*: Holl, Bentham, Schultes, Boissier. *France*: De Candolle, Duby, Delille, Schultz, Vahl, jun., Schimper, Bentham, Soleirol, Baumann, Buchinger, Brunner, Schnitzlein, Schimper. *England* and *Scotland*: Bentham. *Sweden* and *Norway*: Vahl, jun., Fries, Wahlberg, Hornemann, Boek, Clason. *Russia*: Von Ledebour, Richter. *Greece*: Sieber, Frederic Zuccarini, Berger, Sartori, Spruner, Fraas.

About 200 species of the Peninsula of India were communicated by the excellent Dr. Wight, director of cotton culture. From Assam and the Khassia range, the eminent botanist William Griffith presented 290 species; and from Java and the other Sunda Islands, Reinwardt, Siebold, Blume, Korthals, and Müller sent specimens. In this herbarium, as in the general one, the flora of Africa is indifferently represented, with the exception of the Cape of Good Hope. Besides a considerable number of Egyptian plants, there are a few from Madeira, given by Pohl, Hochstetter, and Holl; and a valuable series of Senegambian plants furnished by Thomas Dillinger, by Sieber, and by Baumann. On the other hand, the Cape flora is supplied by more than 2,000 species from Ecklon, Drège, Sieber, Brehm, and Kraus (those of the latter through Henschel, E. Meyer, and Von Schrank).* The Mascarenhan plants, which generally occur but rarely in herbariums, are but few; they were furnished by Sieber and Bojer.

Comparatively speaking, the herbarium is richest in American plants. They were furnished from Sitcha and the adjoining countries,

* With the deserving Mr. C. F. Ecklon I became acquainted during my visit to South Africa, also with the worthy Dr. Pappe (engaged, I learn, in a monography on *Iria*), and Mr. C. L. Zeyher. The infirm health of the first-mentioned botanist deprived me of the advantage of his accompanying me on the tour I was permitted to make, in the first months of 1843, to the western district and mountains of the colony, with my inestimable friend Thomas Maclear, Esq., Astronomer Royal at the Cape of Good Hope. Of the indefatigable Mr. Zeyher there are notices in several portions of Vols. IV. and V. of the preceding series of this Journal. At present he is attached to the Botanic Garden at Cape Town as collector of plants; Mr. M'Gibbon, late of the Royal Gardens at Kew, being the curator. The re-establishment of such an institution, at so important a locality, after having remained in abeyance during very many years, is indeed a matter, not only of local interest and congratulation, but of great consequence to similar institutions in other parts of the globe. It is chiefly to the exertions of the Rev. Dr. James Adamson, whose almost universal knowledge, ardent zeal, and genuine piety entitle him to the respect of all, and in whom I honour a much-valued friend, that we owe this fortunate state of things. Together with Dr. Pappe, and Messrs. Arderne and Clarence, Dr. Adamson constituted a committee, under the orders of Government, to mature and carry into effect the arrangements connected with the garden, which now occupies fourteen acres of land, within the area of the Government grounds; though not all in cultivation as yet. Most of the plants in the beautiful garden of my late friend Baron C. von Ludwig, were purchased as the groundwork of the garden. I may mention here, that there exists a curious account, with a diagram, of the ancient Dutch East India Company's Botanic Garden at Cape Town, in the very useful but exceedingly verbose and voluminous work of Francis Valentyn; together with a list of plants collected by the Danish botanist Henry Bernhard Oldenland, who, it appears, was the curator in 1695. (See *Oud en Nieuw Oost Indien*, vol. v. part ii. pp. 17-30.) —*N. W.*

VOL. III.

by Von Chamisso, Rupprecht, and the Imperial Russian Academy.; from Greenland, by Hornemann, Vahl, jun., and Steetz; from Canada, Labrador, and the United States of North America, there are near 2,000 species, supplied by Hooker, Barth (missionary Heldenberg), Bischoff, Waldmann, Tuckermann, Torrey, Asa Gray, Martens, Lindley, and Bentham (from California); from Mexico, by Schiede and the Baron Karwinski; from the West Indies, by Sieber, Lindley, and Von Barth; from Surinam, by E. Meyer, Baron Römer, Von Martius, and Baumann; from Brazil, by Döllinger, Beyrich, Sellow, Von Langsdorf, Von Martius, Schott, Pohl, and Mikan; from Chile, by Lindley, Cuming, Bertero, Pöppig; and lastly, from Peru, by Endlicher and Pöppig.

From Australia, Sieber, Lindley, Loddiges, and Baron Hügel communicated New Holland specimens; and Hooker, Lindley, and Baron Hügel, Van Diemen's Land, New Zealand, and Norfolk Island plants.[*]

(*To be continued.*)

An Account of the DILPASAND, *a kind of Vegetable Marrow;* by J. ELLERTON STOCKS, M.D., F.L.S., *Assistant-Surgeon on the Bombay Establishment.*

TAB. III.

Citrullus *fistulosus* (J. E. S.); caule petiolisque fistulosis, foliis quinquenervibus quinquelobatis, cirrhis tri-quinquefidis, pepone napiformi concolore.

DESCR. *Stems* diffuse, stout, fistulous, with the cavity stellate in cross section, owing to the prominence towards the centre of five bundles of vessels. *Sap* mucilaginous. Young *shoots* densely villous with long soft spreading hairs, between which is a more minute glandular, viscid, and odorous pubescence, which disappears with age. Older *stems* with scabrous and more scattered hairs, chiefly confined to five prominent and shining bands, between which the surface is

---

[*] In general, this herbarium is richer in phanerogamic than in cryptogamic plants, although it is of much value even in this respect. There is the very important collection of Mosses, extending to 700 species, made by the distinguished anatomist and physiologist Döllinger, during a series of thirty years, and bequeathed by will to Zuccarini. (Here follow some details of the contents of this collection, and of the mosses gathered by Zuccarini himself or presented by his friends and correspondents.)

ACCOUNT OF THE DILPASAND. 75

minutely striated. *Leaves*, when young, green and shining above, with comparatively few lymphatic hairs, and a short glandular pubescence which soon falls off; under surface paler, densely villous on the nerves, and glandular-pubescent on the introvenium. Old leaves with scabrous more scattered hairs, the upper surface dull green, the lower pale. *Leaf* and *cirrhus* (3-4-, more rarely 5-cleft) side by side, and in the joint axil a leaf-branch or leaf-bud, a solitary male or female flower, and a rounded *bract* with induplicate margin. *Petioles* hairy and finally scabrous, furrowed; fistulose, the cavity circular in section. *Blade of leaf* cordate-ovate in general outline, 5-nerved, and cut one-half or one-third part down to the midrib into five, rounded *lobes*, which are themselves wavily lobulated, the margin marked by little white callous terminations of the nerves, between which the parenchyma is puckered and crimped.

MALE:—*Flowers* on a peduncle about half the length of the petiole. *Calyx* villous, the tube spread out nearly flat, crowned abruptly by the short teeth separated by a broad straight sinus. *Corolla* flat, villous outside, smooth and of a sulphur-yellow within; petals 5-nerved, the central nerve running off externally and below the apex into a soft point. *Stamens* of the genus; viz., triadelphous, filaments distinct, and loculus following the windings of the elevated line or ridge of the *connective*. *Disc* filling up the base of the calyx between the stamens. FEMALE:—*Flower* on a thick *peduncle*, lengthening and curving downwards in fruit. *Calyx* quite flat and dish-shaped. *Corolla* as in the male. *Barren filaments* sometimes anther-bearing. *Disc* collar-shaped, round the style. *Ovary* subglobose, with long white soft hairs. *Style* very short, undivided, or more rarely shortly three-cleft towards the apex. *Stigmata* thick, approximated into a round head. *Fruit* at first apple-shaped and hispid, finally quite smooth, light apple-green in colour, and much depressed at base and vertex. *Seeds* black, marked on both sides by an elevated ridge following the outline of the seed, with the margin external to this, narrow, nearly as thick as the body of the seed.

OBS. 1.—Sometimes the leaves are opposite, forming a verticil when taken with the two cirrhi. It is not uncommon to find flowers with six teeth to the calyx, six divisions of the corolla, and the odd stamen which usually has but half an anther with loculi on both sides, as in the two others. This tendency to revert to the ternary division of the floral leaves, and so to agree with ovary and stamens (for I believe the stamens in *Cucurbitaceæ* to be three, not five, in number, although this is masked in many cases by the halving of the filament), is worthy of notice.

OBS. 2.—As a species this is recognized at once from its congeners, the Colocynth and Water-melon (*C. Colocynthis* and *vulgaris*), which are the only others I have to compare, by its much less divided 5-nerved and 5-lobed leaves, not glaucous as in the Water-melon, or hoary as in the Colocynth. Both these last have 3-nerved, 3-lobed leaves, cleft almost to the midrib, with the divisions also deeply lobulated. The tendrils in the Colocynth are generally undivided or rarely bifid, in the Water-melon they are bifid, but here they are generally 3–4- rarely 5-cleft. The fistulous stem and petioles are an absolute distinction. The calyx is here much more flat than in the other two, where it is campanulate at the base. The very short style, the almost globose ovary, the depressed fruit of uniform colour, not striped or speckled in any stage of its growth, are fuller marks of distinction. The seeds differ from the smooth thin seeds of the Colocynth, and resemble more those of the Water-melon, in their surface as it were sculptured by a sharp elevated ridge marking off a centre and a rim, but the ridge is here more conspicuous and more within the margin, so that the rim is broader. Finally, the poisonous Colocynth * and the eatable Water-melon, have associated with them here a cookable vegetable.

OBS. 3.—This species is known in Scinde by the name of *Mého*. In the Punjaub it is called *Hindwana* (the name of the Water-melon in Scinde), and in the Deccan is named *Dilpasand*, or "Delicious," a very appropriate name. I believe it is not known in Bengal proper, and it does not grow in the Concans or on the Malabar coast, but is brought down (when there is a demand) from the more elevated, milder, and drier climate of the Deccan. In Scinde it is cultivated from April to September, generally in the same plot of ground with

---

* Sheep, goats, jackals, and rats eat Colocynth-apples readily, and with no bad effects. Camels refuse them. They are often used as food for horses in Scinde, cut in pieces, boiled, and exposed to the cold winter nights. They are also made into preserves with sugar, having previously been pierced with knives all over, and then boiled in six or seven waters till all the bitterness disappears. The low Gipsy castes eat greedily the kernel of the seed, freed from the seed-skin by a slight roasting. They collect the seeds in large quantities by burying the apples in the sand, and allowing the pulp to rot away.

Common Melons, Luffa, Gourds, and Cucumbers. The fruit is picked when about two-thirds grown, the size and shape of a common field-turnep, two inches and a half high, and three inches and a half across. It is pared, cut in quarters, the seeds extracted, well boiled in water, and finally boiled in a little milk, with salt, black-pepper, and nutmeg. Mussulmans generally cut it into dice, and cook it together with meat in stews or curries. Hindoos fry it in ghee (clarified butter) with split gram-peas (*Cicer arietinum*), and a curry powder of black-pepper, cinnamon, cloves, cardamoms, dried cocoa-nut, turmeric, salt, and last (but not least in their opinion) the never-failing assafœtida. It is sometimes made into a preserve in the usual manner. It is sometimes picked when small, cooked without scraping out the seeds, and regarded a greater delicacy than when more advanced.

In England it might be cultivated and cooked like the Vegetable Marrow (*Cucumis ovifera*), which it much resembles in its qualities.

Kurrachee, Scinde, September 1st, 1850.

TAB. III. Part of a stem:—*natural size*. Fig. 1. Stamens; fig. 2, section of the male flower, showing the disc; fig. 3, ditto of female flower; fig. 4, seed; fig. 5, section of ditto:—*all more or less magnified*; fig. 6, young fruit (downy); and 7, fully-formed fruit:—*nat. size*.

DECADES OF FUNGI; *by the* Rev. M. J. BERKELEY, M.A., F.L.S.

Decade XXXIV.

Sikkim-Himalayan Fungi collected by Dr. Hooker.

331. *B. delphinus*, Hook. fil.; pileo pulvinato glabro luride spadiceo; stipite subæquali sublævi spadiceo sursum rubro apice flavo; poris subdecurrentibus sordide flavis; carne secta multicolore. Hook. fil., No. 76, cum ic.

HAB. On earthy, open places. Darjeeling, 7,500 feet. June. Rare.

Inodorous. Pileus 2 inches or more across, pulvinate, smooth or very minutely tomentose, nearly dry, of a lurid reddish-brown. Stem about 2 inches high, ¼ an inch or more thick, often curved at the base, nearly equal, of the same colour as the pileus, but shaded above with red, and then with yellow, nearly even. Pores somewhat decurrent, dirty-yellow. Flesh of pileus, when cut, instantly changing to blue, especially below; that of the stem exhibiting all the colours of the rainbow. Spores fusiform, yellow-brown when seen by transmitted light.

This species appears just intermediate between Fries's sections *Subtomentosi* and *Calopodes*, and is a more delicate species than any there enumerated. The stem does not seem to be truly reticulated, nor the mouth of the pores to be red, and the change of colour to be far more marked than in any of the former section.

STROBILOMYCES, gen. nov. Hymenophorum ab hymenio prorsus discretum; tubuli ampli, liberi. Sporæ globosæ vel lato-ellipticæ, asperulæ. Fungi e carnoso lenti, ut plurimum verrucis plus minus floccosis quoad pileum ornati. A *Boleto* differt *Strobilomyces* substantia magis lenta, et sporis nequaquam fusiformibus.

332. *S. polypyramis*, Hook. fil.; pileo lato expanso toto areolato verrucoso umbrino carnoso intus rubido : stipite valido lævi fusco-purpureo; poris latis adnexis, ore rubidis. Hook. fil., No. 104, cum ic.

HAB. Jillapahar, 7,500 feet. August. Rare.

Smell strong, ammoniacal, at least in age. Pileus umber-brown, 6–7 inches broad, expanded, slightly convex, covered with large pyramidal soft warts, which disappear almost entirely under pressure; substance thick, firm, and leathery, of a dull red. Stem 5 inches high, 1¼ inch thick, attenuated above, swollen below, nearly smooth. Pores large, shorter towards the stem, not free. Spores globose, dark purple-brown, granulated.

Allied to *Strobilomyces strobilaceus*, but distinguished by its expanded pileus, smoother stem, and less persistent more numerous warts. The spores differ very much from those of true *Boleti*, being more or less globose, without any tendency to be fusiform.

333. *S. montosus*, n. s.; pileo verrucis paucis floccosis pyramidatis montoso nigro-fusco, interstitiis amethystinis; stipite concolore squamoso; poris subadnatis fuscis, ore luteo. Hook. fil., No. 121, cum ic.

HAB. On dead wood, and on the ground. Jillapahar, Darjeeling, 7,500 feet. September.

Inodorous, dry, leathery. Pileus convex, 3 inches broad, rough with large pyramidal flocculent brown warts; their interstices amethyst-coloured. Stem 3–4 inches high, ½–1 inch or more thick, curved, dark brown, scaly; substance brownish. Veil thick, coriaceous, attached to the margin of the pileus. Pores brown within, yellow externally, rounded behind. Spores broadly elliptic, brown, rough.

Very nearly allied to *S. strobilaceus*, but the warts of the pileus, which are not at all imbricated, and the bright colour of their interstices in young specimens, are very remarkable.

334. *Polyporus* (*Mesopus*) *cremoricolor*, n. s.; tener subcarnosus; pileo umbilicato glabro subviscoso, margine nudo repando cremoricolore; stipite centrali deorsum subpubescente sursum glabrescente cum pileo concolore; poris minutis 4–5-gonis niveis, dissepimentis tenuissimis. Hook. fil., No. 107, cum ic.

HAB. On clay banks in Mr. Campbell's garden, attached to decayed wood. Darjeeling, 7,000 feet. August. Rare.

Cream-coloured, with a faint sweet smell. Pileus delicate, of a somewhat fleshy substance, scarcely an inch across, smooth, rather viscid, umbilicate, with the margin arched and naked. Stem $\frac{3}{4}$ of an inch high, central, more than a line thick, slightly pubescent below, becoming naked above, firm, solid. Pores minute, distinctly 4–5-gonal, with extremely thin dissepiments.

Nearly allied to *P. arcularius*, and, like it, approaching closely to the genus *Favolus*. The whole fungus, however, is smoother, and the pores far smaller. In drying it becomes very thin.

335. *P.* (*Mesopus*) *umbilicatus*, n. s.; carnoso-lentus cremoricolor; pileo umbilicato demum subinfundibuliformi tenui rimoso virgato, margine involuto ciliato; stipite curvo gracili sursum deorsumque incrassato fibrilloso subtiliter squamuloso; poris minutis, dissepimentis tenuibus, acie tenui.

HAB. On dead wood. Tonglo and Sinchul, 8,000 feet.

Inodorous. Pileus $1\frac{1}{4}$–3 inches or more broad, at first fleshy, then tough, umbilicate, at length subinfundibuliform, more or less cracked, virgate, snow-white or cream-coloured; margine involute, not ciliate. Stem $1\frac{1}{4}$ inch high, swelling above and below, fibrillose, minutely squamulose, especially towards the base, where it springs from a strigose downy disc. Hymenium white; pores minute, $\frac{1}{50}$ of an inch in diameter, angular; dissepiments thin; edge minutely toothed.

This species has exactly the habit of several *Lentini*, and is most nearly allied to *P. tricholoma*, which is a smaller, thinner, and far more delicate species. *P. ciliatus* also somewhat resembles it, as also *P. arcularius*, and some other neighbouring species, but all differ in essential character. I have a specimen of the same species from Van Diemen's Land, which was given me by Mr. W. Gourlie, of Glasgow. I believe it was bought at Dr. Graham's sale.

* *P. rufescens*, Fr. Ep. p. 438. Hook. fil., No. 26, 27.

HAB. On dead wood. 3–5,000 feet. One specimen springs from the pileus of *P. hirsutus*.

Three distinct forms occur, two from Darjeeling, the third from Jillapahar, differing (I believe according to the height at which they grow) in the substance of the pileus and decomposition of the pores. No. 27 is hard when dry, fuliginous, velvety, with the pores sinuous; No. 26 is rufous, more decidedly tomentose, and has the pores lamellate; a third form has the texture still softer, the down more decided, the colour still more rufous, and the hymenium assuming the appearance of that of a *Lenzites*. English specimens of *Polyporus rufescens* differ greatly in density of texture at the same elevation.

* *P. oblectans*, Berk. in Hook. Lond. Journ. vol. iv. p. 51. Hook. fil., No. 118, cum ic.

HAB. On dead wood. Sikkim, 7,500 feet. Very rare.

A single specimen only occurred, with the hymenium imperfect. The pileus shines more than in the Ceylon or Swan River specimens, but this is apparently the only difference. The pores in this and the allied species vary greatly in form, according to age and circumstance.

* *P. xanthopus*, Fr. Ep. p. 437.

HAB. On dead wood. Darjeeling, 7,000 feet. Below Pimhabania. Hot valleys, 3–4,000 feet.

The specimens have the pores rather larger than usual. They exhibit beautifully the downy disc from which the stem always rises in perfect specimens.

336. *P.* (*Mesopus*) *maculatus*, n. s.; albido-lutescens; pileo coriaceo-carnoso depressiusculo maculato, margine subrepando; stipite centrali deorsum attenuato; poris amplis margine laceratis longe decurrentibus. Hook. fil., No. 3, cum ic.

HAB. On trunks of living trees. Darjeeling, 8,000 feet. April 1848.

Inodorous, coriaceous but brittle. Pileus about 2 inches across, plane or slightly depressed, dry, spotted with obscure adpressed scales; substance thick, white. Spores 1¼ inch high, attenuated below, ¾ of an inch above, smooth at the base. Pores large, 1 line across, running nearly to the base of the stem; edge more or less lacerated, by no means obtuse or entire.

Nearly allied to *P. Michelii*, Fr., but differing in its lacerated not entire pores. *P. lentus*, Berk., which is also nearly allied, and of which *P. coronatus*, Rost., appears to be a synonym, differs in its hispid stem.

* *P. squamosus*, Fr. Ep. p. 438. Hook. fil., No. 8, cum ic.

HAB. On rotten timber. Darjeeling, 8,000 feet. May. Rare.

337. *P.* (*Pleuropus*) *platyporus*, n. s.; inodorus; pileo laterali carnoso-coriaceo pallido, margine inflexo; stipite deorsum nigro-velutino; poris amplis, acie tenui. Hook. fil., No. 39, cum ic.

HAB. On dead timber. Darjeeling, 8,000 feet. May, June.

Pileus 2 inches broad, orbicular, more or less lobed, dull ochraceous, smooth, moist but not viscid; margin incurved; substance rather thick. Stem either a mere disc or much elongated, clothed with very fine velvety down at the base. Pores large, $1-1\frac{1}{4}$ line broad, irregular, edge thin, rather torn.

Most nearly allied to *P. squamosus*, but not only is it on a very reduced scale, but it is destitute of the peculiar smell of that species, and, if at all, is very obscurely spotted. *P. maculatus* is a neater species, with a spotted pileus and smaller pores.

* *P. lucidus*, Fr. Ep. p. 442.

HAB. Trunks of old trees. Darjeeling.

* *P. sanguineus*, Fr. Ep. p. 444.

HAB. On dead wood. Darjeeling, 7,000 feet. Very common.

* *P. flabelliformis*, Klotzsch in Linn. vol. viii. p. 483. Hook. fil., No. 18 (junior), cum ic.

HAB. On dead wood. Darjeeling, 4–8,000 feet. Tonglo. Abundant.

In the young specimens the dissepiments are very thin, and their edge acute. The margin also of the pileus is extremely thin. The stem is sometimes quite smooth and yellow, as in *P. xanthopus*, nor is there in the young individuals any trace of the black or dark bay coat, which is always more or less observable in age. Some of the older smooth specimens are very pretty, and when they have a very slender stem remind one of *P. affinis*.

338. *P.* (*Merisma*) *rubricus*, Berk.; flabelliformis expansus multiplex totus pallide rubricus; pileo lobato subfimbriato glabro demum radiato; stipitibus obesis basi velutinis; poris parvis brevissimis.

HAB. Tonglo, 7,000 feet. May.

One foot or more broad, reddish, tufted, more or less flabelliform, lobed, and somewhat fimbriated, the lobes occasionally confluent and leaving lacunæ. Pileus smooth, at length radiated, fleshy; extremely brittle when dry. Stem more or less developed, obtuse, sometimes 2 inches or more long and 1 inch thick, velvety at the base, sometimes extremely short. Pores very shallow, angular, rather small; dissepiments thin.

A very distinct species, resembling somewhat *P. rufescens*, Fr., but not allied. The pores are so shallow as to resemble those of Dr. Montagne's genus *Glæoporus*.

* *P. intybaceus*, Fr. Ep. p. 446. Hook. fil., No. 98, cum ic.

HAB. On dead wood. Darjeeling, 7–8,000 feet. Rare.

* *P. sulphureus*, Fr. Ep. p. 450.

HAB. Tonglo. Dr. J. D. Hooker.

* *P. crispus*, Fr. Ep. p. 457. *P. armoraceus*, Hook. fil.

HAB. On dead wood. Darjeeling, 7–8,000 feet. Tonglo, 8,000 feet; a single specimen.

The greater part of the specimens have a very dark hymenium and minute pores, but these are connected clearly with one or two exactly according in the hymenium with specimens of *P. crispus* from Switzerland. In fact, most of the specimens are just intermediate between *P. adustus* and *P. crispus*, having a darker hymenium than I have ever seen even in the former.

339. *P.* (*Anodermei*) *ozonioides*, n. s.; pileo reflexo postice effuso lateraliter elongato connatoque fulvo strigoso-stuppeo, contextu pallido; poris amplis; dissepimentis dentatis pallide ochraceis.

HAB. On dead wood. Darjeeling.

Pileus 1–2 inches long, 3–5 inches broad, sessile, reflexed, effused behind, clothed with tawny, rigid, reticulated, branched, towy strigæ; substance nearly white. Hymenium concave, pale ochraceous; pores $\frac{1}{14}$ of an inch broad, dissepiments rather thick, more or less toothed and elongated.

Resembling very much *P. funalis*, of which I have an authentic specimen, and a magnificent tuft, from Sierra Leone. It is, however, less thickly clothed, and the pores are not normally prolonged into Hydnoid teeth; besides which, it is merely reflexed, while that species is much expanded. Occasionally the pores are very much dilated.

* *P. isidioides*, Berk. Lond. Journ. Bot. vol. ii. p. 415.

HAB. On dead wood. Darjeeling.

* *P. licnoides*, Mont. Cuba, p. 401.

HAB. Sinchul, 8,000 feet.

The specimens collected by Dr. Hooker differ slightly from the American individuals which I have from Bahia and Guiana.

The colours, marking, and sculpture of the adult specimens are precisely similar; the young fungi, however, are clothed with a rhubarb-coloured velvety coat, whereas in the American form the early down is rubiginous. The pores, too, are slightly larger, and the whole texture evidently more succulent. These, however, are differences so clearly depending on climate, that I cannot but consider the species the same. The two forms laid side by side, whether on the upper or under side, are so similar when fully grown, that they could not fail to be pronounced the same species by any one at the first glance.

* *P. zonalis*, Berk. Ann. of Nat. Hist. vol. x. p. 375. tab. x. fig. 5.

HAB. On dead wood. Darjeeling, 7,000 feet.

This pretty species was originally found by König, in Ceylon. I believe that it also occurs in Cuba.

* *P. hirsutus*, Fr. Ep. p. 477. Hook. fil., No. 28, cum ic.

HAB. On dead wood. Darjeeling, Mung-durbi, 4,000 feet.

Some of the specimens have a dark margin, exactly as in *P. nigromarginatus*, Schwein., which is a mere variety.

* *P. versicolor*, Fr. Ep. p. 478.

HAB. On wood. Darjeeling, 7,000 feet. Tonglo, 8,000 feet.

The specimens are extremely rigid, the pores larger than usual, and imperfectly formed. The surface of the pileus is just that of the most ordinary European form.

* *P. Neelgherrensis*, Mont. Ann. d. Sc. Nat. Ap. 1842.

HAB. On dead wood. Darjeeling. Not abundant.

* *P. elongatus*, Berk. in Hook. Lond. Journ. Bot. vol. i. p. 149. Hook. fil., No. 30, cum ic.

HAB. On wood, forming large masses. May to the end of the year, till destroyed by the cold. Darjeeling, 7,000 feet, and Jillapahar.

Hymenium, when fresh, of a singularly delicate peach-blossom pink. The Himalaya specimens are much stunted, otherwise they agree exactly with the more expanded form. Sometimes the surface is uniformly zoned, sometimes interrupted with darker fasciæ. A state from the upper Himalayas, sent by Captain Munro, is scarcely zoned at all.

Another form occurs at a lower elevation, externally more like the usual form, but with the hymenium imperfectly developed, as in some specimens of *P. versicolor.* This has been owing, probably, to some sudden check after rapid growth.

340. *Trametes lobata,* n. s.; latissima expansa reniformis lobata, lobis rotundatis; pileo pallido glabro nitidulo zonato rugoso; hymenio ochraceo, poris angulatis, ore elongato dentato demum sinuato; dissepimentis tenuioribus demum lamellosis.

HAB. On dead wood. Mung-durbi, 4,000 feet.

Sessile, but attached by a more or less distinctly marked disc; very broadly expanded, 8 inches broad, $4\frac{1}{4}$ inches long, thin, but coriaceous, rather flexible, reniform, lobed, the lobes rounded, and sometimes assuming the appearance of young pilei; pale wood-coloured, smooth, somewhat shining, rough with radiating more or less strongly marked ridges, repeatedly zoned, the zones rather darker; extreme edge very thin and acute. Hymenium pale ochraceous, uneven; pores angular, $\frac{1}{28}$ of an inch broad; orifice elongated, more or less toothed; dissepiments rather thin, at length broken down into more or less distinct lamellæ.

This species is extremely like *T. læticolor.*

Extracts of Letters from RICHARD SPRUCE, ESQ., *written during a Botanical Mission on the* AMAZON.

(*Continued from* vol. ii. p. 302.)

Santarem, Jan. 29, 1850.

MY DEAR SIR,—It is only five days ago that I received, both together, your letters of the 31st August and 2nd October, and I cannot express to you how welcome they were, for I began to think myself forgotten. The communication between Pará and the Sertao is at best very uncertain. The barque which brought your letters should have reached here several weeks ago, but the captain thrashed one of his Indians at Pará, for which every one of them ran away, and he was obliged to wait until the police captured them again.

When I last wrote to you—which was along with a barrel containing the *Victoria* in spirits—I had some thoughts of ascending to the Barra, and I had a passage promised me in a steamer which the Government had just placed at the disposal of the President; but the next arrival from Pará brought word, that, in consequence of fresh disturbances at Pernambuco, the steamer had been ordered there with troops. There was no other chance for the Barra this season, and I therefore reverted to my original project of visiting Obidos and the Rio Trombétas (the execution of which you will find detailed in the accompanying sheets), and afterwards of making Santarem my head-quarters for the winter. I have now in contemplation two principal excursions during the rainy season; the first a little way up this river, the Tapajoz, and the second to Monte Alegre, in April or May, when the vegetation is said to be most luxuriant.

Santarem is much the pleasantest place on the Amazon, and superior even to Pará in the matter of eatables;—good milk and fresh beef every day—the latter cheap, but the former quite as dear as in London. There are three English residents, from whom we receive much civility, but they are none of them in that flourishing state to enable them to afford us any assistance in the way of diminishing our expenses. In fact, there are (between ourselves) only two Englishmen in the province of Pará—Arch. Campbell and Miller—who have not great difficulty in keeping straight their finances. Poor Captain Hislop complains bitterly of the difficulty he finds in getting any sort of payment for the numerous debts that are owing to him: long credit, and slow or no payment, is the way of doing business here.

I am anxious to learn in what state my sendings from Pará by the George Glen reached you and Mr. Bentham. I took great pains to have them dry and drily packed, but those little bothering vessels of Singlehurst's I am very much afraid of; even when I was by to take care of my goods I had great difficulty to keep them from spoiling.

You will see that I have anticipated some of your queries and wishes. I sent a Mandiocca-strainer,—specimens, &c., of the Pottery-tree,—and some other things you ask about, from Pará. I wanted no incitement to look closely for *Podostemeæ*, but it is only lately that I have come to a country where they can possibly grow: you will see I found two up the Trombétas. I hope to send you painted cuyas, and colours used in painting, from Monte Alegre, which is the famed place.

I expect you will keep your *Victorias*. I fear ere this the seed of the *Victoria* has fallen in this neighbourhood. I commissioned a person to procure some for me during my absence, but he has disappointed me; I intend, however, shortly to revisit the place, if it be only to see for myself whether the plants attain that enormous magnitude they are said to do.

Thanks for the hints about *Arums*: Santarem is not the place for this tribe, any more than for the Ferns. The delta of the Amazon is much richer in Ferns than any place I have since seen, and had I gone to Marajó instead of coming up here, I might have done better in both Ferns and forest-trees; but I was desirous to get into an Orchis country, if such existed. Now, however, I have traversed, from the mouth of the Amazon, a tract extending through from 700 to 800 miles, and I am compelled to conclude that it is the reverse of rich in *Orchideæ*. It is not their utter absence that I can complain of, but their want of variety. Here, up the Tapajoz, are old low trees filled with Orchises, but all of *two* species; in the neighbourhood of Santarem I have seen in all three species. I got a few up the Trombétas, but only two or three that look at all promising. In other tribes of plants I confess to have been disappointed to see the flowers in general so *small*:—in the tropics we look for everything on a gigantic scale, but here the *flowers* are rarely striking from their magnitude, although the plants that bear them are. There are *some* pretty things, but, I fear, very very few which will come up to Mr. Pince's expectations. It is for this reason that I often regret having taken Mr. Pince's £50. I will do the best I can to repay him in plants; and, at any rate, I will hold myself his debtor for the amount advanced, and repay him in one way or other.

I have seen no place yet with a vegetation so varied as that of Santarem, or where I can gather more species in a ramble. The campos, that seemed burnt up in summer, are now assuming a new vegetation; and that not an *annual* one, but of plants whose roots have all the while been buried under the sand. In April and May they are said to be brilliant with flowers. I may hope, too, for a fair proportion of novelty, for the ground must be very imperfectly known. Martius is said to have been sick, from his half-drowning, whilst he remained here, and at Obidos he made no stay at all. From what I have seen, the south side of the Amazon has a much more varied vegetation than the north side; and I was disappointed at Obidos and up the Trombétas, to find the *mass* of the plants quite the same as at Pará.

I wish I was near enough to have your advice as to my next campaign. I suppose it will be up the Rio Negro, as we talked ere I left England, but I am doubtful whether up the main river would not be more profitable. Tabatinga, the frontier town, is in the centre of unknown land, and the eastern slope of the Andes themselves is still imperfectly known. The Rio Negro has *dense* moist forest on each side all the way up, therefore contains few or no Orchises; though, from the absence of mosquitos, &c., it is undoubtedly pleasanter to travel on than the Solimoes. Mr. Bentham told me the Tapajoz had been explored, otherwise my own inclinations would prompt me to ascend it, for it is much the most romantic of the tributaries of the Amazon, its serras extending even to its mouth, at Santarem, while the mountains at its source are the highest in Brazil. Mr. Hislop speaks with enthusiam of the number and abundance of the Ferns near the cachoeiras. Does Mr. Bentham know the particulars of its exploration? Mr. Hislop recollects a party of Germans (among whom was a naturalist) coming rapidly down the Tapajoz; but you know that no one gathers half so much in coming down a river or mountain that he does in going up. I have a capital opportunity of getting up, all the way to Cuyabá if I like, with a merchant who came down lately with a cargo of gold and diamonds.

The Rio Chiquitos is an upper branch of the Madeira, between 15° and 20° S. latitude. I suppose Humboldt means that all the country between the Amazon, the Madeira, and the Andes, is imperfectly known as to its vegetation, which I believe to be the case.

I am glad to hear that *Gramineæ* and *Cyperaceæ* will be acceptable in Europe: the Grasses here are very interesting, those on the sandy campos and volcanic serras being quite a different set from those on the river-banks and low lands. Pará was a better place for *Cyperaceæ*, and now is the best time for them. I shall have a few things for your museum when I send the plants, but as my funds require to be economically dealt with, I cannot purchase many things.

The Mandiocca-plants grow to six or seven feet, and would probably grow higher; I have never seen them *standing* without leaves.

Santarem, April 1850.

When I arrived at Santarem last October, I hired the only house that was vacant (for houses are more scarce here than elsewhere in the province); but it suited me very well, for it has a spacious verandah at the back, where we could work at our plants, and a paved yard, where we could spread our paper, &c., to dry. The adjoining house was tenanted by a single man, and we were very quiet; but when we returned from Obidos we found it tenanted by a family, from several days up the Tapajoz, including amongst them, besides children, several slaves, big and little, numbers of fowls, turkeys, guinea-fowls, goats, dogs, land-tortoises, and other unclean beasts. I should mention that this house claims half our yard, and has a verandah continuous with ours: and then you will understand how on our return we found both yard and verandah befouled and worse than useless to us. A few live plants, that we had left in the verandah, under charge of a slave of Mr. Hislop's, he had the precaution to place in an outhouse which we have, but there, for want of light and air, some of them had died. Since then our live plants have stood in the same outhouse, as near as possible to the window (kept wide open), and would do very well were it not that the niggers and the fowls contrive to enter and play sundry pranks with them,—such as nipping off the leading shoot. What I have for Kew are now all in the case, fastened down, and only half the glass roof remains to be put on. Yesterday morning, when I went to look at them, I found one of my *Sucu-ubas* broken off by the root:—there were two; one from a cutting that I struck last November has grown beautifully, the other was a young plant; it is the latter that is destroyed. I procured another cutting, but I shall be obliged to fasten up the case before I can ascertain whether it has taken root. The other plants are *en bonne santé*, but they will have a severe ordeal to go through between here and Pará, and it is not worth my while going the tedious journey to Pará and back to take care of them: I am very desirous they should reach you alive. The two *Melastomaceæ* are very pretty things, and I send also seeds of both. Will not the "Jará" palm be *Leopoldinia pulchra*, Mart.? I send its fruit and leaves, but the spadices (from which the fruit has fallen), though not very large, are too large to be crammed into Mr. Bentham's boxes.

I send you thirty-five packets of seeds, many of them of species occurring in my dried collection. Some of these seeds and fruits are curious, and you may possibly like to put a portion of such into the museum. I could send many succulent fruits, but I have nothing to put them in; it is impossible to get at once a sufficient number to fill one of our big barrels. Mr. Wallace brought with him, for preserving reptiles, &c., a number of patent cylindrical earthen jars, with tops of the same material encircled by a metallic rim; a half turn fastens these effectually. If you can make out where these jars are to be had, I wish you would send me some—were it only half-a-dozen;—I could fill a part or the whole of these, as materials offered, and then despatch the contents to England in a barrel, reserving the jars for future operations. Bottles here are awfully dear, and often not to be had at any price. Six of the jars, fitting pretty tightly into a box, would be easily moved about from place to place.

My contributions to your museum are this time only insignificant; the only native manufacture at Santarem being baskets from the tender leaves in the young shoots of the Tucumá. I send a small workbasket, and I have two other baskets too large to be sent at present; these, as well as the spadix and spatha of the Bacúba palm, and some other large things, I reserve to be sent after my return from Monte Alegre, whence I hope to bring you some interesting things. I have already commissioned a number of the beautiful painted cuyas, made only at Monte Alegre, through Dr. Campos, whose wife is a native of that place; this is necessary, for they are only made as there is a demand for them. I shall send many more than you require, but I have no doubt purchasers will be found for them, they are so exquisitely done, equalling some of the best Chinese painted articles; and when it is considered that they are done by Indian girls, who make their own brushes and colours, they become still more interesting. I have learnt from Signora Campos what these colours are, and I shall be able to procure specimens of them.

The two or three fruits from the Rio Aripecurú will be hardly worth preserving. No. 27 and 28 are fragments of woods, which may serve until I can procure better. No. 29 is the seeds of the Sapacúya; I have in vain tried to procure the fruit entire, with the lid, and the tree does not grow here.

(*To be continued.*)

Contributions to the Botany of WESTERN INDIA;
by N. A. DALZELL, Esq., M.A.
(*Continued from p.* 39.)

Nat. Ord. LEGUMINOSÆ.

Tribe CÆSALPINIEÆ.

CÆSALPINIA.

C. ? *spicata*; frutex scandens, ramis petiolisque communibus et partialibus aculeis crebris recurvis armatis, foliis bipinnatis, pinnis 5-6-jugis, foliolis 5-6-jugis oblongis obtusis coriaceis supra nitentibus utrinque puberulis, floribus terminalibus spicatis, spicis uni-bipedalibus.

This plant differs from *Cæsalpinia*, to which it appears most nearly allied, by its equal petals, thick margined legume, seeds not compressed, flowers in spikes, and, as the flowers are moreover tubular and connivent, they have a very different appearance to those of a *Cæsalpinia*; it may be called *Wagatea*, and thus defined:—

Calyx coloratus, ad medium 5-fidus, tubo cupuliformi, limbi decidui laciniis æstivatione imbricatis, infima paulo majore concava. *Corollæ petala* 5, *æqualia, uniformia*, unguiculata, calycis summo tubo inserta. *Stamina* 10, cum petalis inserta iisdemque æquilonga, omnia fertilia: *filamenta* subulata, basi pilosa, alternatim breviora. *Ovarium* stipitatum, tomentosum, 4–6-ovulatum. *Stylus filiformis*, filamentis petalisque æquilongus. *Stigma* cavum, bilabiatum, fimbriatum, labio superiore semiorbiculari, inferiore majore cucullato. *Legumen* lineare, acutum, coriaceum, inter semina transversim constrictum, marginibus incrassatum. *Semina* 3–4, obovato-oblonga, transversa, testa crassa dura ossea.

From the absence of any descriptions in the 'Flora Indica' or Hooker and Arnott's 'Prodromus,' this appears to be confined to the Bombay Presidency, and it forms a very conspicuous object in our mountain jungles, by its bright scarlet tapering spikes, generally overtopping the bushes on which it leans for support. It is liable to attract attention in another way not so agreeable to the traveller, who has no sooner got disentangled from the hooks of one of its long trailing branches, than, in turning round to escape, he is caught by another, or, perhaps, he walks out of his difficulty by leaving his hat or part of his coat behind him. From the tubular form of the calyx, the flowers of this plant never open, the apices of its bright orange-yellow petals just appearing above the scarlet calyx; the legumes are much swollen at the seeds, the bases of which are partially imbedded in the thick spongy substance of the suture, resembling those of the *C. resupinata* of Roxburgh. The testa is broken with difficulty, the radicle is obtuse, and at the hilum, which forms the base of the seed's longer axis, the seeds are enveloped in a transparent very tenacious gum, which is capable of being drawn out into long fine threads. No use is made of any part of this shrub, as far as I can learn. Its native name is *Wagatee*, or *Wagaree*, its prehensile character, I suppose, having allusion to that of the tiger (*wag*).

(*To be continued.*)

BOTANICAL INFORMATION.

Professor Reichenbach.

[The following printed circular we received a few months ago, from Leipzig, and are anxious to give it publicity.—Ed.]

"*A Messieurs les Protecteurs, les Professeurs et les Amateurs des Sciences Naturelles.*

"Pendant les jours de terreur à Dresde (mois de Mai de cette année, 1850) il a été incendié plusieurs galeries du Musée d'histoire naturelle, la grande salle d'auditoire, ainsi que le cabinet d'étude de M. le professeur Reichenbach, conseiller du roi de Saxe et directeur du Musée. C'est dans cette salle et dans ce cabinet du pavillon antérieur du 'Zwinger,' vis-à-vis du château, que M. R. croyait avoir sauvé et mis en sureté sa bibliothèque, son herbier et ses collections zoologiques; mais hélas! tandis que sa maison d'habitation, quoique exposée continuellement aux coups des armes-à-feu, resta debout, le pavillon et tout ce qu'il contenait fut dévoré par les flammes. La plus grande partie des ouvrages zoologiques et botaniques, ornés de planches (tels que 16 vol. du Botan. Magazine, 20 vol. Botan. Cabinet, plusieurs ouvrages de M. Lesson, Levaillant, Vieillot, etc.), 1500 paquets d'un herbier, contenant de nombreuses plantes originales, des autographes d'auteurs célèbres, la collection carpologique, 66 cadres remplis de plus de 20,000 exemplaires d'insectes si rarement recueillis; savoir: diptères, hymenoptères, hemiptères, orthoptères, neuroptères et aptères, une collection de petits mammifères et reptiles d'Europe, aussi bien que des objets d'anatomie en esprit de vin, des instruments, des images et des dessins relatifs à l'histoire naturelle, plus d'un millier de planches coloriées, et un plus grand nombre encore en noir de l'ouvrage de M. R. sur les oiseaux, la principale partie de sa correspondance et de ses manuscrits, en un mot, toute cette partie de sa fortune qu'il estimait le plus, a été anéantie.—

"Nous n'aurons pas à dire quelle a été la valeur de ces collections précieuses, auxquelles le possesseur avait, dès sa jeunesse, donné des soins tout particuliers; leur accroissement a été le résultat de son zèle qui ne s'est jamais démenti, et de ses relations non interrompues avec des voyageurs et le monde savant. Nous ne dirons pas non plus, combien il a utilisé ces trésors dans l'intérêt des sciences et de l'enseignement, et personne n'ignore avec quelle complaisance il a prêté des genres et des familles entières, soit de son herbier ou de ses collections zoologiques, aux monographes qui n'ont pas manqué d'en faire une mention honorable et reconnaissante dans ses ouvrages.

"M. R. ayant entrepris, dès sa jeunesse, la publication de travaux en zoologie et en botanique, et les ayant continués jusqu'à l'âge de vigueur auquel il est arrivé, nous avons encore beaucoup à espérer de ses efforts et de ses lumières.

"Au reste, ces collections n'étaient pas seulement la propriété de la science, des hommes de science et du professeur lui-même, mais aussi celle de sa famille, qui y a des droits bien établis, vu qu'elles furent le produit de la publication de nombreux travaux et celui d'un budget qu'elle s'était imposé depuis trente années. L'acquisition de ces trésors scientifiques a été un sacrifice offert par toute la famille, à la science, au professorat et même à l'Etat.

"Il y a donc urgence de restituer et de faire renaître entre ses mains des collections semblables dans l'intérêt de la science, de la justice et de l'humanité.

"D'anciens amis de M. R. se sont proposés d'atteindre ce but, convaincus qu'ils sont, que son zèle et ses lumières contribueront à les compléter pour équivaloir bientôt celles qui ont été détruites; il se dévouera de nouveau à sa famille, à ses amis, aux hommes de science, avec cette tranquillité d'âme propre à embellir un âge plus avancé.

"Les Soussignés prient instamment tous ceux qui voudront bien concourir à cette entreprise d'augmenter leur participation par un appel à leurs amis et en donnant le plus d'extension possible à cette invitation, dont le but est consacré à la science, à un homme de science toujours prêt à obliger les autres et ayant servi sa patrie avec un désintéressement bien rare.

"La somme que nous désirons réaliser le plus tôt possible, par la réunion des divers contingents, servira à l'acquisition de collections. Ceux de Messieurs les intéressés, hors d'état d'y contribuer par des moyens pécuniaires, sont priés d'offrir en nature quelques-uns des objets désignés plus haut.

"Nous nous empresserons de faire connaître le résultat de nos acquisitions à tous ceux qui auront participé à une œuvre si méritoire.

"Adresser l'argent et les paquets à 'Friedrich Hofmeisters Buchhandlung, in Leipzig.'

"Dr. Carl Gustav Carus, à Dresde.
"Dr. Ch. G. Ehrenberg, Prof. à Berlin."

Those who are not disposed to contribute pecuniarily, or in the form of collections, may yet encourage the important labours of Professor Reichenbach by subscribing to some or other of the works in the list announced under the head of " Choix des ouvrages de M. Reichenbach : Botanique."

Flora Germanica (Europæ mediæ) Excursoria.		Les Cupulifères pl. col.	16
		Les Urticacées	17
Iconographia Botanica, s. Plantæ Criticæ Mediæ Europæ. 10 vol. pl. col. 1000 ; plantes environ 3000.		Les Dipsacées et les Valérianées	48
		Les Papaveracées	19
		Les Violacées	23
Icones Floræ Germaniæ (Europæ mediæ) contiennent entre autres :		Les Cistinées	17
		Les Rénunculacées	129
Les Potamogétonées pl. col.	87	Les Euphorbiacées	25
Les Hydrocharidées	18	Les Malvacées	18
Les Graminées	121	Les Géraniacées	16
Les Cypéroidées (Carex, etc.)	126	Les Caryophyllacées	111
Les Iridées	85	Les Tiliacées	14
Les Joncinées	42	Les Linées	17
Les Liliacées	78	Les Hypéricinées	11
(Les Orchidées par. M. Reichb. Fils, sous presse :	100)	Maintenant planches col. 1200 ; plantes plus de 8100.	
Les Conifères	17	La Continuation sous presse.	
Les Salicinées	57	Hortus Botanicus, s. Iconographia Exotica. pl. col. 250.	
Les Betulinées	16		

The above plates and brief notices illustrate a very great number of species common to Great Britain, as well as to middle Europe ; and great pains are taken with the analysis.

Circular from the Association Botanique Française d'Exploration.

M. E. Bourgeau, chargé du voyage, a exploré les environs de Carthagène et de Murcie et une grande partie des Sierras de Alcara et de Segura (Murcie, frontière de Jaen), et a rapporté en nombre près de 600 espèces, qui toutes offrent un intérêt réel, soit à cause de leur nouveauté ou de leur rareté, soit comme constituant le fond de la végétation de contrées jusqu'ici trop peu explorées.

L'Association, comme par le passé, n'a eu qu'à se louer sous tous les rapports de la manière dont M. Bourgeau a rempli sa mission, aussi croyons-nous devoir le charger de poursuivre l'exploration de l'Espagne, qui présente encore un grand nombre de localités qui n'ont pas été visitées par les botanistes, ou qui ne sont qu'imparfaitement connues.*

Le nouveau voyage, qui sera entrepris en 1851, a pour but de compléter l'exploration du sud-est et du sud de l'Espagne.

Le voyageur partira de Paris, le 28 Février, 1851, pour se rendre directement à Almeria.

Le voyage durera environ dix-huit mois, et se divisera en deux parties distinctes :

Dans la première partie (année 1851) seront visités les environs d'Almeria et le littoral, depuis le cap de Gate jusqu'à Malaga et les principales montagnes de la région, spécialement les Sierras de Huescar, de Maria et la Sierra Nevada ; en automne, le littoral sera parcouru une seconde fois, pour recueillir les plantes tardives et les fruits de celles qui n'auraient été prises qu'en fleur.

Dans la seconde partie du voyage (année 1852) seront explorées avec soin les localités les plus riches de la partie méridionale du Royaume de Valence : La Dehesa près Valence, Alzira, Moxente, Tuente-la-Higuera, San Felipe (où M. Léon Dufour a séjourné et dont il a donné la florule, etc.) ; et celles du royaume de Murcie qui n'ont pas encore été visitées, ou que M. Bourgeau pourra revoir utilement dans une saison différente de celle où il les a parcourues en 1850.

Pendant ce voyage, comme on peut le voir d'après l'itinéraire indiqué, M. Bourgeau visitera en général au moins deux fois les mêmes localités, et se trouvera ainsi à même d'avoir presque toujours les plantes en fleur et en fruit ; et il pourra, en hiver et au premier printemps, recueillir une collection nombreuse de monocotylédones bulbeuses qui, par leur préparation irréprochable, presenteront un grand intérêt.

M. Bourgeau espère rapporter de ce voyage environ 800 espèces en nombre.

Toutes les plantes seront déterminées avec soin par des botanistes connus et munies d'étiquettes imprimées ; chaque étiquette portant un numéro d'ordre et le nom du botaniste à qui l'on doit la détermination.

Les conditions de la souscription seront les suivantes :

Le prix de chacune des centuries qui composeront la collection est de 25 francs.

Les botanistes qui désirent avoir droit aux collections les plus com-

---

* These specimens, like all of M. Bourgeau's collecting, are distinguished by their size and perfectness, no less than by their cheapness.—ED.

plètes devront verser entre les mains de M. Bourgeau, avant son départ, une somme de 50 francs au moins. En raison du grand nombre de souscripteurs aux collections ordinaires, M. Bourgeau ne peut recueillir de collections complètes que pour les personnes qui auront effectué ce premier versement.—Toutes les collections seront d'ailleurs réparties d'après l'ordre d'inscription sur la liste de souscription.

Vers la fin de la présente année, il ne sera distribué que deux centuries environ à chaque souscripteur ; et les 50 francs versés à l'avance ne seront déduits que sur le prix de la partie principale de la collection dont la livraison aura lieu peu de temps après le retour de M. Bourgeau.

Vous êtes instamment prié de vouloir bien faire parvenir votre réponse, dans le plus bref délai possible, à M. Ernest Cosson, à Paris, rue du Grand Chantier, 12, ou à M. Bourgeau, rue des Blancs Manteaux, 11.

Paris, 4 Février, 1851.

[A printed circular, of which the following is a translation, has been communicated to us by Dr. Nees von Esenbeck.—Ed.]

"It is hereby announced to sympathizing friends, and to all who interest themselves in the matter, that by a ministerial proceeding I have been for the present suspended from my function as professor at the University of Breslau, and the branches of duties connected therewith. I make this announcement in the consciousness, that those who really know me will not connect this act of the Government with any dishonouring guilt on my part. Busy papers of the day, deriving their information from secret sources, will not permit much delay to take place in the disclosure of the affair; and that will be the time for me to take it up in continuation of the present notice. Only thus much will I add to-day: my guilt is of a personal nature, and such, that the letter of the law can only award an inadequate punishment; inasmuch as the real punishment which the law must and will award, depends not on her physical power, but on the judgment, which will be pronounced by contemporaries, according to their humane opinion of the accused individual. Meantime, I will submit to the pains and penalties inflicted in the name of the law; while I calmly await the real judgment, which will be pronounced by the heart and mouth of my contemporaries, as soon as they shall have been properly informed of the case, and have heard also what I have to say.

"Dr. Nees von Esenbeck.

"Breslau, Feb. 1, 1851."

SALICORNIA.

We have received the following from the author of 'The Tourist's Flora':—

"You have attributed to me some opinions in the number for February of your 'Journal of Botany' which I am sorry to see go out to the world as mine. I think hardly any two species are more clearly distinct than *S. herbacea* and *S. radicans*, and I enclose the concluding paragraph of my paper, that you may see distinctly what my notions are on this subject.—JOSEPH WOODS."

"If I were to sum up the result of my observations of this year on the genus *Salicornia*, I should say that *S. procumbens* is a distinct species. *That* S. radicans *and* S. lignosa *are certainly specifically distinct from* S. herbacea, but whether they are so from each other, and whether, if that be the case, *S. lignosa* ought not to be considered as a variety of *S. fruticosa* of Linn., and the plant with tubercled seeds to be called *megastachya*, I do not feel competent to decide."

NOTICES OF BOOKS.

MUSEUM BOTANICUM LUGDUNO-BATAVUM; *sive Stirpium Exoticarum Novarum ex vivis aut siccis brevis Expositio, additis figuris.* Scripsit C. L. BLUME. Leyden, 1849.

Under this title we have received nine numbers, in large 8vo, each of sixteen pages, and one plate illustrative of new or rare plants. The descriptions are drawn up, and the figures made, chiefly from the author's collections made in the Malay Islands, and in part from the "Herbarium regium Neerlandicum," and from other public and private sources. Although the species introduced into this work are of a very miscellaneous character, yet a good many are brought together under one and the same Natural Order, and there is a great deal of illustration of certain families; such, for example, as the *Melastomaceæ*, *Pangieæ*, *Bignoniaceæ*, *Gnetaceæ*, *Orchideæ*, *Phytocreneæ* (including *Miguelia*, Meisn.), *Asclepiadeæ*, the genus *Sarcosiphon* in *Rhizantheæ* (?), *Myrtaceæ*, *Hydrilla* in *Hydrocharideæ*, *Lythrarieæ*, *Apocyneæ*, *Legnotideæ*, *Rhizophoreæ*, and *Fœtidia* in "*Combretaceis affine*." Each plate has three compartments, and as many Genera, extremely well executed, —chiefly explanatory of genera in *Melastomaceæ*, *Phytocreneæ*, *Asclepiadeæ*, and *Orchideæ*. We are prevented from noticing as we could wish, the splendid "*Rumphia*" of the same author (Blume), in consequence of our being unable hitherto to complete our copy.

Characters of some GNAPHALIOID COMPOSITÆ *of the Division* ANGIANTHEÆ; *by* ASA GRAY.

Some Australian *Compositæ*, chiefly of the subtribe *Gnaphalieæ*, having been placed in my hands by Sir William Hooker, for examination, I here offer the characters of those which belong to De Candolle's division *Angiantheæ*. Figures of the more interesting, now in preparation, will appear in the forthcoming volume of the 'Icones Plantarum.' Mr. James Drummond's collections in South-western Australia furnish so many plants of this groupe which I am obliged to consider as the types of new genera, that it becomes needful to preface the account of them with an arrangement of the whole Division.

This groupe is not so perfectly limited in nature as it is by the character assigned to it by De Candolle. *Calocephalus*, Br., and *Pycnosorus*, Benth., want the general involucre altogether; in *Chamæsphærion*, &c., it is represented merely by a circle of radical leaves; in several genera the capitula are pedicellate, as in *Craspedia*, where there is merely a subtending bract for each; and in the plants which I have referred to the obscure genus *Crossolepis*, Less., the compound glomerule is merely subtended by the more crowded uppermost leaves, making an evident transition to the *Helichryseæ*, as do *Chamæsphærion* and *Chthonocephalus* to the *Cassinieæ*.

The capitula are homogamous in all the known *Angiantheæ*, although some of the flowers, otherwise perfect, are occasionally sterile by the abortion of the ovary, or, in *Pycnosorus*, according to Bentham, one of them is neutral.

CONSPECTUS GENERUM DIV. ANGIANTHEARUM.

§ I. *Involucrum generale cyathiforme, bivalve; partiale nullum nisi palea cujusque floris.*

1. DITHYROSTEGIA, nov. gen. Receptaculum villis prælongis lanatum.

§ II. *Involucrum generale e squamis foliisve discretis plerumque imbricatis constans, aut nullum; partiale imbricatum.*

* *Capitula 1-2-flora.*

† Pappus nullus, vel (in *Hyalolepide*) unisetosus.

‡ Receptaculum generale nudum, planum.

2. HYALOLEPIS, DC. Pappus e seta unica tenerrima.

‡‡ Receptaculum generale paleaceum, planum vel conicum, involucro pluriseriali brevius.

3. HYALOCHLAMYS, nov. gen. Involucrum partiale 1-florum, 4-phyllum, squamis biformibus; ext. 2 navicularibus axi incrassatis corneis cum paleis receptaculi orbiculatis persistentibus, int. minimis hyalinis. Involucrum generale e squamis orbiculatis tenuissime hyalinis. Glomerulum radicale.

4. SKIRŔOPHORUS, DC. (Pogonolepis, *Steetz*). Involucrum partiale 2-1-florum, 3-4-phyllum; squamæ scariosæ, paleis squamisque interioribus involucri generalis similibus et cum eis deciduæ. Corollæ tubo post anthesin basi skirroso-incrassato.

5. NEMATOPUS, nov. gen. . Involucrum partiale 2-florum, biseriale, 10-phyllum; squamæ, paleæ et squamæ involucri generalis consimiles, angustæ, apice lamina parva petaloidea appendiculatæ.

‡‡‡ Receptaculum generale filiforme, involucrum oligophyllum squamaceum longe superans.

6. CHRYSOCORYNE, Endl. Glomerulum amentaceum. Capitula biflora, inferiora sæpe uniflora.

†† Pappus e paleis parvis 2-5 discretis setam longam gerentibus.

7. ANGIANTHUS, Wendl. (Cylindrosorus, *Benth*.) Glomerulum cylindricum, involucro parvo paleaceo subtensum. Pappi setæ nudæ.

8. PHYLLOCALYMMA, Benth. Glomerulum subglobosum, foliis involucratum. Pappi setæ superne subplumosæ.

††† Pappus e paleis 5 fere discretis exaristatis ciliato-fimbriatis.

9. STYLONCERUS, Spreng. Glomerulum ovatum vel cylindraceum, foliis involucratum. Achænium papillosum.

†††† Pappus calyculato-coroniformis, nudus.

10. CEPHALOSORUS, nov. gen. Glomerulum globosum, involucro aut majusculo foliaceo aut paleaceo parvo cinctum. Capitula uniflora.

††††† Pappus e setis paucis tenuibus basi concretis laxe plumosis.

11. BLENNOSPORA, nov. gen. Glomerulum ovatum, oligocephalum; capitulis 2-floris, breviter pedicellatis.

12. ANTHEIDOSORUS, nov. gen. Glomerulum hemisphæricum; capitulis 1-floris arcte sessilibus, centralibus sterilibus. Involucrum generale radiatiforme.

** *Capitula 3–5-flora, in glomerulum globosum conferta, arcte sessilia. Receptaculum generale simplicissimum; partiale parvum, nudum.*

† Pappus nullus, vel e setis 2–3 tenerrimis nudis. Capitula 4–5-flora.

13. MYRIOCEPHALUS, Benth. Glomerulum depressum, involucro pluriseriali squamarum petaloideo-appendiculatarum cinctum. Achænia angusta.

†† Pappus e setis 4–12 plumosis vel plumoso-penicillatis. Capitula 3-flora.

14. LEUCOPHYTA, R. Br. Glomerulum sphæricum, squamis parvis uniseriatis involucratum, paleis consimilibus inter capitula.

15. CALOCEPHALUS, R. Br. Glomerulum ovoideo-globosum, exinvolucratum, epaleatum.

*** *Capitula 5-flora, in glomerulum globosum exinvolucratum aut parce folioso-bracteatum congesta. Receptaculum partiale paleis hyalinis inter flores onustum. Pappus e setis filiformibus plumosis.*

16. PYCNOSORUS, Benth. Capitula in glomerulum densissimum exinvolucratum aggregata, arcte sessilia. Flos centralis interdum sterilis vel neuter.

17. CRASPEDIA, Forst. Capitula in receptaculum generale cylindraceum indivisum pedicellata, singula bractea foliaceo-scariosa fulcrata; involucri squamis omnino hyalinis.

**** *Capitula 3–9-flora, pl. m. pedicellata, secus receptaculum cylindricum indivisum spicata, paleis latis scariosis bracteata, in glomerulum obovatum basi attenuatum paleis vacuis imbricatis involucratum dense aggregata. Involucrum partiale duplex, squamis interioribus petaloideo-appendiculatis. Receptaculum partiale nudum.*

18. GNEPHOSIS, Cass. Capitula 3–4-flora. "Pappus minimus, coroniformis, lacerus, caducissimus."

19. PACHYSURUS, Steetz (Actinobole, *Fenzl?*) Capitula 6–9-flora. Pappus caducus, e paleis 4–5 tenuibus brevissimis, basi in discum patellæformem hyalinum coroniformi-concretis, apice in fila totidem flaccida summo apice longiuscule ramificantia et ibi lanam intricatam referentia desinentibus.

***** *Capitula 5–15-flora, pl. m. pedicellata, folioso-bracteata, capitato-glomerata ; bracteis extimis (ut in* Craspedia) *glomerulum iuvolucrantibus. Axis, seu receptaculum generale, pl. m. divisum, lanatum. Receptaculum partiale nudum.* Transitus facile ad *Helichryseas.*

20. CROSSOLEPIS, Less.? Pappus nullus. Involucri squamæ inappendiculatæ, fimbriato-laceræ.

****** *Capitula 6–20-flora, in glomerulum exscapum humi adpressum, foliis radicalibus rosulatis involucratum, dense aggregata, sessilia. Receptaculum partiale paleis scariosis persistentibus flores fulcrantibus onustum. Pappus nullus vel coroniformis.*—Transitus ad *Cassinieas.*

21. CHAMÆSPHÆRION, nov. gen. Pappus coroniformis, lacerus, caducus. Capitula 5-7-flora: receptaculum planum.

22. CHTHONOCEPHALUS, Steetz. Pappus nullus. Capitula pluriflora: receptaculum conicum.

DITHYROSTEGIA, nov. gen.

Capitula pauca, pauciflora, in glomerulum involucro generali cyathiformi subspathaceo bivalvi herbaceo cinctum dense aggregata, singulis exinvolucratis. *Receptacula* generalia et partialia parva, angusta, angulata, villis prælongis flores subæquantibus onusta. *Flores* hermaphroditi, singuli palea lata hyalina tenuissima enervi superne in villis tenerrimis soluta obvoluti, pauci centrales ovario inani sæpius steriles. *Corollæ* tubulosæ, glabræ ; tubo gracili, limbo cyathiformi 5-partito. *Antheræ* basi caudatæ, appendice angusta superatæ. *Styli* rami apice capitellati, penicillati. *Achænia* fertilia ovoideo-fusiformia, hispidissima, suberostria; sterilia linearia, glabra. *Pappus* coronula setularum pilis achænii consimilium et breviorum.—*Herba* digitalis, annua, simplex, glaberrima; caule exili paucifoliato cum ramis raris glomerulo solitario terminato; foliis alternis squamæformibus oblongis ovatisve amplexicaulibus, supremo majore sæpe glomerulo approximato ; involucro e foliis duobus, basi vel ad medium usque coalitis. Flores lutei; lana candida. (Nomen ex δίθυρος, *bivalvis*, et στέγη, *tectum*, compositum, ad involucrum glomeruli bivalve alludens.)

D. *amplexicaulis.* (Ic. Pl. tab. ined.) South-western Australia, *Drummond,* 1850.—This singular little plant, of the division *Angian-*

theæ, is remarkable, not only for its two-lobed gamophyllous general involucre, and densely villous receptacle, but for the total want of partial involucral scales, and for the reduction of the palea which subtends each flower to an excessively delicate and diaphanous membrane. The whole might well be taken for a simple capitulum, with the receptacle obscurely branched. Some of the partial axes, or capitula, if they must be so called, bear six or eight, others two or three, or the central only single, flowers. The radical or primordial leaves are verticillate, about four in number, very small.

HYALOCHLAMYS, nov. gen.

Capitula uniflora, homogama, in glomerulum sphæricum involucro generali cinctum conglobata. *Involucrum generale* pluriseriale, e squamis latissimis tenuissime hyalinis orbiculatis retusis didymobiconcavis, nempe axi opaco valido subintruso, apice spathulæformi herbaceo. *Axis* seu *receptaculum generale* convexum, paleis persistentibus involucri squamas referentibus, sed sensim angustioribus atque nervo basi incrassato indurato, onustum, singulis capitulum unicum fulcrantibus. *Involucrum partiale* (capituli) 4-phyllum, lanigerum, corolla brevius; squamis 2 exterioribus (palea fulcrante contrariis) navicularibus, equitantibus, axi incrassato corneo, marginibus hyalinis, persistentibus; 2 interioribus alternis, minimis, totum hyalinis, deciduis. *Flos* hermaphroditus. *Corolla* filiformi-tubulosa, basi dilatata indurata et apicem ovarii tegens, apice leviter ampliata 5-dentata. *Antheræ* basi ecaudatæ. *Styli* rami apice capitellato-truncati. *Achænium* pyriforme, glabrum, basi attenuatum, calvum.—*Herba* pygmæa, annua, furfuraceo-pubescens, rhizocephala, nempe glomerulo inter folia radicalia spathulata (interiora basi seu petiolo hyalino-alato) sessili, seu flagellari-prolifera. Squamæ et paleæ glomeruli lanigeræ.

H. *globifera*. (Ic. Pl. tab. ined.) Swan River, *Drummond*.—The globose glomerules of this curious little plant vary from one to three lines in diameter, and the leaves are about the same length. Those which immediately subtend the glomerules, and form a sort of exterior involucrum for them, especially the proliferous ones, have a very short spathulate lamina, below which they are winged with broad hyaline margins; thus making a transition to the scales themselves, in which the green lamina, or tip, barely projects, or, in the inner ones, is within the retuse apex. These broad and silvery hyaline scales, of extreme thinness and delicacy, are externally, as it were, *didymous*, from the introflexion of their thickened axis or costa. In the inner paleæ the base becomes more and more thickened and corneous, like the outer squamæ of the partial involucres, this corneous axis making a strong contrast with the diaphanous margins. This genus should stand between *Skirrophorus*, from which it differs by its very remarkable and persistent squamæ and paleæ (both of the glomerule and of the one-flowered capitula), as well as in its whole habit, and *Hyalolepis*,* from which it is distinguished by the paleaceous general receptacle, the glabrous pyriform achænia, and the total absence of pappus.

(*To be continued.*)

Das KÖNIGLICHE HERBARIUM *zu* MÜNCHEN *geschildert.* (*Sketch of the* ROYAL HERBARIUM *at Munich.*) *By Dr. C. F. Ph. von Martius. Translated from the German in the* Gelehrten Anzeigen, Bd. xxxi. No. 89-93 [*separately printed at Munich,* 1850]; *by* N. WALLICH, M.D., F.R.S., V.P.L.S.

(*Continued from p.* 74.)

Zuccarini estimated the amount of species in his herbarium at 30,000; among which were many specimens received from Botanic Gardens, such as those of Erlangen, Munich, Berlin, Vienna, Paris, Padua, Verona, Milan, Kew, Löwen, Naples, Bonn, Montpellier, and St. Petersburg. Owing to his lamented early death, there was only found a numerical list of the genera, but none of the species, of this very considerable botanical treasure.†

* The examination of Cunningham's specimens of *Hyalolepis*, DC., in the Hookerian Herbarium, confirms, in most respects, the characters given by De Candolle, of the correctness of which, owing to the state of the specimen, he was uncertain. The narrow hyaline scales of the general involucre are entire, and the greenish costa is evanescent long below the summit. The general receptacle is flat and epaleate. The involucre of the capitula, which are sometimes two-flowered, is formed of three similar lanceolate-spathulate scarious scales, with a thin costa, firmly coalescent below the flower into a short and indurated terete base, or partial receptacle, which separates at maturity from the general receptacle. The corolla is 5-toothed, and the anthers are manifestly bicaudate. The very immature achænia are linear, and some of them apparently want the single delicate bristle of the pappus.

† We extract from this list a very general sketch, arranged according to Endlicher's method, as it may be of interest to know the numerical extent of an herbarium, raised by a continental botanist, during the last decenniums, by means of that system of

The following is a MS. estimate by Zuccarini, of the extent of his herbarium, with reference to the quarters of the globe :—

Flora Europæa	6500	species.
Flora Asiatica	6750	,,
Flora Africana	3380	,,
Flora Americana	7860	,,
Flora Australasica	1700	,,
	26,190	,,

Zuccarini added to this, besides, 5,000 plants of culture, from various gardens.

---

mutual exchange which we have pointed out; and may likewise serve as a sort of key, wherewith to compute the relative numerical proportions of the grand divisions of the Vegetable Kingdom.

1. Algæ	306		22. Principes	18		43. Rhœades	854
2. Lichenes	165		23. Coniferæ	139		44. Nelumbia	20
3. Fungi	204		24. Piperitæ	57		45. Parietales	350
4. Hepaticæ	101		25. Aquaticæ	9		46. Peponiferæ	60
5. Musci	767		26. Juliflorae	404		47. Opuntiæ	11
6. Equiseta	12		27. Oleraceæ	443		48. Caryophyllinæ	585
7. Filices	560		28. Thymeleæ	489		49. Columniferæ	455
8. Hydropterides	8		29. Serpentarieæ	34		50. Guttiferæ	200
9. Selagines	54		30. Plumbagines	149		51. Hesperides	38
10. Zamieæ	4		31. Aggregatæ	2817		52. Acera	218
11. Rhizantheæ	4		32. Campanulinæ	296		53. Polygalinæ	175
12. Glumaceæ	1813		33. Caprifoliaceæ	620		54. Frangulaceæ	254
13. Enantioblastæ	152		34. Contortæ	520		55. Tricoccæ	370
14. Helobieæ	22		35. Nuculiferæ	1312		56. Terebinthaceæ	317
15. Coronariæ	597		36. Tubifloræ	471		57. Gruinales	387
16. Artorhizæ	28		37. Personatæ	1090		58. Calyciflorae	360
17. Ensatæ	363		38. Petalanthæ	270		59. Myrtifloræ	505
18. Gynandræ	385		39. Bicornes	439		60. Rosifloræ	560
19. Scilomeneæ	84		40. Discanthæ	780		61. Leguminosæ	2325
20. Fluviales	46		41. Corniculatæ	346			
21. Spadicifloræ	60		42. Polycarpicæ	561			26,230

It may be interesting, also, in a literary point of view, to enumerate all the botanists from whom Zuccarini received direct or indirect contributions from the year 1819 to 1848; we accordingly subjoin an alphabetical list of them. Albers, Andrzeiowsky, Balsamo, Baumann, Bentham, Berger, Bertero, Bertoloni, Besser, Beyrich, Biasoletto, Bieberstein, Bischoff, Boissier, Bongard, Boek, Bracht, A. Braun, Brehm, R. Brown, Bruch, Brunner, Bunge, Chamisso, Chesney, Cumming, Czyhak, De Candolle, sen. et jun., Delille, D'Herigoyen, Diesing, Döllinger, sen. et jun., Drege, Duby, Duval, Ecklon, Erhardt, Einsele, Elsmann, Endlicher, Endress, Erdl, Eschweiler, Eschenloher, Eubel, Fenzl, Fischer (Petrop.), Fleischer, Fraas, Frank, Fries, Frölich, Fürnrohr, Funck, Gay, Graf, Gray, Hargasser, Haenke, Henschel, Heuffel, Hinterhuber, Hohenacker, Holl, Hooker, Hoppe, Hornemann, Hügel, Karwinski, Köberlin, Koch, Krauer, Krämer, Krauss, Kröber, Kummer, Landerer, Lang, Langsdorff, von Ledebour, Lehmann, Le Maire, Leo, Lessing, Leydoldt, Lindlachner, Lindley, Loddiges, Lucas, Lunz, Martens, von Martius, C. A. Meyer, E. Meyer, Meisner, Mertens, jun., Mettingh, Michahelles, Mikan, Mohl, Noë, Nolte, de Notaris, Ohmüller,

The specimens are of smaller size than those of the Academical Herbarium, and, as yet, are placed in papers of different qualities. They will be incorporated into the general collection, as soon as this shall have been made completely accessible to the public.

Herbarium Boicum.—By a resolution of his Majesty, the botanical examination of the kingdom was especially delegated to the Royal Academy of Sciences, and Dr. Otto Sendtner, assistant in the botanical conservatorium, was sent, during three years, in summer and autumn, to the southern parts, in order to collect phytogeographical and other data. It now became necessary to establish an herbarium, destined to comprise all the plants found in Bavaria, and communicated to the Royal Academy, together with all geographical information having reference to them, as the principal means to attain the object in contemplation. Besides the native plants of Bavaria, found by Sendtner, both on its hills and table-lands, this division of the Royal Herbarium received highly valuable additions from M. von Spitzel, Dr. Einsele at Berchtesgaden, Dr. Kummer, custos of the Royal Herbarium, the Rev. Mr. Ohmüller, M. von Krempelhuber at Mittewald, Professor Fürnrohr at Ratisbon, the Botanical Union at Augsburg, &c.

It has been a chief object in founding this Herbarium Boicum, that each species should be represented as completely as possible by instructive specimens, gathered in the different localities of its extension ; and that the remarks added to it, should point out the places of growth according to the degree of longitude, latitude, height above the sea, as well as the properties of the soil, and other data calculated to establish the vegetable geography and statistics of the Bavarian kingdom. With the view of enriching this collection of plants and of phytogeographical facts as rapidly as possible, the Botanical Conservatorium has set on foot an extensive communication with patriotic and zealous botanists in the kingdom, aided especially by the Royal Botanical Society

---

Otto, Panzer, Petter, Plaschnik, Pohl, Pöppig, Pritzl, Putterlick, Raab, Reinward, Reisseck, A. Richter, Rochel, Römer, Röper, Roth, Sadler, Salzmann, Sartori, Sauter, Savi, Schenk, Schiede, C. Schimper, N. Schimper, W. Schimper, von Schlechtendal, Schleicher, Schnitzlein, sen., Schott, Schubert, Schübler, Schuch, Schultes, Schultz, sen. et jun., Seitz, Sellow, Sendtner, Seringe, Sieber, von Siebold, Soleirol, Steven, Szowitz, Talbot, Tenore, Thomae, Trachsel, Tretzel, Treviranus, Trinius, Tschernjajew, Turczaninow, Unger, Vahl, jun., Visiani, Viviani, Voit, Wahlberg, Waldmann, Wallich, von Welden, Wiest, Waltl, Wolf, Frederic Zuccarini (brother of the collector of this herbarium).

of Bavaria; and it will furnish periodical reports of the results of all these combined researches.

Such, then, are the principal elements of the treasure of dried plants, which the Royal Herbarium has acquired since its foundation to the year 1850, consisting, at present, on computation, of 42,000 species. Its arrangement is that of Endlicher's 'Genera Plantarum' (Vindobonæ, 1836–40, 2 vols. 8vo). It is placed partly in high presses with glazed doors; partly in low cabinets. Each specimen is fixed by means of glued strips of paper on half a sheet of royal folio paper, having the labels of the authors or contributors attached. The autographs are carefully preserved, in order to enhance the historical value of the collection. Specimens belonging to the same species are all put within a common sheet of greyish paper, having, in a similar manner, the specific name attached at the lower end on the left side. All the species of one genus are similarly placed in a wrapper of blue paper, bearing the systematic generic name; and the genera are kept in portfolios, closed by means of tape. As a security against insects, a great number of papers drenched with *styrax liquida* are scattered among the packets; and the specimens which are known to be especially subject to attacks from insects, are powdered with *mercurius dulcis*, at the time when they are first attached to the paper, or subsequently from time to time. For security's sake, part of the collection is annually examined during summer.

With regard to the catalogue, the arrangement is likewise that of Endlicher, according to the natural families and genera; but the species are in alphabetical order. Up to the present period, the general catalogue is finished as far as *Leguminosæ*, inclusive.

The index of species in Zuccarini's herbarium was so far advanced, that he was able to finish the acotyledons and monocotyledons. There is a similar index of the specimens belonging to the Royal Ludwig-Maximilian University; they have already been incorporated, from time to time, with the general herbarium. A regular entry is separately made of each accession to the herbarium. It is the intention of the conservator to furnish the sheets of the general herbarium with a current number, as soon as the business of gluing the specimens to papers of uniform size shall have been accomplished; which, being added to the catalogue of species, will serve as a check on the general number of specimens.

VOL. III.

Other parts of the Collection.—The Royal Herbarium possesses, likewise, a considerable number of fruits, seeds, and flowers, preserved in spirits, specimens of wood, drugs, and other interesting productions of the vegetable kingdom. Several objects of this description, namely, a considerable series of specimens of woods, were contained in the Schreberian collection. Still more important is the series of fruits and seeds brought from Brazil by Spix and Von Martius, and likewise specimens of the different sorts of Peruvian barks, which they obtained from the son of Hippolyt Ruiz, author of the 'Quinologia;' besides other specimens of drugs. But owing to want of room and the requisite receptacles, these articles have not hitherto been scientifically arranged, or made accessible to the public. The series of woods, augmented by various specimens from the Royal Garden, possessing anatomical and physiological interest, has been placed in the open passage to the Royal Herbarium, in glazed cases.

There is likewise a small botanical library attached to the herbarium, for the most part transferred, with the Academy of Mannheim, from the botanical garden there. Of late years the most necessary handbooks for scientific gardening purposes have been acquired by the funds of the Royal Garden, where they are kept, and where is preserved, likewise, a collection of drawings of remarkable plants from the garden, with their analyses, executed in water colours, by Mr. Joseph Prestele, under direction of the conservator.

Locality.—So long as the Royal Herbarium consisted of the Schreberian collection only, it was placed in one single room (now one of the offices) of the Royal Academy. On the accession of the harvest of Spix and Von Martius's Brazilian Voyage, the locale was extended to the three apartments, in which the academical sessions are at present held. In 1839 the locality in the eastern wing of the Wilhelmina building, formerly occupied by the collection of engravings, became that of the herbarium, and to it were subsequently added, on building the new public library, two more apartments, which had hitherto been under the control of the directory of the library. At present the locality occupies an area of 2,755 square feet (Quadratschuhen); 1,767 being the extent of the six apartments, and 988, of the saloon which divides them. This locality is, however, destined for the mineralogical cabinets, as soon as the new buildings for the herbarium at the Royal Botanic Garden, which has already received the royal sanction, have been completed. It is only in this last most appropriate vicinity, that the interesting objects of the Royal Herbarium can attain their due development, for the benefit of botanical science and the good of the Royal Garden.

Publications which have emanated from the Royal Herbarium.

The conservators of the herbarium have, by means of its treasures, been enabled to publish the following larger and smaller works:—

I. VON FRANZ VON PAULA VON SCHRANK.

Anmerkungen, &c. (Notes on Panzer's List of twenty-five plants of the Bavarian Flora); in the fourth annual report of the math. and phys. class of the Royal Bavarian Acad. of Sciences, 1812, p. 236.

Omphalodes, a restored genus of plants. Memoirs of the R. Acad. for 1811 and 1812, p. 217.

Three rare Bavarian Plants. Ibid. 1813, p. 313.

Botanical Observations. Munich Memoirs, 1813, vol. v. p. 57.

Botanical Observations. Mem. of the Royal Bavar. Bot. Soc. at Ratisbon, vol. i. 1815, p. 104.

Porella, not a distinct genus. Nürnberg, 'Magazin für Nutzen und Vergnügen,' 1816.

Anacis, a new genus of plants. Mem. of the Munich Acad., 1817, Class ii. p. 1.

Aufzählung, &c. (Enumeration of some plants of Labrador, with observations). Mem. of the Ratisbon Bot. Soc., vol. i. part 2, 1817, p. 1.

Plants from Sarepta. Ibid. p. 157.

Four new Plants. Bot. Zeitung, 1819.

Commentatio de rarioribus quibusdam maximam partem Arabicis plantis in amplissima Schreberi collectione repertis. Munich Memoirs, vol. vi. 1817, p. 161.

Neue Beyträge, &c. (New Contributions to the Flora of Bavaria), vol. vii. 1820, p. 41.

Observationes in P. Leandri de Sacramento nova genera plantarum. Ibid. p. 239.

De plantis quibusdam Africanis commentariolus in Sylloge plantarum novarum vel minus cognitarum. Ratisbonæ, vol. i. 1824, p. 41.

Plantæ Ucranicæ. Bot. Zeit. 1822, vol. ii. p. 641.

Quatuor nova genera plantarum in Sylloge, &c., vol. i. 1824. p. 85.

Commentariolus in Irideas Capenses. Mem. of the Bot. Soc. at Ratisbon, vol. ii. part 1, 1822, p. 165.

Commentarius in Gnaphaloideas Capenses. Munich Mem., vol. viii. p. 141.

Botanische Beobachtungen (Botanical Observations). Botan. Zeit. 1824, vol. ii. append. p. 1-46.

Plantæ novæ aut minus cognitæ in Sylloge plantar. novar. Ratisbonæ, vol. i. p. 189, vol. ii. 1828, p. 55.

Spergula laricina restituta. Munich Mem. vol. x. 1832, p. 171.

II. C. Fr. Ph. von Martius.

Fasciculus Plantarum herbarii academici. Mem. of the Munich Acad. vol. vi. 1816, 1817.

Polygalæ quatuor novæ. Mem. Bot. Soc. Ratisb. vol. i. 1815, p. 183.

De Plantis nonnullis antediluvianis ope specierum inter tropicos viventium illustrandis. Ibid. vol. ii. 1822, p. 121, t. 2, 3.

Lychnophora, novum plantarum genus. Ibid. p. 148, t. 4-10.

Beytrag, &c. (Contribution to the Flora of Brazil, by Max. Prince of Wied, with descriptions by Nees von Esenbeck and Martius). Nova Acta phys. med. Acad. Cæs. Leop. Car. vol. xi. part 1, 1823, p. 1-88, t. 1-6; and vol. xii. 1824, p. 1-54, t. 1-8.

Fraxinellæ, plantarum familia naturalis definita; auct. Nees ab Esenbeck et Martio. Ibid. p. 147-190, t. 18-31.

Specimen Materiæ Medicæ Brasiliensis. Munich Mem. vol. ix. 1823, with nine plates.

Beytrag, &c. (Contribution towards a knowledge of the natural family of *Amarantaceæ*). Nova Acta, vol. xiii. 1826, p. 209-322.

Beyträge, &c. (Contributions towards a knowledge of the genus *Erythroxylon*). Munich Mem. 1840 (vol. xvi. part 2.), p. 281, with nine plates.

Fridericia, novum plantarum genus. Nov. Act. Acad. C. Leop. vol. xiii. part 2, 1827, p. 4-12, c. tab. 2.

The *Eriocauleæ* set forth and illustrated as a distinct family of plants. Ibid. vol. xvii. 1835, p. 1-72, t. 1-5.

Plantæ aliquot Brasilienses descriptæ. Mem. of the Ratisb. Bot. Soc. vol. iii. 1841, p. 295, with four plates.

Herbarium Floræ Brasiliensis. Flora, or Allg. Bot. Zeitung, 1837, vol. ii. (Suppl. numbers) and in the succeeding volumes.

The following are the more extensive works which have been founded on the specimens preserved in the Royal Herbarium:—

Nova Genera et Species Plantar. Brasil. Three vols., small fol., 1823-1830.

Historia naturalis Palmarum. Fol. max. three vols., 1823–1850.

Icones selectæ plantarum cryptogamicarum Brasiliæ. Small fol. 1826–1831.

Flora Brasiliensis, s. Enumeratio plantarum in Brasilia provenientium. Two vols. 8vo. Vol. I. *Cryptogama, auctor. Martio, Nees ab Esenbeck et Eschweiler.* Vol. II. *Agrostologia, auctore Nees.*

Flora Brasiliensis, ediderunt Steph. Endlicher et Martius. Nine numbers have been hitherto published. 1840–1847. Fol. c. tabulis.

III. JOS. GERH. ZUCCARINI.

Bemerkungen, &c. (Observations on Aug. de St. Hilaire's Monography on the genera *Sauvagesia* and *Lavradia*). Bot. Zeit. 1825.

Monographie, &c. (Monography on the American species of *Oxalis*). Munich Mem. vol. ix. 1823, p. 125, with nine plates.

Nachtrag, &c. (Addition to the above). Ibid. vol. x. 1829, p. 177.

Plantarum novarum vel minus cognitarum descriptio. Fasc. I. Ibid. p. 287. II. Vol. xiii. 1831, p. 309, with ten plates. III. Ibid. p. 597, with five plates. IV. Vol. xvi. part 1, 1837, p. 219, with nine plates. V. Vol. xix. part 2, 1845, p. 1, with six plates.

Plantarum quas in Japonia collegit Dr. Ph. Fr. de Siebold, genera nova. Fasc. I. Munich Mem. vol. xvi. 1843, p. 717, with five plates.

Zuccarini et Siebold, Floræ Japonicæ familiæ naturales. Sectio I. Munich Mem. vol. xix. part. 2, 1845, p. 111, with two plates. Sect. II. Ibid. part 3, 1846, p. 125, with one plate.

Administration and wants.—It will have been seen, from the preceding history of the rise and enlargement of the establishment, that everything was to be remodelled. The materials which gradually raised the herbarium to its present magnitude were multifarious, the access to them not properly regulated, and generally they were of such a nature, that it was only after a long scientific inquiry, and the systematic determination of many thousand plants, that their exhibition and indexing could be thought of. The want of a proper locality at first, of the requisite presses and the like, and the small sum allowed for the herbarium, were great obstacles. Until the year 1827–28, it was under no specific board of administrators, and the most necessary funds for its support had to be taken from those of the Garden, whose directors could not conscientiously devote any considerable sum for that purpose. It was only after the present spacious locality had been made available, in 1839–40, that the work of systematic arrangement could be properly undertaken; the business of determining and naming the plants having already been attended to. On this occasion it became manifest, how urgently necessary it was, that the conservators, who were constantly engaged in their duties at the Garden or in the lecture-room of the Academy, should have the aid of a proper assistant. To this office Dr. Ferd. Kummer was accordingly appointed that year. The herbarium has at present 300 florins annually for its maintenance and augmentation; a sum wholly inadequate for those purposes. It has been fully proved that 700 florins will be required for the purposes named. The following plain statement of expenses, incurred since the herbarium was first placed under a proper management, from 1827 to 1850, will establish this last-mentioned fact.

	Fl.	Kr.
Furniture for the plants and officers	1033	87
Freight and carriage	567	2
Paper for the specimens, catalogue, &c.	1872	54
Purchase of plants (rendered possible, in part, only by extraordinary grants)	1654	59
Allowances to scientific assistants, for copying, &c.	1466	12¼
Bookbinding	54	89
Contingencies, such as labels, glasses, preparation of the woods, spirits of wine, boxes, twine and tape, tinfoil, bladder, gum, washing, baskets, &c.	262	31
Gluing of specimens	745	26
Cleaning rooms, &c.	86	31
Total	7243	51¼*

It is clear from the above, that no more than 1,939 fl. 56 kr. have been expended during a series of 33 years, in the purchase and transport of new acquisitions, averaging less than 60 florins per annum. It is quite unnecessary to demonstrate the inefficiency of such a sum to maintain a public collection in a condition corresponding to the actual

* 7243 florins expended on the Royal Herbarium of Munich in 33 years;—equal to £579, or £17 sterling per annum! We could, without much difficulty, point to private herbaria, both in England and on the Continent, where we are sure that four times that sum (of £17) is spent annually in the purchase of *paper alone*: so that however thankful we may be to the excellent Von Martius for the interesting history of the formation of the Royal Herbarium, we cannot congratulate the country on the liberality of the Government in its encouragement of this department of the sciences in the capital of Bavaria.—Ed.

state and progress of the science, and capable of representing the aggregate materials for such a purpose. We will content ourselves with pointing out the numerous voyages undertaken during the last forty years, partly by private individuals, or by persons forming unions for supporting such expeditions, and partly at the public expense, whereby the treasures of regions hitherto unknown in botanical respects, have acquired such an importance, that the herbariums imported from thence have become objects of commerce. Such a state of things could not otherwise than exercise a powerful influence on our means of becoming acquainted with the forms and geographical distribution of plants; and it has, as it were, compelled leading public institutions, devoted to the extension and spread of botanical knowledge, not to limit their collections to a partial augmentation only, but to endeavour constantly to obtain additions from all zones and countries.

But, independent of these considerations, the present state and extent of the collections comprised in the herbarium of Munich, render it quite impossible that the yearly sum of 300 florins should suffice for making the herbarium generally available for scientific purposes. We mention here, especially, the laborious and tedious work of attaching the specimens to paper; for which objects alone, as well as for other purely manual labours, a fit bookbinder, with two pupils from Mr. Meyer's celebrated establishment for crippled boys, are employed.

It may reasonably be expected from the munificent head of an enlightened government, that a suitable increase of our means will be granted in time, to advance the grand work to such a state, that the herbarium when ready for being placed in the contemplated building at the Royal Garden, may be of general use to the public at large.

Second Report on Mr. Spruce's *Collections of Dried Plants from* North Brazil; *by* George Bentham, *Esq.*
(*See First Report,* vol. ii. of this Journal, pp. 209, 233.)

The collections forming the subject of the present report were made chiefly in the neighbourhood of Santarem, during Mr. Spruce's residence there in October and November 1849, and from January to October 1850, and during an excursion made from thence to Obidos, to the Rio

Trombétas and up one of its affluents, the Rio Aripecurú, as far as the first *cachoeiras*, or cataracts, in the months of November and December 1849. The detailed account of the localities visited will be found in Mr. Spruce's letters, inserted in this Journal, Vol. II. pp. 173, 193, 225, 266, and 299, and at p. 82 of the present volume, as well as in some further extracts about to appear. The collections made up to the end of March 1850 were received in this country in the end of July, and distributed to the subscribers early in October. Another portion, despatched from Santarem in July 1850, reached Pontrilas in the end of December; it is preparing for distribution as soon as the remainder, sent off on Mr. Spruce's leaving Santarem for Barra do Rio Negro, on the 7th of October, shall have arrived. This last despatch, now daily expected, will complete the Santarem collections, and will, it is hoped, be in time to be included in the following memoranda.

The subscribers will no doubt have observed that the specimens maintain well their character for excellence in drying and preservation. Mr. Spruce's residence at Santarem enabled him to supply many species in different states as to flower and fruit, including many new or rare ones of considerable interest, more especially among those from Obidos and from the Rio Aripecurú. It is only to be regretted, that, owing partly to his not having at the time been aware of the increased list of his subscribers, and more especially, as to the Obidos plants, owing to the difficulty of carrying with him a large stock of paper, there should still be so many species of which the number of specimens are insufficient to supply the whole of the sets. They are, however, much more numerous in the despatch last received than in the preceding ones.

The arrangement followed in the following observations is the same as that of the first report, the Obidos and Rio Trombétas and Aripecurú collections being included with those made at Santarem.

Among the *Dilleniaceæ* are specimens of the common *Curatella Americana*, Linn., whose geographical range extends from the Tierra Caliente of Mexico, over the whole of tropical America west of the Andes, to South Brazil; for the *C. çambaiba*, A. St. Hil., is surely, as already pointed out by some botanists, a mere variety rather more downy in some parts than the more common form. Mr. Spruce found it in open campos near Santarem, where it forms a low spreading tree, with the habit of an old oak. There have also been generally distributed specimens of a *Davila*, which appear to be a variety rather more hairy than usual of *D. asperrima*, Splitg. (Pl. Nov. Surinam., 1842, p. 1), published also by Miquel (Linnæa, vol. xviii. p. 611) under the name of *D. Surinamensis*. It is an intermediate form between the old *D. rugosa*, Poir., figured in Delessert's 'Icones' under De Candolle's name of *D. Brasiliana*, and the *D. radula*, Mart., Herb. Fl. Bras. no. 239; all of which may possibly prove to be varieties of one species extending over tropical Brazil and Guiana, and perhaps also into Columbia and some of the West Indian islands. Mr. Spruce describes the first single specimen he gathered as from a low procumbent shrub, and those more generally distributed as from a shrubby climber attaining about fifteen feet in height, with pale green flowers, strongly smelling of ripe peaches, and usually apetalous; there are also a few specimens of a smoother variety, with more constantly petaliferous flowers. All were from the campos near Santarem.

The *Anonaceæ* include the widely diffused *Xylopia grandiflora*, St. Hil., the *Xylopia barbata*, Mart., apparently confined to this part of Brazil, a new *Anona*, described below, and two Guiana species of *Guatteria*. The one, *G. Ouregou*, DC., has already been indicated in North Brazil; the other, *G. Schomburgkiana*, Mart. (Fl. Bras. *Anon.* p. 36), agrees in every respect with Schomburgk's specimens, except that the pedicels are perhaps rather longer. The same species appears again among Hostmann's Surinam plants (no. 1221), and has been published by Steudel (Flora, 1843, p. 754) under the name of *Anona Hostmanni*. Mr. Spruce found it on gravelly hills near Santarem. The new *Anona*, above mentioned, was from the Serra d'Escamas, near Obidos, where it formed a dwarf shrub of about three feet high, allied in other respects to the *A. tenuiflora*, Mart. (Fl. Bras. *Anon.* p. 10. t. 3). There were, unfortunately, but three or four specimens received. It may be thus described :—

Anona *humilis*, sp. n.; trunco humili frutescente, ramulis pedunculis venisque foliorum subtiliter ferrugineo-strigillosis, foliis ellipticis v. obovato-oblongis cuspidatis membranaceis præter venas glabris, pedunculis lateralibus solitariis minute bracteolatis, sepalis minimis acutis, petalis subæqualibus rhomboideis acuminatis basi demum in unguem contractis tenuibus subtiliter pubentibus.—*Frutex* 3-pedalis. *Folia* breviter petiolata, 4–6-pollicaria, latiora quam in icone *A. tenuifloræ* Mart. depicta, cæterum iis subsimilia. *Pedunculi* semipollicares v. paulo longiores. *Sepala* linea breviora, angusta, acuta, sæpe decidua (v. deficientia ?). *Petala* alba, quam in *A. tenuiflora* minora et latiora, vix 6 lin. longa, distincte acuminata et in unguem basi contracta, medio late rhomboidea et concava. *Stamina* numerosissima.

The *Menispermaceæ* have been examined by Mr. Miers, who has recognized three new species of *Cissampelos*, of which one only, *C. Amazonica*, Miers, has been distributed into most sets under the name of "Cissampelos, sp. n., affinis C. ovalifoliæ," the two others having been only in single, or in very few, specimens. Another plant inserted in a very few sets under the name of "Cocculus affinis C. lævigatæ, Mart.," is the *Anelasma Spruceanum*, Miers. Besides these, a very curious plant has been rather more generally distributed as "Genus novum affine Menispermaceis:" it is the *Aptandra Spruceanum*, Miers, whose very doubtful affinities will be more fully discussed by Mr. Miers on a future occasion. In the meantime, that gentleman has kindly communicated to me the following characters for the above-mentioned new species:—

1. Cissampelos *assimilis*, Miers, n. sp.; ramulis striatis tomentosis, foliis orbiculato-ovatis undulato-sinuatis apice retusis opacis supra pubescentibus subtus cinereo-tomentosis (junioribus acutis utrinque fulvo-tomentosis) longiuscule petiolatis, petiolo erecto, limbo subrefracto, racemis paniculatis ternis axillaribus petiolo subæquilongis tenerrimis, sepalis spathulatis extus pilosis, corolla 4-loba glabra, anthera 4-loba.—Prope Santarem.

This closely approaches *C. crenata*, DC., but the leaves are more glabrous above; the under surface not so densely tomentose. The petiole is half the length of the blade (while in *C. crenata* the leaves are almost sessile), the panicles are three (not two) in each axil, the peduncle and pedicel are far more slender, and the anther is 4- (not 6-) lobed.

2. Cissampelos *Amazonica*, Miers, n. sp.; foliis ovalibus subacutis crenato-sinuatis mucronatis coriaceis supra glabris nitidis reticulatis subtus fusco-tomentosis 5–7-nerviis sublonge petiolatis, racemis fœm. (ramulis novellis) binis folio brevioribus bracteatis, bracteis conspicuis orbicularibus mucronatis utrinque tomentosis, floribus minutis pedicellatis in axillas 6–7-fasciculatis, fructu piloso.—Prope Santarem.

This plant may be confounded with *C. ovalifolia*, DC., which it greatly resembles in the shape and texture of its leaves, but their upper surface is far more shining; the petiole is about one-fourth the length of the blade, while in *C. ovalifolia* it is scarcely more than one-twelfth or one-fifteenth of that length; the racemes (or rather young floriferous branchlets) are also much longer, and in axillary pairs (not single); the bracts or young leaflets are more membranaceous (while in the other they are thick and opake), with six or seven (instead of four) flowers in each axillary fascicle.

3. Cissampelos *denudata*, Miers, n. sp.; volubilis, caulibus striatis tuberculatis junioribus retrorsum ferrugineo-pubescentibus delapsu foliorum cito nudis et ex axillis hinc deinde floriferis, foliis (marium) orbiculato-cordatis obtuse acuminatis mucronatis membranaceis subtus cinereo-glaucis utrinque parce pubescentibus 7–9-nerviis petiolo gracillimo griseo-pubescente æquilongis, racemis masc. binis longissimis gracilibus paniculatis bracteisque minutis linearibus caducissimis pubescentibus, fœm. 2–4 simplicibus crassioribus multo brevioribus, pedicellis 5–6 fasciculatis 1-floris, fasciculis bractea unica minuta lineari donatis, baccis glabris?

Found by Mr. Spruce near Obidos, and on the lake Quiriquiry, near the river Trombétas: it is evidently a long straggling climber, known to the natives by the Caribbean name of *Ambóa Rembó*. The leaves (at least in the male plant) are broadly heart-shaped, tapering obtusely to the summit, where it is terminated by a pubescent mucronate point; they are $2\frac{1}{2}$ inches long, or $2\frac{1}{4}$ inches from the insertion of the petiole, and $2\frac{1}{4}$ inches broad; the petiole of marginal insertion is slender, reflexed at its origin, fulvo-pubescent, and $1\frac{1}{4}$ inch long. The male racemes are very slender, 6–8 inches long, pubescent, and throw out at intervals of every $\frac{1}{4}$ or $\frac{1}{2}$ inch, three fasciculate pedicels $\frac{1}{4}$ to $\frac{1}{2}$ inch in length, which are again dichotomously divided, bearing numerous minute florets; at each branchlet there is a very minute linear tomentose bract, so that the inflorescence appears almost ebracteate: the ultimate pedicels are smooth, the flowers quite glabrous, the sepals spathulately linear and obtuse, the corolla deeply bell-shaped, with an obsoletely 4-lobed margin, and the annular anther is 4-celled. About three racemes grow out of the bare axils of the female plant, each having a much stouter peduncle, about 2 inches long, producing, at intervals of one-eighth of an inch, about five or six fasciculate pedicels, each bearing a solitary minute flower · at the base of each fascicle is a minute and very caducous bract, so that the female racemes are apparently ebracteate: the sepal and petal are both glabrous, the former obovate, the latter much shorter and broader : the ovarium is sericeous, crowned by three reflexed pointed stigmata. It appears that in both the male and female plants, the branches soon lose their leaves, or probably do not renew them on the second year, when the bare axils, at about four inches apart, throw out their racemes of flowers.

4. Anelasma *Spruceanum*, Miers, n. sp.; caule tereti nitido vix striato, axillis nodosis, foliis late ellipticis basi cuneatis apice repente attenuatis mucronatis margine reflexis glaberrimis subtus pallidis 5-nerviis, nervibus 3 prominulis rubescentibus, lateralibus tenuibus fere marginalibus et mox evanidis, venis transversis conspicuis et anastomosantibus, petiolo canaliculato apice basique valde incrassato 3-plo longioribus, racemis masc. binis subpaniculatis supra-axillaribus gracilibus petiolo longioribus glabris, bracteis membranaceis ciliatis, sepalis exterioribus extus pubescentibus.—Ad fluv. Amazon., inter Santarem et Obidos.

Near A. *Sellowianum*, but with larger and broader leaves, which are not so thick and coriaceous, and with shorter petioles; the racemes are not pubescent, as in that species. The leaves are here 5 inches long, $2\frac{1}{4}$ inches broad, with a petiole $1\frac{1}{4}$ inch long; the racemes are 2 inches in length.

APTANDRA, gen. nov. *Calyx* brevissimus, patelliformis, 4-dentatus, 4-sulcatus, carnosus, persistens et demum augescens. *Petala* 4, æqualia, calycis lobis alterna, carnosula, lineari-lingulæformia, summo parum latiora et concava, apiculo inflexo, æstivatione valvata, demum spiraliter reflexa. *Squamæ* petaloideæ 4, liberæ, petalis alternæ, inter eadem et tubum staminalem sitæ, crassæ, rotundatæ. *Stamen* integrum (forsan 4 coalita), longitudine fere corollæ, tubo tereti, cylindraceo, carnoso, pistillum cingente; *anthera* ex loculis 8 oblongis, æqualibus, arcte in annulum extrorsum dispositis, et in connectivum fere globulare crasso-carnosum summo pervium immersis, singulatim valvula exteriore membranacea ab apice ad basin valvatim soluta et hinc omnino reflexa. *Pollen* subfarinaceum, granulis amplis, cruciato-lobatis. *Ovarium* substipitatum, conico-oblongum, subcompressum, 2-sulcatum, 2-loculare; *ovula* in loculis solitaria, anatropa, obovata, facie ventrali ad dissepimentum funiculo brevissimo utrinque affixa. *Stylus* erectus, longitudine staminis. *Stigma* oblongum, compressum, obtusum, inclusum. *Fructus* ignotus.—Arbor *biorgyalis Amazonicus*; folia *alterna, elliptica, penninervia, reticulata, petiolata, exstipulata.* Inflorescentia *paniculata, axillaris, gracilis;* flores *minimi.*

1. Aptandra *Spruceanum*, Miers; foliis ellipticis subreflexis apice subito-attenuatis utrinque glabris subtus punctis minutissimis lentiginosis et pellucidis notatis, rachi nervisque rubentibus, paniculis folio tertio brevioribus, pedicellis gracilibus subfasciculatis in fructu valde elongatis et crassioribus, bractea e dichotomiis minuta caducissima.—Prope Obidos. (J. Miers.)

Among *Nymphæaceæ*, besides the *Victoria*, were a few specimens of a *Nymphæa*, probably a variety of *N. ampla*, DC.; but the leaves are rather bluntly sinuated than sharply toothed, and on that account Salzmann, who gathered the same form near Bahia, gave it the MS. name of *N. sinuata.* The *Cabomba* distributed as *C. aquatica*, is the variety or species described by Gardner as *C. Piauhiensis*, and figured Hook. Ic. t. 641.

The *Capparideæ* contain single specimens of *Cleome villosa*, Gardn., and of *C. Hostmanni*, Miq., both from Obidos; the latter, perhaps, a mere aculeate form of *C. latifolia*, Vahl. Also a few specimens of *Physostemon intermedium*, Moric., from Lake Quiriquiry; of a variety of *Capparis lineata*, Pers., from Obidos, with much less of down on the leaves than in the ordinary Brazilian forms; and lastly, of a species of *Capparis*, from Santarem, very nearly allied to, if not identical with, *C. Vellosiana*, Mart.

To the *Bixaceæ* belongs a new species of *Lindackeria*, from Obidos, nearly allied to the *L. Maynensis*, Pöpp. (Nov. Gen. et Spec. vol. iii. p. 63, t. 270), and distributed under the name of *L. latifolia*. The genus is, perhaps, too closely allied to Aublet's *Mayna*, and especially to the three species which I described from Schomburgk's collection, and which Walpers has referred to *Carpotroche*, for what reason I am unable to guess, as they have not the peculiar fruit, nor the other characters, by which *Carpotroche* has been distinguished from *Mayna*. As the original species of *Mayna* is as yet imperfectly known, and as Mr. Spruce's plant is certainly a congener of Pöppig's, and probably of Presl's *Lindackeriæ*, I here describe it under the name I gave it in distributing. I have, however, little doubt that further investigation will cause to be united under *Mayna*, not only the three species I formerly described, but all the *Lindackeriæ* and two or three as yet unpublished species I have in my herbarium; whilst Pöppig's *Mayna longifolia* (Nov. Gen. et. Spec. vol. iii. p. 64, t. 271.) is said to have plaited wings to the fruit, and would, therefore, prove most probably to be a second species of *Carpotroche*.

Lindackeria *latifolia*, sp. n.; glabra, foliis longe petiolatis late ovatis acuminatis integerrimis, racemis petiolo brevioribus plurifloris, sepalis ovalibus, petalis (9) calycem subæquantibus caducissimis, filamentis hirtellis anthera longioribus.—*Frutex* 15-pedalis, ramis foliisque siccitate pallentibus. *Folia* 6-8-pollicaria, 4-5 poll. lata, petiolo 2-3-pollicari, tenuiter coriacea, concoloria, penninervia et reticulato-venosa. *Racemi* axillares, axi brevi a basi florifera. *Bracteæ* minutæ. *Pedicelli* pollicares, ad 1½ lin. a basi articulati. *Alabastra* clavata. *Sepala* 3, ovali-oblonga, concava, alba, patentia v. reflexa, diu persistentia, 5-6 lin. longa. *Petala* in alabastro a me aperto novem vidi, in floribus apertis fere omnia jam delapsa; alba, tenuia, oblique oblonga, obtusa, sepalis subbreviora. *Stamina* numerosa, toro dilatato inserta, in floribus masculis totum torum vestientia. *Filamenta* erecta, arcte approximata, columnam simulantia at non vere cohærentia, 2-2¼ lin. longa, pube brevi villosula. *Antheræ* lineares, 1-1½ lin. longæ, loculis summo apice discretis breviter bifidæ, basi et apice hirtellæ et ad rimas loculorum plus minus ciliolatæ. *Ovarium* in floribus hermaphroditis sessile, ovoideum, verrucosum et mox echinatum, glabrum, intus uniloculare, apice desinens in stylum flexuosum staminibus longiorem leviter puberulum summo apice expansum in stigma minute 6-lobum. *Placentæ* in ovario tres, parietales, ovulis numerosis. *Fructus* junior globosus, aculeis longis mollibus pyramidatis undique echinatus; maturum non vidi.

Among the Pará plants was a single specimen of the following additional and very distinct large-flowered species of the same genus: it was gathered at Tanaü in September 1849 :—

Lindackeria *pauciflora*, sp. n.; glabriuscula, foliis breviter petiolatis ovali- v. oblongo-ellipticis acuminatis, racemis 2-3-floris, pedicellis petiolum longe superantibus, sepalis orbiculatis, petalis (9) calycem superantibus, filamentis anthera multo brevioribus.—*Frutex* glaber v. partibus novellis minute puberulis. *Folia* majora, semipedalia, 2¼ poll. lata, acumine abrupto angusto longiusculo, petiolo rarius semipollicem excedente. *Racemorum* axis brevis; pedicelli ultrapollicares, prope basin articulati. *Sepala* alba, 5 lin. longa. *Petala* alba, anguste et oblique oblonga, pleraque acutiuscula, 6 lin. longa.

Stamina numerosissima, in floribus masculis totum torum tegentia, petala æquantia; filamentis abbreviatis vix puberulis. *Antheræ* elongatæ, apice hirtellæ. *Ovarium* in floribus hermaphroditis et fructus junior omnino *L. latifoliæ*.

Both the above species are polygamous; in both of them hermaphrodite and male flowers are to be found in one and the same raceme, as is probably the case in all *Lindackeriæ* and true *Maynæ*.

The following *Flacourtia*, from Santarem, is also new; the specimens have all male flowers only.

Flacourtia *nitida*, sp. n.; inermis glabra, foliis breviter petiolatis ovatis breviter acuminatis basi cuneatis margine calloso-serratis venosis supra nitidis, floribus (masculis) glomeratis, pedicellis glabris, calycis foliolis ciliatis, disco 8-lobo.—*Frutex* biorgyalis, ramosissimus, ramulis teretibus glabris punctis albis notatis. *Folia* 1½–2 poll. v. rarius 3 poll. longa, circa pollicem lata, petiolo 1–2 lin. longo, rigidule chartacea, venis prominulis, faciebus concoloribus, superiore nitida, inferiore opaca, punctis pellucidis paucis notata. *Racemuli* brevissimi, glomerulos 6–10-floros formantes. *Bracteæ* ad basin pedicellorum oblongæ, cymbæformes, pedicellis breviores. *Pedicelli* vix lineam longi. *Flores* parvi. *Calycis* foliola 4, ovali-oblonga, obtusa, concava, margine ciliolata, semilineam longa. *Discus* annularis, fere usque ad basin divisus in glandulas 8 truncatas v. retusas. *Stamina* circa 20, calyce paulo breviora, receptaculum totum vestientia, ovarii rudimento nullo. *Stirps* fœminea mihi ignota.

There are four *Samydeæ* from the neighbourhood of Santarem, viz.:—

1. Casearia (Iroucana) *ramiflora*, Vahl, var.? This is the large-leaved form, or true *Iroucana Guianensis* of Aublet, referred to in Hook. Journ. Bot. vol. iv. p. 110, which, if really different from the small-leaved West Indian form, might take the name of *C. Iroucana*. Blanchet's n. 3119, referred to in the same paper as a new species, has been since published by Miquel ('Linnæa,' vol. xxii. p. 801) under the name of *C. Blanchetiana*. He describes it as decandrous, but the flowers in my specimen are certainly 5-fid, with 8 stamina, as in all other species of the same section.

2. Caseariæ (Pitumba) *spec.*; perhaps new, but closely allied to *C. stipularis*, Vahl, from which it only appears to differ in its smaller and less toothed leaves, and smaller stipules and flowers. It is, also,

very near to *C. Benthamiana*, Miq., 'Linnæa,' vol. xviii. p. 737, and *C. lanceolata*, Miq., l. c., p. 753 (two species scarcely distinct from each other, if my specimens are correctly determined), but differs from them chiefly in the leaves being hoary underneath. The plant described by Miquel in the 'Annales des Sciences Naturelles' (Ser. 3, vol. i. p. 38), inadvertently under the same name of *C. Benthamiana*, appears to be a very different species, belonging to the section *Crateria*.

3. Casearia (Crateria) *sylvestris*, Sw.? (DC. Prod. vol. ii. p. 49); foliis ovatis oblongisve acuminatis integerrimis v. obsolete crenato-serratis subcoriaceis nitidulis subtus pallentibus crebre punctatis ramulisque glabris, umbellis sessilibus, floribus parce pubentibus 5-fidis, staminibus 10 calyce paulo brevioribus, stylo trifido.—*Arbor* triorgyalis, trunco tenui, coma ramosissima. *Folia* 2-3-pollicaria, acumine longo obtuso, basi rotundata v. cuneata, parum inæquilatera v. rarius æquilatera, tenuiter penninervia et reticulato-venosa, venulis in pagina superiore vix conspicuis, petiolo vix 2 lin. longo. *Nodi* floriferi axillares, tomentosi, multiflori. *Pedicelli* lineam longi. *Flores* magnitudine *C. parviflora*, albi, calyce turbinato, laciniis oblongis. *Ovarium* glabrum, et fructus fere *C. parviflora*.

4. Casearia (Piparea) *Javitensis*, Humb. et Kunth; a species very generally diffused over Columbia, Guiana, and North Brazil. Two specimens separately gathered in the forest near the Tapajoz, differ slightly in their nearly sessile flowers, but are probably not specifically distinct.

(*To be continued.*)

Contributions to the Botany of WESTERN INDIA;
by N. A. DALZELL, ESQ., M.A.

(*Continued from p.* 90.)

Nat. Ord. LABIATÆ.

DYSOPHYLLA.

D. *rupestris*; suffruticosa 3-4-pedalis, radice crassa lignosa, caule erecto parum ramoso, ramis pilis patulis dense velutino-tomentosis, foliis ternis quaternisque lineari-lanceolatis acutis serratis basi angustatis utrinque molliter pubescentibus, floralibus lineari-spathulatis flores subæquantibus, *calycis obconici* dentibus *æquilateraliter triangularibus obtusis* pilis albis rigidis dense ciliatis.

Folia 10-20 lin. longa, 2-3 lin. lata, utrinque glandulis albis conspersa. *Calyx* 1 lin. longus, pilis albis rigidulis hirsutus, dentibus tubo 3-plo brevioribus. *Spicæ* 3-4 poll. longæ, graciles, densifloræ, floribus minutis. *Anthemidem nobilem* redolet.—Crescit in rupibus prov. Malwan.

I do not give this so much as a new plant, as with a view to elucidate a seeming inconsistency with respect to Roxburgh's *Mentha quadrifolia* and Bentham's *Dys. quadrifolia* being one and the same plant. I gathered the plant now described on dry rocks, where no water could lodge, and this seems to me to point to a plant having a very different habit and constitution to the "herba aquatica" of Bentham. I believe, therefore, that Roxburgh's *M. quadrifolia*, which he describes as "growing among rocks," and for which, he adds, "common garden-soil is too moist," is the plant above described, and not Bentham's *D. quadrifolia*, which is said to grow naturally in stagnant water. The shape of the calyx-teeth is also very different.

Nat. Ord. MEMECYLEÆ.

MEMECYLON.

M. *terminale*; frutex 2-3-pedalis, ramis dichotomis teretibus gracilibus, foliis sessilibus lanceolatis acuminatis, pedunculis axillaribus terminalibusque solitariis semipollicaribus, floribus umbellatis pedicellatis, pedicellis pedunculo 2-plo brevioribus.

Folia obscure penninervia, 2-2¼ poll. longa, infra medium 9-10 lin. lata, basi rotundata. *Fructus* globosus, siccus, unilocularis, pisi majoris magnitudine. *Semen* unicum: *cotyledones* reniformes; *radicula* longiuscula, compressa, duobus lateribus alato-marginata.—Crescit in umbrosis montanis prov. Canara; fruct. Mart. Apr.

This very distinct species I found in Canara, in localities very far from each other, so that it is probable that it is common throughout that province. It is, perhaps, the smallest of our Indian species, as I have never seen it higher than three feet, and very slender. It is sufficiently distinguished by its having always terminal as well as axillary inflorescence, though the axillary is often wanting. It resembles *Mem. Heyneanum* most in the shape of its leaves, but differs by the leaves being sessile and the peduncles solitary.

Nat. Ord. RUBIACEÆ.

IXORA.

I. *pedunculata*; fruticosa, foliis breve petiolatis ellipticis coriaceis glabris, stipulis triangularibus breve cuspidatis, panicula terminali trichotoma parva laxa *pedunculum semipedalem nudum terminante*, inflorescentiæ ramis bracteis subulatis suffultis.

Folia 5 poll. longa, 2 poll. lata. *Calyx* basi bibracteolatus, puberulus, dentibu striangularibus obtusiusculis, post anthesin patentibus. *Corollæ tubus* $3\frac{1}{2}$ lin. longus; limbi segmenta ovato-oblonga, acuta, $1\frac{1}{4}$ lin. longa, sub anthesi reflexa. *Filamenta* crassa, brevissima; *antheræ* basi sagittatæ, apice acuminatæ, laciniis corollinis breviores. *Stigma* claviforme, exsertum. *Fructus* glaber, leviter bilobatus.—Crescit in montibus Syhadree, prope Parwarghat; floret Feb.

Readily distinguished from all other species, by having the inflorescence on a long bare peduncle, twice the length of the last pair of leaves; the flowers are pretty numerous, small, and pink.

Nat. Ord. EUPHORBIACEÆ.

ROTTLERA.

1. R. *Mappoides*; dioica, foliis amplis longe petiolatis cordato-triangularibus acuminatis peltatis, adultis supra glabris subtus tomento stellato albo-ferrugineo densissime vestitis, fol. junioribus utrinque tomentosis, floribus masculis spicatis, spicis axillaribus compositis, floribus aggregatis 5-6-nis nudis.

Calyx 3-partitus, extus stellato-tomentosus, utrinque glandulis flaviæ conspersus. *Stamina* plurima, libera, filamentorum apices glanduloso-graniformes. *Spicæ* fœmineæ terminales, 4-5-pollicares, densissime ferrugineo-tomentosæ. *Calyx* 4-partitus, laciniis oblongis v. ovatis, extus densissime pilis stellatis ferrugineis farinosis vestitus. *Capsula* 3-4-locularis. *Stigmata* 3-4, plumosa. Tota præter folii adulti paginam superiorem densissime ferrugineo-tomentosa.—Crescit in montibus Syhadree; floret tempore frigido. R. *peltatæ*, Rox., valde affinis.

2. R. *aureopunctata*; dioica, foliis oppositis oblongo-obovatis acuminatis basin versus attenuatis breve petiolatis supra glabris nitidis subtus pilis raris stellatis squamulisque crebris aureo-nitentibus conspersis, stipulis linearibus ferrugineis caducis, floribus racemosis, racemis axillaribus terminalibusque simplicibus folio brevioribus, pedicellis brevibus bracteatis, floribus masculis numerosis in rachi 5-nis fasciculatis, calycis 4-partiti laciniis late ovatis reflexis, staminibus plurimis receptaculo nudo insertis, exterioribus brevioribus.

Antherarum loculi, connectivo in glandulam apice bilobatam producto, utrinque adnati. *Racemi* fœm. 4-5-flori, calycis laciniis in spatham uno latere fissam coalitis. *Stylus* longus, crassus, apice trifidus, ramis plumosis recurvis simplicibus. *Capsula* diametro 9 lin., setis mollibus pilosis vestita.

CROTON.

C. *hypoleucos*; monoica, ramis petiolis rachique ferrugineo-tomentosis, foliis ellipticis utrinque acutis integris longiuscule petiolatis, squamulis argenteis stellatis supra sparse subtus densissime obsitis, lamina basi glandulis 2-4 stipitatis prædita, floribus racemosis, racemis axillaribus terminalibusque folia subæquantibus, floribus masculis numerosis geminis superioribus, fœmineis inferioribus distantibus solitariis paucis (1-2).

FL. MASC. *Calyx* 5-fidus, segmentis ovatis stellato-tomentosis. *Petala* 5, lineari-oblonga, dense albo-lanata, calyce paulo longiora. *Stamina* 15 : *filamenta* libera, filiformia, glabra, in receptaculo villoso inserta; *antheræ* adnatæ, muticæ. FL. FŒM. *Calyx* 4-5-fidus, laciniis foliaceis, lanceolatis, obtusis, utrinque pilis stellatis conspersis. *Styli* 3, ramis stigmaticis capitatis. *Ovarium* pilis stellatis ferrugineis densissime obtectum.—*Frutex* 3-4-pedalis, in sylvis umbrosis cum præcedente. Fl. temp. frigido.—*Crot. bicolori*, Rox., valde affinis : an differt ?

Nat. Ord. AMPELIDEÆ.

VITIS.

V. *Canarensis*; tota pubescens, caule herbaceo, foliis trifoliolatis longe petiolatis, foliolis oblongis subito acuminatis basi cuneatis margine anteriore serrulatis membranaceis utrinque pubescentibus.

Petala 4, distincta, extus pubescentia. *Filamenta* subulata, sub disci plani margine inserta. *Stylus* brevis : *stigma* acutum. *Petiolus* $3-3\frac{1}{2}$ poll. longus ; *foliola* 4 poll. longa, 2 poll. lata, lateralia basi inæqualia, brevius petiolulata.—Crescit in Canara ; fl. Aprili. *Cissus hirtella*, Blume, an eadem ?

Nat. Ord. MALVACEÆ.

ABELMOSCHUS.

A. *Warreensis*; caule setulis bulbosis erectis exasperato, foliis late cordatis acuminatis grosse crenatis utrinque pilis rigidis simplicibus vel stellatis conspersis, calyce spathaceo scarioso pubescente apice bifido, involucro 4-*partito persistente cum fructu* increscente, laciniis late ovatis extus hispidulis intus pubescentibus, floribus in ramorum brevium axillarium apice fasciculatis vel terminalibus.

Fructus lineari-oblongus, rostratus, pentagonus, pilis albis longis patulis rigidis vestitus. *Pedicelli* ½–1 poll. longi, basi bracteis 2 subulatis suffulti. *Folia* cum petiolo 3-pollicari 6–7 poll. longa, 3¼ poll. lata. *Flores* lutei, fundo purpurei.—Crescit in regno Warreensi; fl. et fruct. Januario.

The four-divided involucre, which is persistent, and increases in size with the fruit, sufficiently distinguishes this species, although it is not the only species which has a four-leaved involucre; the *Hibiscus tetraphyllus* of Roxburgh, which also belongs to this genus, has a similar involucre, but very different leaves.

(*To be continued.*)

Figure and Description of a new species of RANUNCULUS, *from the Rocky Mountains; by* SIR W. J. HOOKER, D.C.L., F.R.S.A.

(TAB. IV.)

RANUNCULUS (§ *Ranunculastrum*) DIGITATUS.

Humilis, glaberrimus, radice grumosa, foliis paucis petiolatis digitato-subpedatifidis laciniis oblongo-spathulatis, floribus 2–3 terminalibus, sepalis 5 patenti-reflexis, petalis 7–11 oblongo-cuneatis, capitulis subglobosis, carpellis ovatis glabris stylo subulato paululum recurvato terminatis.

HAB. Rocky Mountains, near Fort Hall. *Mr. Burke.*

Planta digitalis, glaberrima. *Radix* e tuberibus paucis (3–5) clavatis fasciculatis, una cum fibris 2–3 crassiusculis e summitate tuberum. *Caulis* brevis, simplex, parce foliosus. *Folia* digitata, laciniis 3–4, oblongo-spathulatis, exterioribus nunc bifidis bipartitis vix radicalibus sublonge petiolatis, caulinis (sub-2) approximatis subsessilibus. *Flores* 2–3, terminales, pedunculati, magnitudine variant. *Calycis sepala* 5, membranacea potius quam herbacea, ovata, patenti-reflexa, petalis duplo breviora. *Petala* flava, 7–11, cuneato-oblonga, obtusissima, ungue squama nectarifera. *Stamina* numerosa. *Antheræ* ovatæ, adnatæ. *Pistilla* numerosa, in capitulum subglobosum congesta. *Ovaria* ovato-subrotunda, compressa. *Stylus* ovario longior, late subulatus, paululum recurvus.

With a general habit resembling the arctic species of *Ranunculus*, *R. nivalis* and *R. Eschscholtzii*, this species differs very much in other respects from them, having a grumose root (not much unlike that of our *Ranunculus Ficaria*), and in that character according with the *Ranunculastrum* groupe of De Candolle, and in the numerous narrow petals it approaches several of the South American species.

Tab. IV. Fig. 1. Flower; fig. 2, petal with nectary; fig. 3, stamen; fig. 4, capitulum of carpels; fig. 6, single carpel :—*magnified*.

BOTANICAL INFORMATION.

Mr. N. Plant's *Natural History Journey in South America.*

Mr. Plant, late Curator of the Leicester Museum of Natural History, has issued circulars explanatory of his intended visit as a collector to South America, the Sandwich Islands, &c. His route is so extremely interesting, that we gladly lay before our readers a notice of it.

He intends going first to Rio Grande, and thence to the La Plata and Paraguay, he will then cross to Chili, and work northwards, examining the western slopes of the Chilian and Peruvian Andes; from Peru, he will make for the Sandwich Islands, and carefully examine that groupe, and proceed thence to Vancouver's Island and several adjoining districts of the North American continent; here he will turn his steps homewards, and visit the East India Islands, and devote as much time as possible in exploring these little-known but rich localities.

Mr. Plant will make collections of birds, insects, shells, dried plants, and other objects of natural history, and he has no doubt of being able to send home many interesting and valuable specimens during the four or five years which will be necessary to accomplish the journey sketched out.

Mr. Plant is encouraged to undertake this exploration with the earnest and devoted spirit of a true naturalist, believing that sufficient support will be afforded him in the profitable sale of his collections, when consigned to his agent (Mr. Stevens) in London, to enable him to defray the great expenses attending the labours of a naturalist abroad; he, therefore, respectfully solicits every promise of support in this respect, and will feel greatly obliged for any orders and communications which naturalists and collectors will favour him with, relative to the specific objects to which they would direct his attention during this journey.—Letters may be addressed to Mr. N. Plant, Museum, Leicester; Mr. John Plant, Royal Museum, Salford; and Mr. Samuel Stevens, Natural History Agent, 24, Bloomsbury Street, London.

We understand Mr. Plant intends to embark in April or early in May of the present year.

NOTICES OF BOOKS.

Nederlandsch Kruidkundig Archief. (Dutch Botanical Archives.) Uitgegeven door W. H. DE VRIESE, F. DOZY, en J. H. MOLKENBOER. Leyden.

This is a periodical work, of which two volumes, each of four parts, or stout numbers, have appeared. It is gratifying to find so many of our botanical neighbours in Holland zealously devoting themselves at this time to the publication of the vast treasures they possess from their oriental possessions in the Malay Archipelago. The works of Blume, Korthals, Miquel, and De Vriese, are among the most important. The latter, united with Messrs. E. Dozy and J. H. Molkenboer, gentlemen especially devoted to the study of the Mosses, have established a journal under the above title, to which we cannot but wish every success. We only now call attention to the publication, as containing a great deal on the subject of the vegetation of the Dutch East Indian possessions and other countries, reserving further notice for another opportunity.

PLANTÆ JUNGHUNIANÆ. *Enumeratio Plantarum, quas, in insulis Java et Sumatra, detexit* F. JUNGHUN. 8vo. Fasc. I. Leyden, 1850.

M. Junghun appears to have made a very extensive collection of plants in the Dutch Islands of Java and Sumatra, and is known to science by his writings in the scientific journals of his country, and by his excellent memoir on *Balanophoreæ* in the 'Nova Acta Academiæ Curiosorum,' &c.

The collection of this gentleman is about to be published in the work now mentioned, of which the first fasciculus of 106 pages has just appeared. It begins with *Coniferæ*, by Dr. Miquel. One true Pine is shown to be a native of Sumatra, *Pinus Merkusii*, Jungh. et De Vriese, Pl. Ind. Or. fasc. i. tab. 1 :—" probably the *P. Finlaysoniana*" of Dr. Wallich, Cat. n. 6062, from Cochin-China. *Cupuliferæ*, by Miquel, has four new Oaks ; and a new genus, *Callæocarpus*, Miq., allied to *Castanea*. *Piperaceæ* (Miquel) has three new species. *Urticeæ* (Miquel), besides several new species, includes three new genera, *Dendrocnide*, *Leucocnide*, and *Oreocnide*. In *Moreæ* (Miquel) is one new species. In *Artocarpeæ* (by Miquel), *Conocephalus*, Blume, has a new species, *C. gratus*, Miq., and a new genus, *Stenochasma*, Miq. The *Ficeæ* embrace many new species, especially *Ficus* proper, and *Urostigma*. *Parasponia*, Miq., is a new genus in *Celtideæ*. *Ranunculaceæ*, by De Vriese, has only one new species—in *Anemone*, *A. Sumatrana*, De Vriese ; but many observations are given on little-known species. *Dipterocarpeæ*, by De Vriese, contains a full botanical history of *Dryobalanops Camphora*, and he adds, " Quæ de camphoræ sede et colligendi ratione sunt ad hunc usque diem relata, ea plerumque erronea esse, aliis locis demonstrabimus, utpote minus hujus loci existimanda."* *Leucopogon Javanicus* (in *Epacrideæ*), *Anacyclodon pungens*, Jungh. in Nat. en Gen. Archief voor Nederl. Indië. Batav. 1845, II. Jaarg. bl. 49–51, is fully described. *Primulaceæ*, by De Vriese, affords a new genus in the splendid *Primula imperialis*, Jungh. MSS. and in Tijdschrift voor Nat. Gesch. en Phys. vol. vii. p. 298, and was also last year fully described by De Vriese, with an excellent plate, in 'Jaarb. der Koninkl. Nederl. Maatsch. van Tuinb.' 1850, bl. 29, pl. 1. The scapes of this attain a height of three feet, with leaves in proportion, and numerous golden flowers

* Dr. De Vriese is so good as to assure us that he is preparing a translation of his Memoir on this interesting subject, for insertion in our Journal.—ED.

arranged in whorls.* *Cankrienia chrysantha* (not *Cancrinia chryso-cephala*, Karel. et Kiril.). *Umbelliferæ* are undertaken by Molkenboer, and it is curious that of the sixteen species noticed, only one, the remarkable *Horsfieldia*, should exhibit any unusual form: all the other fifteen belong to well-known genera common to Europe; viz., *Hydrocotyle*, *Sanicula*, *Falcaria*, *Pimpinella*, *Fœniculum*, and *Coriandrum*. *Aroideæ* are elaborated by De Vriese, and present three new species in the genus *Pothos*.

RHODODENDRONS OF SIKKIM-HIMALAYA, *by* Dr. JOSEPH HOOKER; Fasc. II. Imperial folio.

Messrs. Reeve and Benham have recently published a second fasciculus of this work, containing ten species, quite equal in beauty to those of the first number, and the plates are nearly finished, of a third and last fasciculus, of ten more species.

The VICTORIA REGIA, *beautifully illustrated with four coloured plates, by* MR. FITCH; *the descriptive portion by* SIR W. J. HOOKER.

This is another splendid work, in elephant folio, which has issued from the press of Messrs. Reeve and Benham. The figures are splendidly executed by Mr. Fitch, from drawings made from the living plants partly at Sion and partly at Kew.

Icones Plantarum; by SIR W. J. HOOKER.

This work, which has extended to eight volumes, each of 100 plates of rare or little-known plants in the author's herbarium, and which has been discontinued for a period of more than twelve months, owing to circumstances over which the author had no control, will now be published by Messrs. Reeve and Benham, and will appear in monthly numbers, each number containing eight plates. The first of them will appear on the first day of next month.

* Among the numerous drawings recently sent home by Dr. Hooker, from Sikkim-Himalaya, is one of a yellow *Primula*, that vies with this in size (the flowers, indeed, are larger, but not whorled), of which Dr. Hooker says, "The pride of all the alpine *Primulas*: inhabits wet, boggy places at elevations of from 12–17,000 feet, at Lachen and Lachoong, covering acres with a golden carpet, in May and June."

On the Character of the SOUTH AUSTRALIAN FLORA *in general; by* DR. H. BEHR. (*Translated from the German in Schlechtendal's* 'Linnæa,' Bd. xx. Heft 5; *by* RICHARD KIPPIST, Libr. L. S.)

The Flora of South Australia, and with it the physical character, may be divided into two widely separated forms, that of the *Grass-land* and that of the *Scrub*. In the hill-country, and the plains lying to the westward of it, the grass-land prevails; yet so that extensive tracts, as well as small portions, of the other form of vegetation, likewise occur commonly enough. In the east, the Scrub predominates to an extent that is only interrupted by the formation of the fertile grass-land in the flats (pastures) of the Murray, and in the marshes of the lower part of its course, together with those of its estuary. The grass-land resembles for the most part European pastures in its physical character, as do also the individual plants which constitute its turf resemble corresponding European ones. A rather thick meadow-carpet is the essential characteristic of these regions, with which is associated in most cases a light park-like forest of gigantic *Eucalypti*, whose crowns, however, are never in contact with each other. Their smooth stems, robbed of their outer bark, stand at measured, and often very regular intervals; so that the idea involuntarily presents itself, that the whole must be the park of some landed proprietor, enthusiastically devoted to the quincunx of Cicero. Where the soil is poorer, *Casuarinas* make their appearance here and there, whose brown-green crowns contrast strangely in spring with the sap-green of the turf. They reach the height of twenty, or at most of thirty feet, and stand like twigs among the *Eucalypti*. The gummiferous *Acacias*, *retinoides* and *pycnantha*, belong, likewise, to this form of vegetation. *Acacia retinoides* reaches the height of the *Casuarinas*, and grows in a more isolated manner; *Acacia pycnantha* is usually little above the height of a man, yet of a very decided arborescent growth. It forms an umbrella-like crown, and often constitutes little forests. But few shrubby species occur, and only where the poorer soil forms a transition to the Scrub vegetation. Among the commonest plants is *Bursaria*.

One variety of the grass-land is the pit-land ("Bay-of-Biscay-land"), consisting of undulating plains or gently inclined slopes, which resemble a sea suddenly frozen during the beating of the waves. The depressions are pit-formed, and surrounded by circular elevations; yet, even in the most strongly-marked form, the distance between the bottom of the cavity and the level of the surrounding ring scarcely amounts to five feet. The continual change of level produces a very broken surface, which, however, becomes effaced in the course of a few years under the plough. The Flora of these districts has some peculiarities. Whilst in other tracts which I have visited, grass-land destitute of trees is comparatively rare, these districts show a decided aversion to the elsewhere almost universally prevalent *Eucalyptus*, which here seldom occurs except as a border to the ravine-like water-courses, and even then only as a less robust species, *Eucalyptus odorata*, Schldl. *Casuarina* is a more common shrub, but the commonest is *Acacia pycnantha*, which *here* evinces an unusual tendency to unite in groves. *Bursaria* is also characteristic of these localities, together with the creeping bushes of a few *Grevilleas*. The generally treeless ground is peculiarly rich in Syngenesious plants; but, with the exception of the Grasses, poor in Monocotyledons. *Orchideæ* it does not produce at all.

A second variety of the vegetation of the grass-land is afforded by the beds of rivers when dried up in summer. The stems of the *Eucalypti* on the banks attain here to incredible dimensions; trunks of eight feet diameter being by no means uncommon. Crowded together in the actual bed of the stream, is a Flora of principally European forms, which, hitherto retarded by the water flowing over them, first develope their flowers when all others are withered. The bed is likewise often filled up with bushes of *Melaleuca* or *Leptospermum*. Reeds and brushwood, hanging from the tops of these bushes, then mark the height to which the water rises in winter. This form affords the transition to another—that of those shady ravines, which, during the whole year, are more or less supplied with water. Here is found a vegetation whose herbaceous representatives generally remind us of Europe, even more strongly than those of the dried-up creeks and of the remaining grass-land; but whose arborescent and fruticose forms assume the habit of the Scrub-land. But as this Flora occurs for the most part in the transverse valleys of the upper course, seeds, rhizomes, &c., are carried down during the winter rains, and enrich the originally different Flora of the ill-watered lower course.

The Scrub differs from the above-described forms of vegetation by the utter want of a turf; at least, I do not think that the most imaginative colonist could construct a turf out of the few scattered *Stipas* and *Neurachnes*. This almost entire want of herbaceous plants is compensated, however, by an endless profusion of bushes and small trees. Here is the especial source of those plants which for some dozen years past have been the ornament of our green-houses. The general impression produced by this district is nevertheless not an agreeable one. Heath-like foliage, or vertically-placed leaves, crowd about mossy compactly grown spherical bushes, or but sparingly conceal the nakedness of long rods which jut out from hideously lanky bushes. The prevailing colour of the leaves is a dead blue-green, but in this respect Nature lays little restriction upon herself, the *Rhagodia* bearing white leaves, other bushes brownish-red; the most conspicuous, because in such company the most unnatural, being the lively May-green of *Cassia* and *Santalum*. Pinnate or otherwise divided foliage is rare; the only example that I can remember is a species of *Cassia*. In other respects, the greatest possible variety is found among the rigid leaves, from rotundo-ovate, through the lanceolate form, to the mere bristle; from the most dense crowding, through every possible shade, to the bare leafless twig. Moreover, plants belonging to very different families coincide so completely in habit, that only flowers or fruit afford a safe criterion. The bushes and trees of the Scrub regions are of very different heights; many species of *Eucalyptus* rivalling those of the fertile land.

One variety of these forest districts is distinguished by the colonists under the name of "Pine-forests." Except the frequent recurrence in such places of the *Callitris* ("Pine" of the colonists), I should scarcely be able to point out any characteristic which would distinguish the Flora of the Pine-forests from that of other Scrubs. The *Callitris* itself never forms a wood; it grows always singly, intermixed with the Scrub, and I have never met with it as the predominant tree.

The "Sand-plains" are more evidently distinct from the true Scrub. The brush-wood of these districts does not reach to the height of a man, and although differing but little in habit from the other Scrub districts, it nevertheless continually afforded me new species. In the hills and in the western plains such tracts are very rare; in the east they form a principal constituent of the Murray Scrub.

It will readily be understood that transitions are to be found between the two forms of vegetation—that of the Grass-land and that of the Scrub; for example, as above stated, there are found intermixed the forms usually occurring in the vegetation of dried-up water-courses, as the seeds and rhizomes of different regions are here brought together by means of the winter rains. In most cases, however, the differences are sufficiently well-marked to strike the most inexperienced cockney.

The same dichotomy which prevails locally in the approximating Floras of the country, manifests itself also with respect to time, in the change of seasons. The names of the European seasons have, indeed, been imported to South Australia, but there are in truth only two, a dry and a wet, that can readily be distinguished. The gradual cessation of the rainy season can as little be compared with our spring, as its tardy entrance with our autumn.

The beginning of the winter rains, which in most districts may be assumed as April, elicits from the soil, changed by the dryness of the summer months to ashes, its earliest green, which glimmers forth from under the dead stems, and is only to be readily perceived on the lands which have been wasted by fire. Except the flowers of a few *Eucalypti*, which are developed at this late season, the grass-land shows only the stems, and the hard leathery leaves of its trees and few bushes, its yellow haulms, or the ground wasted by fire. Soon, however, under the influence of the winter rains, the soil becomes covered with fresh juicy turf, interrupted by larger or smaller pools of water. Beautiful *Droseras* and the dwarf *Oxalis cognata* form the vanguard of a host of lovely flowers, which in the course of a few weeks emerge rapidly from the soil. The bright sunny days which now and then interrupt the rainy season, towards the end of August become more and more frequent, and a profusion of flowers, which in many places almost conceals the turf, is rapidly developed upon the land, strengthened by the summer rest, and fertilized by continued rains. The *Orchideæ, Melanthaceæ*, and *Asphodeleæ*, in equal variety and beauty, and often disposed in figures resembling parterres, shine forth pre-eminently from the smiling green. *Stackhousias* fill the mild spring air with their honey-like scent, and creeping *Kennedias*, with glowing red blossoms, lurk beneath the overarching culms, above which the fine-stalked Bell-flowers swing, and Buttercups wave their yellow heads. A crowd of European kinds entwine themselves in the series of genuine Australian forms, just as the entire surface of the country, with its light park-like groves of *Eucalyptus*, reminds us of the meadows at home. New forms are now developed in quick succession; every week offers fresh flowers. The pools of water are dried up; but clear rivulets and brooks yet meander through the land, fed by the rain which at this season still falls occasionally. The turf becomes a luxuriant meadow, in which Syngenesious plants of many different species are produced, and form, as with us, the last act of the beautiful drama. The turf, of which the *Stipaceæ* form no small proportion, now ripens its seeds, which, in conjunction with the prickly fruit of *Acæna*, put him, whom business or inclination leads through the tall-grown meadow-grass, to no trifling inconvenience. The ground, which a little earlier had been of a rich green, now resembles a ripe, but very thinly-sown corn-field; and the number of plants in flower diminishes daily, till at last all vegetable life is reduced to the peculiar form of vegetation of the now dried-up rivers and brooks. This epoch arrives at different times in different localities, but, as far as my observations reach, never before the end of November, and never after the beginning of February. From this time I have never found herbaceous plants in flower on the grass-land, with the exception of *Lobelia gibbosa*, which, although wild only in a few districts, springs from the dried-up soil with its leafless, fleshy stems.

On these forms of vegetation, altitude seems to have but little influence. Mount Barker (2,000 feet above the level of the sea) is regarded as the highest mountain of the colony. I have climbed to its summit, aud found nothing there, which I have not met with, either before or since, at the foot of the mountain. *Xanthorrhœas* and *Epacrideæ* certainly seem to prefer the mountain; yet rather on account of the stony soil than of the elevation, since I found nearly all the alpine species on the coarse gravelly soil of the plains also. The species of the western plains, on the contrary, and those of the eastern (Murray Scrub), are almost always different. The fertile land in the Murray valley also possesses many peculiarities, which, however, without entering too much into detail, I am unable to notice further here. But the character of its vegetation differs in no respect from the corresponding districts of the west. A peculiar vegetation occurs only in the immediate neighbourhood of the sea, in the woods upon the beach overflowed by the tide, which consists of a *Rhizophorea*—I believe, *Ceriops*. This tropical form has a very distinct boundary-line on the side of the Scrub, commencing with a shrubby *Salicornia*, which has taken up its position on those parts of the strand farthest removed from the sea.

A much greater influence than that of elevation above the sea-level, is exercised by the neighbourhood of man, especially that of a cattle-dealing population. Annual plants seem to be among the first to yield before the foreign influx. I was informed of a pretty flower, from the description probably an *Argyrophanes*, which had formerly covered whole plains of the upper district of the Oncaparinga, and had now entirely disappeared. The *Anthistiria*, which in the hilly lands formed pastures, is now supplanted in many places by new grasses; the original vegetation of the more cultivable Scrub-districts, in the neighbourhood of the town, lurks timidly and secretly about the hedges which separate it from the reclaimed districts, and views with terror the destructive progress of the merciless intruders.

The Australian Flora has been but little enriched by any European plants, except cultivated ones. We find, indeed, here many that are identical with European plants, but the native home of most of them is a very critical point; while the Australian burgher-right of others is beyond all doubt. As might naturally be expected, my researches upon this point have not led to any certain results; the following plants, however, appear to me unquestionable immigrants, viz., *Lolium temulentum*, *Centaurea Cyanus* (rare), *Capsella Bursa*.

Contributions to the Botany of WESTERN INDIA;
by N. A. DALZELL, Esq., M.A.

(*Continued from* p. 124.)

Nat. Ord. URTICEÆ.

POUZOLZIA.

P. *integrifolia*; suffruticosa 3–4-pedalis, radice crassa, caule compresso utrinque linea pilosa instructo, foliis oppositis sessilibus e basi cordata lanceolato-acuminatis 3-nerviis *integris* supra et in nerviis subtus pubescentibus 3–4 poll. longis, floribus masculis et fœmineis in foliorum axillis glomeratis breve pedicellatis, pedicellis apice articulatis, floribus masculis plurimis fœmineis paucis, perianthio masculino 4-partito, filamentis complanatis, fœmineo fructifero bi- interdum trialato inter alas 8–10-costato, alis ciliatis, involucro oligophyllo membranaceo.

Semina ovata, acuta, brunnea, nitida.—Crescit in montibus Syhadree, prope Phondaghat; fl. Septembre.

Perhaps the above is one of the continental species of *Pousolzia* alluded to by Bennett, in Horsfield's Plantæ Jav. Rar.; but as I believe two of them are yet undescribed, those only who may have specimens from Dr. Wight will be able to determine.

Nat. Ord. LEGUMINOSÆ.

SMITHIA.

S. *hirsuta*; annua erecta sesquipedalis parce ramosa, caule ramis pedunculisque pilis fulvis patulis hirsutis, foliolis 3-4-jugis obovato-cuneatis basi inæquilateris margine ciliatis subtus interdum parce pilosis, floribus breve pedicellatis capitato-racemosis, pedunculis folio longioribus 10-floris, calyce pilis longis parce vestito, *segmento superiore cuneato apice truncato* emarginato, *inferiore cuneato-obovato integerrimo*, vexillo amplo orbiculari, legumine 6-7-articulato, articulis reticulato-venosis etuberculatis.

Flores flavi.—Crescit prope Phondaghat; fl. Sept.

A very distinct species, easily recognized by the calyx. This plant is very rare: I have never met with it but in one locality.

Nat. Ord. COMMELYNEÆ.

Of all the genera included in this family, none presents so many different forms in this Presidency as the genus *Aneilema*. It may be naturally divided into three sections, according to the inflorescence:—1. *flowers fasciculate*; 2. *flowers paniculate*; 3. *flowers cymoso-racemose*. The genus is said to have deciduous petals (inner sepals), but in no instance have I been able to verify this statement. Their delicate and beautiful flowers expand about ten A. M.; they last from three to four hours, after which they curl up, gradually dissolve into a pulpy mass, and thus terminate their brief existence. Many Indian species are known hitherto only by name, or by the name having "Wall. Cat." attached to it.

Sect. 1. *Floribus fasciculatis*.

1. Aneilema *ochraceum*; caulibus erectis simplicibus teretibus glabris foliosis, foliis vaginatis alternis, inferioribus ovato-oblongis, superioribus cordato-ovatis acutis minoribus, pedicellis axillaribus terminalibusque pluribus (6-7) fasciculatis medio articulatis parte inferiore

hirtellis, calycis (sep. exter.) laciniis glabris, corollae petalis rotundatis *ochraceis*, antherarum fertilium filamentis barbatis.

Stamina 6, quorum tria sepalis interioribus opposita effœta, glandulis bilobatis instructa; tria iisque alterna fertilia. *Stylus* brevis; *stigma* simplex. *Ovarium* oblongum, trigonum. *Capsula* cartilaginea, glabra, loculicide dehiscens, trilocularis, polysperma; semina in quoque loculo 7–8, *biserialia*, valde inter se compressa, hinc latere interiore angulata, lateribus omnibus impressione umbiliciformi.—Crescit in locis saxosis humidis Concani australioris; fl. tempore pluviali.

2. A. *versicolor*; ramosum, ramis erectis teretibus striatis *patenti-hispidulis* foliosis, foliis alternis bifariis distantibus lanceolato-acuminatis glabris amplexicaulibus subtus 5–7-nerviis supra striatis vaginis longiusculis hispidulis sulcato-striatis, pedicellis axillaribus fasciculatis 3–4-nis folio longioribus medio articulatis parte superiore glabris inferiore pubescentibus, sepalis exterioribus oblongis obtusis concavis, interioribus petaloideis semiorbiculatis brevissime unguiculatis *ochraceis* diametro 3-lin.

Stamina fertilia 3, patentia : *filamenta* subulata, parte inferiore barbata; *antheræ* lineari-oblongæ, intense purpureæ, medio dorso affixæ, versatiles. *Stamina* glandulifera 3, fertilibus breviora, *filamentis* parce barbatis, *stylo* erecto longioribus, *glandulis* albis 3-lobatis. *Petala* in alabastro *rosea*, marcescentia cyanea. *Capsula* linearis, trigona, sepalis exterioribus vix longior, loculis 7-spermis, semina *uniserialia*.—Crescit cum præcedente.

These two species are distinguished from all others (as far as I am aware) by their yellow flowers; and perhaps it has not hitherto been known that flowers of that colour exist in the genus. These two species are closely allied, but the flowers in the latter are twice as large, and the hispid stems, the much more acute leaves, and the seeds in a single series, form a ready mark of distinction. The latter species appears later in the rainy season.

3. A. *pauciflorum*; bipedale, tota planta (vaginarum ore excepto) glabra, basi ramosa, ramis strictis erectis teretibus striatis, foliis linearibus acuminatis nitidis planis 8 poll. longis 3 lin. latis internodiis duplo longioribus, inferiorum vaginis fissis, superioribus subito in bracteis floralibus brevibus vaginantibus transeuntibus, pedicellis ex bractearum axillis solitariis geminis vel ternis semipollicaribus unifloris bractea subduplo longioribus medio bis articulatis apicem versus

puberulis, ovulis in loculis solitariis.—*Calycis laciniæ* exteriores ovato-oblongæ, puberulæ, fructu paulo longiores. *Stamina* 2 fertilia; *antheræ* aurantiacæ. *Stamina* effœta glanduliformia 4, *glandulæ* flavæ. *Stylus* cæruleus. *Capsula* obtuse trigona, rotundata, glabra, nitida; *ovula* in quoque loculo, solitaria; *semina* tuberculata.—A. *vaginato*, Br., affine, sed folia multo longiora, et capsulæ loculi 1-spermi.—Crescit in Concano australiore; fl. tempore pluviali.

Sect. 2. *Floribus paniculatis.*

4. A. *elatum*; 3–4-pedale, caule erecto tereti glabro folioso, foliis lineari-lanceolatis acutis glabris planis albo-marginatis marginibus undulatis 6–8 poll. longis 2 poll. latis, vaginis integris pollicaribus ore puberulis vel subglabris, pedunculis terminalibus dichotomo-ramosis, ramis ramulisque distantibus paucifloris basi bracteis integris vaginantibus suffultis, floribus ternis.

Sepala exteriora profunde concava, ovalia, obtusa, interiora petaloidea, obovato-cuneata, concava, sub anthesi reflexa. *Stamina fertilia* 3, filamentis longis subulatis, antheris flavis. *Stam. glanduliformia* 3 breviora, glandulæ luteæ, cordato-bilobæ: *filamenta omnia* infra medium barbata. *Capsula* sepalis exterioribus inclusa, ovalis, obtuse trigona, nitida, loculis 3-spermis.—Crescit in sylvis umbrosis regni Warreensis; fl. Augusto. A. *giganteo*, R. Br., et *A. elato*, Kunth, Enum. vol. iv. p. 70, valde affine; a priore differt staminibus omnibus barbatis; ab *A.elato*, Kunth, sepalis exterioribus non linearibus.

This remarkable species has cylindric fleshy, or rather tuberous, roots, and blue metallic bead-like fruit; it is very probably the *Commelyna elata* of Vahl, and that there are errors in the short description; I have, therefore, kept the specific name.

5. A. *canaliculatum*; 6–7-pollicare, radice fibrosa, caule simpliciter ramoso erecto striato uno latere linea pubescente alternatim notato, foliis inferioribus late lineari-lanceolatis, superioribus cordato-oblongis, omnibus amplexicaulibus glabris 10–18 lin. longis 4–5 lin. latis medio *canaliculatis*, pedunculis terminalibus et ex foliorum superiorum axillis solitariis vel geminis dichotomo-ramosis paucifloris, floribus longiuscule pedicellatis bifariis distantibus, pedicellis fructu longioribus bracteis minutis persistentibus oppositis.

Sepala exteriora sub anthesi *patentia*, interiora majora, petaloidea, rotundata. *Stamina* perfecta 3 fertilia, antheris apice apiculatis, cyaneis;

glanduliformia 3 breviora, stylum æquantia, glandulis 3-lobatis albis; *filamenta omnia barbata*. *Capsula* oblonga, acute trigona, sepalis exterioribus dimidio longior, loculis 4–5-spermis; *semina* uniserialia. *Flores* cyanei. Crescit in prov. Malwan; fl. tempore pluviali.

6. A. *dimorphum*; totum (vaginarum ore ciliolato excepto) glabrum pedale, radice fibrosa, basi parce ramosa, ramis erectis teretibus striatis, internodiis uno latere linea pubescente notatis, foliis inferioribus lineari-acuminatis, superioribus lanceolatis acutis, omnibus amplexicaulibus, floribus terminalibus dichotomo-paniculatis paucis, paniculæ ramis pedicellisque basi bracteatis, bracteis minutis rotundato-cucullatis.

Sepala exteriora ovata, obtusa, concava, sub anthesi *reflexa*, interiora rotundata, basi cuneata, petaloidea. *Stamina* fertilia 3, antheris purpureis, filamentis medio barbatis stylo 4-plo longioribus; glanduliformia 3 breviora, *filamentis glabris*, glandulis albis 3-lobatis. *Ovarii* loculi 3-*ovulati*. *Stylus* brevissimus, strictus, subulatus. *Stigma* simplex. *Capsula* sep. exter. longior, oblonga, acute trigona. *Flores* cyanei.—Crescit in prov. Malwan; fl. tempore pluviali.

When specimens of this species are gathered on a stony soil, the internodes are so much shortened as to give the plant a very different appearance, the leaves appearing all radical.

7. A. *semiteres*; caule erecto simplici tereti 2–5-pollicari glabro, foliis paucis *subulatis carnosis semiteretibus*, vaginis integris, floribus terminalibus et ex folii supremi axilla dichotomo-paniculatis, paniculis paucifloris, pedunculis pedicellisque rubris, vaginis floralibus truncatis unidentatis.

Stamina fertilia 3, incumbentia, sep. exter. opposita, antheris fuscis, filamentis glabris; sterilia 3, glanduliformia, glandulis 3-lobatis albis, *filamentis omnibus ima basi coalitis*. *Ovarium* ovale, stylo stricto paulo brevius, 3-loculare; *ovula* in quoque loculo 6, biserialia, latere interiore angulata. *Petala* obovata, cuneata, apice mucronata, irregulariter dentata. *Flores* cyanei.—Crescit in Concano utroque, in locis saxosis; fl. tempore pluviali.

A very distinct species.

Sect. 3. *Floribus cymoso-racemosis*.

8. A. *compressum*; basi ramosum radicans, ramis adscendentibus *compressis* simplicibus glabris, foliis bifariis brevibus 2–3 poll. longis

4–5 lin. latis *ensiformibus obtusiusculis* subcomplicatis, superioribus minoribus, *vaginis undique hispidis*, pedunculis terminalibus et in bractearum vaginantium axillis pollicaribus plurifloris, floribus racemosis breve pedicellatis, sepalis exterioribus oblongis obtusis glabris, interioribus petaloideis rotundatis *roseis*, ovulis in loculis ternis.

Antheræ fertiles 2 albæ, *filamentis* albis, pilis roseis barbatis. *Stamen* 1, depauperatum barbatum; glanduliformia 3 parce barbata, glandulis 3-lobis. *Capsula* oblonga, sepalis exterioribus persistentibus *duplo longior*, 9-sperma; *capsula* cartilaginea.—Crescit in prov. Malwan; fl. tempore pluviali.

Very like *A. nudiflorum*; the leaves, however, in this species are shorter, more fleshy, darker in colour, and there are more seeds in the capsule. The anthers in *A. nudiflorum* are blue and the glands bilobed. This species may be easily distinguished by the *compressed stems*.

CYANOTIS.

C. *hispida*; 4–5-uncialis annua tota hispida basin versus parce ramosa, *caulibus erectis* teretibus striatis rubris, foliis lineari-eusiformibus carnosis subplanis 1–2 poll. longis 3–6 lin. latis, floribus terminalibus in bracteis falcato-semicordatis sessilibus capitato-congestis paucis (4–5).

Calycis laciniæ exteriores lineari-lanceolatæ, dorso marginibusque ciliatæ, 3-nerviæ, ima basi connatæ, interiores exterioribus paulo longiores, 3-lineares, in tubum petaloideum obconicum apice 3-lobatum connatæ; *lobi* rotundati, læte punicei, tubo albo breviores. *Stamina* 6 fertilia, longe exserta, corollam duplo superantia, stylum æquantia: *antheræ intense violaceæ; filamenta* superne pilis cæruleis secundis barbata, sub apice incrassata. *Capsula* inclusa, oblonga, obtusa, apicem versus pilosa, loculis dispermis.—Crescit in rupibus prov. Malwan; fl. Aug. et Sept.

(*To be continued.*)

Extracts of Letters from RICHARD SPRUCE, ESQ., *written during a Botanical Mission on the* AMAZON.

(*Continued from p.* 89.)

Santarem, June 28, 1850.

I wrote to you last when just recovering from a rather severe attack of fever. I was beginning to get out again when I had another seizure, and it was at least two months before I regained my usual health, and could bear that exposure to sun and weather to which a field-botanist in the tropics is accustomed. At the same time illness of a similar kind was almost universally prevalent in Santarem, and proved fatal to a great number; while in the villages up the Tapajoz ague of the worst kind was rife, and above four hundred people fell victims to it. These maladies were attributed, I believe justly, to the unprecedentedly rapid rise of the rivers, and the consequent inundation of all the lower grounds. Last year the waters attained their maximum elevation on the 12th of June, at Santarem; but this year they were higher on the 15th of April than ever last year; and from that date they continued at the same average height—now rising, now falling a few inches—until the early part of the present month, when they began to subside. Nearly all the cacoals between Monte Alegre and Obidos were flooded, and the people who resided on them driven into the towns, in the outskirts of which they erected temporary habitations of palm-leaves. Our neighbour and countryman, Mr. Jeffreys, had a plot of mandiocca at his sitio on the Rio Arripixuna, and being alarmed by the sudden rise of the waters, he set all his hands to work to get it up, dress it, and roast it. It was near midnight of their last day when they withdrew from the furnace the last batch of farinha; next morning the furnace (generally elevated two or three feet from the ground) and the whole of the mandiocca plot were laid completely under water! We suffered also in the matter of provisions: the milch-cows were flooded out of their pastures, and strayed away into the forest, so that often no milk was forthcoming at our breakfast—a great privation. The rich low meadows on the opposite side of the river, on which numerous oxen were fattened for the Santarem market, were transformed into a complete lake; the poor cattle were some starved, some drowned, and not a few of the younger ones fell victims to the jacarés. These rapacious monsters thread their way in the water, concealed by the gigantic grasses, and thus approach, unperceived, their unconscious victim, whom they first stun with a blow of their tail, and then speedily crush in their enormous jaws.

The city of Pará had a still more serious visitation than Santarem, in the yellow fever. Out of a population of 25,000, above 13,000, it is said, were ill at one time; though, perhaps, not all of the fever. Many people of distinction fell victims to that dreadful malady, including Her Britannic Majesty's Consul, Richard Ryan, Esq. The yellow fever had never before visited the shores of the Amazon, and great was the alarm it created, even at Santarem. The good people of Santarem are not ordinarily remarkable for attention to religious observances, except at Christmas and other festivals, when there is a pious display of rockets, crackers, and balloons, and of processions of a very dramatic character; but when we were in daily fear of the dreaded fever reaching us, we had vespers every night in the church, and those families who were happy enough to possess a rude daub of some saint, assembled round it on their knees at stated times, and recited a number of prayers taken *ad libitum* from the breviary. The most amusing process was the dragging a couple of field-pieces through the streets, and discharging them at short intervals, with the object of clearing the atmosphere, and so preventing the entrance of the threatened "pesta"! With the same intention lumps of "brêo branco" were fastened on poles, stuck up at the crossings of streets, and set fire to after sundown; thus illuminating the whole town, and emitting a perfume by no means disagreeable. But the most efficacious precaution of all was considered to be the kissing a small wooden figure of St. Sebastian, which was nightly exposed at the foot of the altar, during the "novenas" of Whitsuntide, to receive the homage of such as feared the pest and trusted to secure the saint's intercession against it; including every man, woman and child in the church, with the exception of the "estrangeiro," whose neglect did not fail to be remarked on, though, as he added his mite to the contributions towards the expense of the festa, his crime was considered venial.

At intervals during my illness, I was able to take short excursions, sometimes on foot, at others in a canoe. Perhaps the most interesting was across the Tapajoz to the low meadows of the tongue of land terminating in the Ponta Negra, at the junction of the Tapajoz and Amazon. From the hill behind Santarem these meadows seem covered with a rich green herbage, but on approaching them they are seen to be under water to the depth of from three to five feet, with nearly as great a depth of mud below that. In summer they are traversable on foot, though very muddy, and they contain two small lakes, in which it was reported the "Forno" had been seen. To ascertain if this report was correct was the principal object of my visit, and I was well pleased to find a plant of the *Victoria* in each lake.

Numerous leaves rose from the roots, and each plant bore a single flower, but I found no fruit, as I had hoped to do. To attain these lakes our men had to push the canoe through a thick grove of grasses, which stood out of the water to a height of from two to five feet, besides the length of stem buried in water and mud, and they were fortunately all in excellent flower. They belong chiefly to the tribes *Oryzeæ*, *Chlorideæ*, and *Paniceæ*, and the most abundant is the *Pirimembéca*, celebrated everywhere on the Amazon as the richest of meadow-grasses. The same grasses constitute the mass of the numerous floating islands on the Amazon (called "Ilhas de Capim"), as I have ascertained by repeated examinations. These islands are sometimes acres in extent, and from five to eight feet of their thickness is under water; from this, some idea may be formed of the shock with which they would meet a canoe sailing against the furious current of the Amazon, and instances are not rare of large vessels being swamped or half-buried in the floating mass. During the force of the rainy season no vessel anchors in the Amazon; the least evil that could result from such imprudence would be the dragging of her anchor by the onslaught of an Ilha de Capim. To return to the Ponta Negra. In the lakes and among the tall grasses were several small floating plants, including a *Salvinia*, a pretty *Riccia*, a very curious plant with the aspect of a *Salvinia*, but proving to be a *Euphorbiacea*, of which I can find no description in Endlicher, though it approaches *Phyllanthus* in technical character; and some others. It was strange to see quantities of a floating sensitive-plant, a *Neptunia*, whose slender tubular stems were coated with a cottony felt of an inch in thickness, as buoyant as cork, while the delicate leaves and flowers were by this means sustained completely out of water.

In the middle of May, though still not very strong, I began to think of visiting Monte Alegre, and I applied to the Commandante for men, my friend Dr. Campos having lent me an excellent boat. Santarem is noted at all times of the year for a scarcity in men and unoccupied houses; but now, when many vessels are making up their cargoes for Pará, all the hands that can be got are required to man them. The Commandante, however, sent up the Tapajoz in quest of Tapuyas, but he has himself been very ill for some time, and whether from this cause he has not taken care that his commands were more promptly attended to, or whether there are really no men to be had, I cannot say, but nearly five weeks have elapsed and I as yet hear no tidings of my crew. A singular accident that happened to myself and Mr. King, about ten days ago, would have effectually prevented our visiting Monte Alegre, had even the means been forthcoming. Since the rise of the waters we had been unable to get into the Serras, on account of an intervening stream (the Igarapé d'Irurá) having widely overflowed its banks; but when the rivers began plainly to ebb, I was desirous to see how the igarapé was affected. We visited it one day with this intent, and were well satisfied to find it passable by wading up to the middle. The ground on the opposite side, though still plashy, was not impassable, and we saw that the foot of the serras could be reached without difficulty. On a slightly rising ground a little beyond the igarapé are the ruins of a cottage, half of the walls and roof of which have fallen, and are now so overgrown with rank grasses as quite to hide the beams and rafters from the eye. In passing over this place Mr. King had the misfortune to tread upon a nail, and having, like myself, only India-rubber shoes on, which are a protection against nought but wet, he was severely wounded in the broad part of the foot. As the wound was very painful, I thought it better that he should return to the igarapé and wash it, and then await my return, as I wished to penetrate a little farther. Having gone far enough to satisfy myself that there was no obstruction from water, I was retracing my steps and expected I had already passed the dangerous ground, when I felt myself pierced in the left foot, and was immediately thrown forward with violence. On withdrawing my shoe my foot was bathed in blood; a nail had entered the narrow part of the sole and pierced through a little below the ancle. How we reached Santarem I hardly know, and I shudder at the remembrance of the excruciating pain. We cut sticks wherewith to assist our faltering steps, but we were every now and then obliged to lie down on the ground. The distance is nearly three miles, and we were three hours traversing it. On reaching home I had poultices applied to our wounded and swollen feet, and as I know rest to be the best of all remedies in such a case, we did not attempt to leave our hammocks for three days. Mr. King's wound is by this time nearly healed; and though my own wounds are not yet closed, the pain and inflammation have nearly subsided.

It is a curious coincidence that the builder of the cottage at Irurá came to his death by a nail. This man, a Portuguese, was pursuing a runaway slave along a narrow track in the forest; the slave, who was armed with a musket, ascended a tree, and, as his master passed underneath it, shot him in the forehead with a nail.

From the causes just stated, I find myself compelled to forego the visit to Monte Alegre, on which I had quite set my heart, or at least to postpone it until my return down the river. We are now preparing to ascend to the Barra do Rio Negro, with a countryman, Mr. Bradley, who is settled there. He descended the Amazon to Pará with two laden vessels about two months ago, and we are in daily expectation of his reaching Santarem on his way up. If I miss this opportunity, I may have to wait many months for another, as there is rarely any direct communication between Santarem and Barra, and few vessels ascending from Pará call here, on account of their having to undergo three or four days of quarantine—a serious sacrifice of time.

On a separate sheet I have added a few notes respecting the articles I am sending for your museum, especially as to the use of the Guaraná. I would gladly have visited the Guaraná country, which is six days' journey or more from here, but it would seem to be not very promising to a botanist, and the Guaraná plant is already perfectly well known. Respecting the bow I send you, I may add that the manufacture of such a one occupies an Indian *three months*;—not exactly of continuous labour; but it must be borne in mind that it is made of the intensely hard heart-wood of the *Pao d'Arco*, and that his only tool is a shell. The wood which he intends to fashion into a bow is first smeared with oil, to soften it; he then scrapes it down with his shell as far as the oil has penetrated, when he anoints it anew, and betakes himself to the chase. Returned from hunting, he again falls to work to scrape his bow; and so on, until it is completed; and no joiner can make one so symmetrical, so nicely poised, as these which are made by the Memdrucú and Mauhé Indians. The price of a good bow in Santarem is five or six patacas. I send an arrow, such as is used at Santarem for killing fish, such as Pirarucú, Tucunaré, &c.; the one-barbed head is called, in *lingoa géral*, " taçu-umba."

July 26.—Since writing the above, we have been looking out daily for Mr. Bradley's arrival, but hitherto in vain, and we have lately learnt that his partner has been dangerously ill of the yellow fever in Pará, and that until he regains his strength they cannot set out. At the worst we shall have a chance of getting up to the Barra in about a month, with a gentleman of Santarem, who goes there on business.

As I have held myself in constant readiness to embark with Mr. Bradley, I have been unable to take any long excursions; but I have made several shorter ones, both by land and water, and have found more abundant work than at any former period. The Amazon began to fall in the first days of June; about the same time the forest-trees, and especially those of the river-margins, commenced pushing their new leaves, and the process is not yet completed with all. The trees of the "gapó,"—that is, of the land adjoining the river, which has been inundated during winter,—begin to flower as the water leaves them, and according to my experience of last year, will continue to blossom in their turn until the end of September. After that period, few trees are in flower at a time; yet all the year round some tree or other is bursting into flower, and a botanist who should suspend his operations during any one month in the year would infallibly miss several trees.

As the waters leave the sandy shores of the Amazon and Tapajoz, and especially of certain small lakes connected with them, several small annuals spring up. I may almost call them *ephemerals*, so brief is their existence. They start up from the sand, flower, and ripen their seeds, and by the time the sand is quite dry—that is, in a few days—they are withered away. Amongst them are an *Alisma*, resembling *A. ranunculoides* in miniature, two or three *Eriocauleæ*, a *Xyris*, and some minute *Cyperaceæ*.

After making several attempts to procure the flowers and fruit of the Itaüba, I have at length succeeded. The nearest place in which I could obtain information of its growing was in the forest beyond Matricá, an Indian village about four miles down the Amazon; and in a visit I paid to them by water in March last, I found the flower-buds of the Itaüba just appearing. My illness prevented me from visiting the same place again until a long time afterwards, and in an attempt which Mr. King made to reach it alone, over land, he did not succeed, on account of the quantity of water in the low grounds. In another excursion, the trees we met with were all sterile. At length, in the early part of the present month, we were fortunate in falling in with a tree laden with fruit. The only way to obtain the fruit was to cut down the tree; but our *trésados*, which generally suffice for this purpose, made no impression on the hard wood of the Itaüba. In this emergency, Mr. King made his way to an Indian cottage which we had passed a few minutes before, and soon returned with a heavy American felling-axe. With this he succeeded in severing the trunk, but not until he had well blistered his hands. The drupes resemble in size and colour our small black grapes, only they are more elongated, and they hang in small panicles. The Brazilians compare them, and justly, to the small variety of olive which is imported in great quantities from Portugal. They have a slight bloom on them, and the pellicle is studded with pallid glandular dots. The pulp is about the eighth of an inch in thickness; it is good eating, though with a strong resinous flavour, much resembling that of an edible myrtle frequent on the campos, and a wine is made from it in the same way as that of the Assaí Palm. The testa is horny and very thin; albumen none; cotyledons amygdaloid, rose-coloured on the inner face; embryo pendulous from a little below the apex of the seed. The 6-cleft calyx is persistent, but not enlarged in fruit as in most of the *Lauraceæ* I have seen on the Amazon.

I had long suspected the dioicity of the Itaüba; I have now confirmed it; and I find that I gathered male flowers on the 30th of April, though at the time I did not recognize the tree, which was small and young, and grew in a part of the forest quite near to Santarem, which had been cut down some dozen years ago. On revisiting the place within these few days, I found two or three female trees, of the same size, growing near, and laden with unripe fruit. The male inflorescence is of minute yellowish-green flowers, arranged in small umbels on a raceme. Perianth 6-cleft, in two series. Stamens 3, fleshy, with 2 anther-cells (rarely 3) imbedded in their substance, and opening *outwardly* by an orbicular operculum. These characters seem to indicate a genus hitherto undescribed, and certainly prove the Itaüba to be distinct from the Greenheart of Demerara (*Nectandra Rodiæi*), with which some of the English settlers here have supposed it identical.

As I have before informed you, the Itaüba is the most valuable timber for shipbuilding which the Amazon affords. Its range seems to be from the mouth of the Tapajoz to that of the Rio Negro, and it is most abundant on the Rio Trombétas. It prefers gravelly or stony rising ground, and is never found in marshes.

(To be continued.)

Characters of some GNAPHALIOID COMPOSITÆ *of the Division* ANGIANTHEÆ; *by* ASA GRAY.

(*Continued from p.* 102.)

SKIRROPHORUS, DC. (Skirrophorus et Pogonolepis, *Steetz.* Eriocladium, *Lindl.*)

Char. Reform. Capitula biflora v. uniflora, homogama, in glomerulum obovatum vel globosum dense aggregata. *Involucrum generale* pauciseriale, duplex; squamis exterioribus foliaceis seu herbaceis persistentibus, interioribus scariosis cum paleis consimilibus capitula fulcrantibus plus minusve deciduis. *Axis* seu *receptaculum generale* conicum, convexum, vel planum, stipitibus capitulorum minimis onustum. *Involucrum partiale* 3–5-phyllum; squamis scariosis planiusculis (costa tenui), cum floribus deciduis. *Flores* hermaphroditi. *Corolla* tubulosa, sursum ampliata, 5-dentata, raro 4-dentata, tubo basi *post anthesin* in tuberculum skirrosum sæpius dilatato. *Antheræ* basi breviter caudatæ vel ecaudatæ. *Stylus* basi nunc bulbosus; rami apice capitellati, papillosi seu penicillati. *Achænium* obconicum vel obovatum, basi attenuatum, glabrum, apice basi skirroso corollæ tectum, calvum, vel in *S. Preissiano* coronula obsoletissima superatum.—Herbæ habitus diversi.

The *Skirrophorus Preissianus* of Steetz is, as this author remarks, very different in aspect from the original *S. Cunninghami*, DC., yet much too closely resembling it in the whole structure, to admit of a generic separation. Besides a third species closely allied to *S. Preissianus* (*S. eriocephalus*, Hook. fil. ined.), I have before me two or three others, differing still more widely in habit from the original species, and manifestly connecting *Pogonolepis*, Steetz, with *Skirrophorus*, unless, indeed, we establish nearly as many genera as there are known species. The common receptacle is paleaceous, the paleæ resembling the proper involucral scales, which they subtend, and deciduous with them, except perhaps in *S. demissus*. The species may be disposed as follows:—

§ 1. SKIRROPHORUS, DC., Steetz. Involucrum partiale fere semper biflorum, 4–5-phyllum; squamis hyalinis concavis obovatis v. spathulatis obtusissimis corollisque undique glabris.

* Involucrum generale exterius glomerulo sphæroideo brevius, quasi truncatum, squamis æqualibus lana intricata inter se

conjunctis. Capitula cum paleis hyalinis cito decidua. Receptaculum generale conicum.—Herba cano-lanata, caulibus adscendentibus basi suffruticosis, capitulis citrinis.

1. *S. Cunninghami*, DC., Prodr. vol. vi. p. 150; Deless. Ic. Sel. vol. iv. p. 51; Steetz in Pl. Preiss. vol. i. p. 438. *Eriocladium pyramidatum*, Lindl.! Swan Riv. Bot. p. 24.

Sands of the coast, Dirk Hartog's Island, Western Australia, *A. Cunningham*. Swan River, *Fraser, Preiss*.

** Involucri generalis exterioris squamæ foliaceæ, inæquales, discretæ (glabræ v. glabrescentes), glomerulum obovatum subsuperantes. Receptaculum generale convexiusculum. Corolla tubo brevi, limbo 4-dentato.—Herbæ pusillæ annuæ, caulibus simplicibus vel e basi ramosis, glomerulo solitario terminalis, capitulis albido-fuscis paleisque tardius deciduis.

2. *S. Preissianus*, Steetz, Pl. Preiss. vol. i. p. 439. Swan River, *Drummond, Preiss*.

3. *S. eriocephalus*, Hook. fil. in Herb. Hook. Georgetown, Van Diemen's Land, *Gunn*.—This has a narrower glomerule than the preceding, which it much resembles, and the external involucre, filiform stems, &c., are clothed with an arenose wool, which is, however, deciduous. The achænium does not show the minute border of that of *S. Preissianus*, which Steetz calls a very minute coroniform pappus; but the whole achænium is covered with a granulated cellular pellicle.

*** Involucrum generale præcedentis sed laxum, glomerulum depresso-globosum subæquans. Receptaculum generale convexiusculum. Corolla 5-dentata.

4. *S. pygmæus*, n. sp.: annuus; glomerulis primariis inter folia radicalia linearia glabra sessilibus, sequentibus ramos breves proliferos terminantibus humi adpressis; involucro generali exteriore laxe lanato, squamis interioribus cum paleis et squamis capitulorum albido-scariosis costa viridula obovatis et oblongo-spathulatis inappendiculatis.

South-western Australia, *Drummond*.—This little plant has the habit of *Hyalochlamys*, &c. The full-grown glomerules are nearly three lines broad. Leaves about the same length, apiculate, narrowly linear, or the larger linear-spatulate, dilated at the sessile base. The base of the corolla is not thickened in the specimens examined, which are young. There are no full-grown achænia.

§ 2. POGONOLEPIS. Glomerulum obovatum. Involucrum partiale uniflorum, 3-phyllum; squamis linearibus subconduplicatis apice rotundato pube brevi barbatis. Paleæ et squamæ interiores glomeruli squamis capitulorum consimiles et cum iis deciduæ. Receptaculum generale planum. Involucrum generale exterius squamis foliaceis subsquarrosis apice pungentibus glomerulum subsuperantibus. Corolla gracilis, glabra.—Herba annua, laxe lanata, caulibus erectis simpliciusculis, ramis brevibus rigidis glomerulis 1–3 terminatis.

5. *S. strictus.*—*Pogonolepis stricta*, Steetz, Pl. Preiss. vol. i. p. 440. Swan River, *Preiss, Drummond.*

§ 3. PSEUDOPAPPUS. Glomerulum obovatum. Involucrum partiale uniflorum (raro biflorum?), 3–4-phyllum; squamis lineari-lanceolatis carinatis v. subconduplicatis apice acuminato tenui lanigero-barbatis. Palæ et squamæ interiores glomeruli squamis capitulorum consimiles, subpersistentes. Receptaculum gen. convexum. Involucrum generale *Pogonolepidis.* Corolla persistens, tubo gracili, ima basi lana longa tenuissima, pappum pilosum mentiente, instructo!—Herba pygmæa depressa, annua.

6. *S. demissus*, n. sp.: caulibus e radice exili ramosis depressis; foliis subulatis cuspidatis basi dilatata amplexicaulibus cito glabris; glomerulis junioribus dense lanatis folioso-bracteatis. (Hook. Ic. Pl., tab. ined.)

South-western Australia, *Drummond,* 1850.—Stems one or two inches long, spreading on the ground, soon glabrous. Glomerule turbinate, 2 lines long. The narrow paleate scales of the general receptacle appear to persist after the capituli they subtend fall away. Tube of the corolla longer than the infundibular-campanulate limb, its base much dilated, and covering the apex of the ovary and achænium, thickened, and just at its insertion bearing a ring of very long and delicate simple woolly hairs, more than half the length of the corolla, and exactly imitating a fine pilose pappus. The attenuate base of the turbinate achænium is tipped with a small callus, nearly as in *S. Preissianus.* I think the ring of hairs belongs to the base of the corolla, and not to the achænium (as it does in *Pachysurus,* Steetz), but the two are so firmly coalescent that, when pulled asunder, the vertex of the achænium usually comes off with the thickened base of the corolla, like a lid, opening the pericarp by a clean circumscission, and exposing the seed!

NEMATOPUS, nov. gen.

Capitula biflora, homogama, in glomerulum obconico-hemisphæricum densissime aggregata. *Involucrum generale* pauciseriale, glomerulo paulo brevius; squamis appressis, oblongo-linearibus, disco paulo brevioribus, inter se et cum paleis receptaculi consimilibus (extimis 2-3 magis herbaceis), apice viridulo lamina parva hyalino-petaloidea aurea rotundata abrupte appendiculatis. *Receptaculum generale* convexum; *partialia* brevissima, arcte sessilia. *Involucrum partiale* biseriale; squamis cujusque seriei circiter 5, conformibus, spathulato-linearibus vel anguste oblongis, concaviusculis, marginibus hyalinis, apice parce lanato lamina inflexa aurea (ut squam. glomeruli) appendiculatis, exterioribus ut videtur persistentibus, non raro capitulis duobus, uno circ. 5-phyllo, altero 7-10-phyllo, amplectentibus. *Flores* hermaphroditi. *Corolla* 5-dentata; fauce cylindracea, tubo brevissimo glanduligero. *Antheræ* breviter caudatæ. *Styli* rami breves, apice capitellati. *Achænia*, potius *ovaria*, brevia, teretia, glaberrima. *Pappus* plane nullus.—Herba annua gracili; caule erecto corymboso-ramosissimo ramisque filiformibus cito glaberrimis; glomerulis flavis solitariis ramulos capillares divergentes nudos terminantibus; foliis lineari-filiformibus alternis laxe lanatis glabratis.

N. effusus. Swan River, *Drummond*.—Plant between a span and a foot in height. Glomerules a line and a half in diameter; the scales sparingly lanate. Scales of the glomerule, paleæ, and proper involucral scales all alike, except the interior are successively thinner and more scarious or hyaline, all abruptly tipped with the same short and rounded yellow petaloid appendage.—The plant appears to constitute a very distinct genus, between *Skirrophorus* and *Gnephosis*. It has the habit of the latter; but the scales, &c., are all narrow and homogeneous, the axis is convex (instead of cylindrical), the partial receptacles sessile (not stipitate), and there is no trace of pappus.—The specimen of *Gnephosis tenuissima* in the Hookerian herbarium has lost all the flowers, and although some achænia remain loose in the scarious heads, I can find no pappus, &c., to compare with the curious and nearly-allied *Pachysurus* of Steetz, which was abundantly gathered by Drummond. There is another allied plant in Drummond's collection (No. 201 of the coll. 1848, also in the earlier collection), which has the habit of *Gnephosis*, but with strict and very flexuous branches, a more compound glomerule, the axes of the few-flowered capitula all arising from the depressed summit of the general axis, and clothed with numerous series of scarious squamæ; but none of the flowers are developed so that their structure can be made out.

CHRYSOCORYNE, *Endl. Gen. Suppl.* vol. iii. p. 70.

Char. Emend. *Capitula* biflora (inferiora nunc uniflora), homogama, singula sub palea scariosa dilatata orbiculata concava recondita, in spicam amentaceam secus receptaculum filiforme dense aggregata. *Involucrum generale* e paleis infimis vacuis parvis constans. *Involucrum partiale* e squamis 2–6 hyalinis, exterioribus dilatatis conduplicato-navicularibus. *Flores* hermaphroditi. *Corolla* gracilis, apice infundibulari-ampliata, 5-dentata vel 3-dentata. *Antheræ* basi bimucronulatæ. *Styli* rami capitellato-truncati. *Achænium* glabrum, sæpe atomis resinosis conspersum, calvum.—Herbæ pusillæ annuæ, furfuraceo-pubentes; foliis alternis sessilibus sublinearibus; spicis terminalibus solitariis subternisve, cylindricis vel clavatis, glabris, aureo-fuscis. Capitulum imæ basi crassiori paleæ fulcrantis pl. m. adnatum.

In *Plantæ Hugelianæ*, Mr. Bentham briefly characterized a plant under the name of *Crossolepis? pusilla*, supposing it might possibly belong to Lessing's obscure genus *Crossolepis*. Afterwards Sir William Hooker figured under this name, in the *Icones Plantarum*, a different species; and from this figure (and not from the Hugelian plant) Endlicher recently established the genus *Chrysocoryne*. I have not seen any original specimen of Hugel's, indeed; but Steetz, who has, describes the proper involucre of the capitulum as consisting of six hyaline scales. There are only two in the species figured by Hooker, and also in a third, lately gathered by Mr. Drummond, which is distinguished by its very long spike and 5-toothed corolla. The three species must be disposed in two sections, as follows:—

§ 1. Involucri squamæ 6, inferne parce lanato-ciliatæ, intimis minoribus planiusculis. Paleæ receptaculi basi laxe lanatæ.

1. *C. Hugelii:* caule gracili superne corymbosi-ramoso; foliis inferioribus linearibus, summis parvis ovatis; spicis clavatis paniculatis. —*Crossolepis? pusilla*, Benth. Pl. Hugel. p. 61, non Hook. *Chrysocoryne pusilla*, Steetz, Pl. Preiss. vol. i. p. 441, non Endl.

Swan River, *Hugel, Preiss, Drummond.* The larger specimens, from Drummond, are five inches high, and copiously branched above, the branches and branchlets filiform; and the spikes, as described by Bentham, are only three or four lines long. The lowermost capitula are principally one-flowered.

§ 2. Involucri squamæ tantum 2, oppositæ, naviculares, marginibus superne parce setigero-denticulatæ v. fimbriatæ, cum palea fulcrante inferne glabræ. Radix multicaulis.

2. *C. Drummondii*: spicis cylindricis breviusculis; corolla tridentata.
—*Crossolepis? pusilla*, Hook. Ic. Pl. t. 413. *Chrysocoryne*, Endl. l. c. ex char.

Swan River, *Drummond*.—Stems 1 to 2 inches high. Spikes half an inch long, straw-colour or ferruginous.

3. *C. myosuroides*, n. sp.: spicis cylindricis lineari-elongatis; corolla 5-dentata exserta.

Swan River, *Drummond*, 1845.—Stems and foliage as in the foregoing; the orange-fuscous spikes an inch or more in length, narrowed at the base, much resembling the fructification of *Myosurus*. The two squamæ of the partial involucre are much narrower than in *C. Drummondii*, and less setigerous-fimbriate above.

CEPHALOSORUS, nov. gen.

Capitula uniflora, in glomerulum sphæroideum dense aggregata. *Involucrum generale* 2–3-seriale; squamis foliaceis vel subherbaceis inappendiculatis. *Receptaculum generale* aut latum convexum aut cylindraceum, paleis linearibus scariosis inter capitula et cum iis plus minus deciduis onustum. *Involucrum partiale* e squamis scariosis concavis exappendiculatis 4–6, exterioribus oblongis, interioribus latioribus florem invicem involventibus. *Flos* hermaphroditus. *Corolla* tubulosa infundibuliformis, limbo 5-fido. *Antheræ* basi subcaudatæ. *Styli* rami apice capitellati, breviter penicillati. *Achænium* obovatum, aut glabrum pellicula diaphana humectate tumefacta gelatinosa, aut hirsutum, pappo calyciformi conspicuo tenuiter scarioso deciduo coronatum.—Herbæ annuæ; caule virgato simplici vel parce ramoso, apice subnudo glomerulum solitarium gerente; foliis alternis. *Flores* flavidi. (Species duæ quoad involucrum et receptaculum glomeruli plane diversæ, nec tamen meo sensu divellendæ.)

1. *C. phyllocephalus*, n. sp.: foliis spathulatis cauleque tomento floccoso deciduo lanatis; glomerulo depressi-globoso involucro foliaceo (squamis ovatis) cincto; receptaculo generale depresso; achænio glabro;

pappo inæqualiter lobato corollæ tubum subæquante. (Ic. Pl. tab. ined.)

Swan River, *Drummond*, 1846, 1848.—Stem a foot or more in height. Glomerule from one-half to two-thirds of an inch in diameter, its conspicuous foliaceous involucre shorter than the disc. Capitula very densely crowded, closely sessile on the much dilated convex receptacle, the scales nearly glabrous, as long as the corolla, their tips yellowish. Corolla much thickened and indurated at the base after anthesis, and the base of the style bulbous, as in *Skirrophorus*. The achænia have a thick, cellular, diaphanous pellicle, which, when moistened, swells into an extremely thick gelatinous covering.

2. *C. gymnocephalus*, n. sp.: foliis lineari-filiformibus cauleque gracillimo cito glabris; glomerulo ovoideo squamis brevissimis inconspicuis lanceolatis margine scariosis paleis fere similibus cincto; receptaculo angusto cylindrico; achænio hirsuto; pappo brevi margine subintegro. (Ic. Pl. tab. ined.)

Swan River, *Drummond*, 1848.—Stems a foot or eighteen inches high; the leaves an inch or more in length, very narrow. Glomerule 4 or 5 lines in diameter, globose, or at maturity ovoid, apparently naked; the general involucre so small and similar to the paleæ that it might escape notice. Except in this respect, and in the different shape of the receptacle, however, the floral structure accords with the preceding species. The two form a genus well distinguished among the *Angiantheæ* by the calyciform pappus, and of a different habit from any other, except *Pycnosorus*, Benth., which has no involucre, more than one-flowered heterogamous capitula, and a very different pappus.

BOTANICAL INFORMATION.

Sale of a great Herbarium and extensive collection of Drugs.

The *Herbarium* and the collection of *Drugs* of the late Dr. Lucæ, Apothecary at Berlin, are offered for sale. The first consists of 36–40,000 species, and contains, beginning from the collections published by Sieber, almost all which have been distributed during the last thirty years, besides those of several travellers which could not be procured by purchase. It includes about 150 different collections, and is, the latest acquisitions excepted, arranged according to the natural system of De Candolle. The specimens of all phænogamic plants, save the *Glumaceæ*, are poisoned, and therefore in a state of perfect preservation. The price, 6,000 Prussian dollars.

The collection of *Drugs* contains 1,604 objects, and is one of the most complete on the Continent. There are in it, for example, forty-four sorts of "China," thirty sorts of *Sanguis Draconis*, &c. It is distinguished by exquisite specimens and many rare objects. Its price, 2,500 Prussian dollars. The cabinets for both collections, included in the above sums, are elegant and commodious. For particulars apply, by post-paid letters, to R. F. Hohenacker, Esslingen, near Stuttgart.

NOTICES OF BOOKS.

The RHODODENDRONS *of the* SIKKIM-HIMALAYA; *being an Account of the Rhododendrons recently discovered on the mountains of Eastern Himalaya.* By JOSEPH DALTON HOOKER, M.D., F.R.S. *Edited by* Sir W. J. HOOKER, K.H., F.R.S. Part I., 1849. Part II., 1851. Imperial folio.

The progress of our acquaintance with East Indian *Ericeæ* (taking the family in its widest extent, so as to comprise the *Cranberries*) presents some curious features, which it may not be uninteresting to place before the reader. I will, therefore, preface my few remarks on the above important work with a condensed chronological summary of authors and their publications on the family, giving the abstract number of genera and species of each, as far as practicable, but without any analysis or reduction; their total amount having accordingly no direct reference to the actual state of our knowledge, since several of the genera have been abandoned, and many species are mere repetitions or else synonyms.

1712. Kæmpfer, *Amœnitates Exoticæ*: Rhododendron, 1; Azalea, 2; Vaccinium, 2: of all of which there are specimens in his original herbarium, preserved in the Sloanean collection at the British Museum.

1768. N. L. Burmann, *Flora Indica*: Rhododendron, 1; Azalea, 1.

1780. C. Thunberg, *Flora Japonica*: Rhododendron, 1; Andromeda, 1; Vaccinium, 3.

1789. R. Saunders, *Some account of the vegetable and mineral productions of Bootan and Thibet*, in Philosophical Transactions, vol. lxxix. (republished in 1800, as an appendix to Turner's 'Embassy to

Thibet in 1783'): simple allusion to Rhododendron, 1; Cranberries, 2; Vaccinium, many species; Arbutus, 1.

1790. J. Loureiro, *Flora Cochinchinensis*: Enkianthus, 2; Acosta, 1 (Agapetes).

1799. T. Hardwicke, *Narrative of a Journey to Sirinagur* (in 1796) in 'Asiatic Researches,' vol. vi.: Rhododendron, 1; Andromeda, 1.

1804. J. E. Smith, *Exotic Botany*: Rhododendron, 1.

1814. W. Roxburgh, *Hortus Bengalensis*: Rhododendron, 1; Ceratostema, 2.

1819. F. Hamilton, *Account of the Kingdom of Nipal*: Rhododendron, 1.

—— N. Wallich, *Letter to Dr. F. Hamilton*, in Edinburgh Philos. Journal, vol. i.: simple allusion to Rhododendron, Andromeda, and Gaultheria.

1820. N. Wallich, *Description of some rare Indian Plants*, in 'Asiatic Researches,' vol. xiii.: Andromeda, 4; Gaultheria, 1.

1821. D. Don, *Description of several new plants from the Kingdom of Nepal*, in 'Memoirs of the Wernerian Society of Natural History,' vol. iii.: Rhododendron, 3; Andromeda, 1.

1822. W. Jack, *Descriptions of Malayan Plants*, in 'Malayan Miscellany,' vol. ii. n. 3: Rhododendron, 1; Vaccinium, 1.

1825. C. Blume, *Bijdragen tot de Flora van Nederlandsch Indie*: Vireya (Rhododendron), 5; Vaccinium, 1; Azalea, 4; Gaultheria, 3; Diplycosia, 3; Thibaudia, 8; Gaylussacia, 1; Hymenanthes, 1; Clethra, 1.

—— D. Don, *Prodromus Floræ Nepalensis*: Rhododendron, 4; Andromeda, 4; Gaultheria, 1.

1828–1832. N. Wallich, *Lithographic Catalogue of the East India Company's Herbarium*: Rhododendron, 8; Andromeda, 5; Gaultheria, 3; Thibaudia, 7; Vaccinium, 1.

1831, 1832. N. Wallich, *Plantæ Asiaticæ Rariores*, vol. ii. and iii.: Rhododendron, 2; Andromeda, 1.

1832. W. Roxburgh, *Flora Indica*, vol. ii.: Rhododendron, 1; Ceratostema, 2. (It is proper to remark here, that this posthumous work was published very many years after the author wrote his descriptions; and this applies to some extent also to the 'Hortus Bengalensis;' see above, under 1814.)

1834. G. Don, *A General System of Gardening and Botany*, vol. iii.: Andromeda, 1; Pieris, 4; Enkianthus, 2; Diplycosia, 3; Gaultheria, 5; Clethra, 1; Vireya, 1; Rhododendron, 23; Hymenanthes, 1; Vaccinium, 3; Gaylussacia, 1; Agapetes, 16.

1835. J. C. Zenker, *Plantæ Indicæ quas in montibus Nilgherry dictis legit B. Schmid*: Rhododendron, 1.

1838. J. J. Bennett, *Plantæ Javanicæ Rariores quas legit T. Horsfield*: Rhododendron, 2.

1839. J. Royle, *Illustrations of the Botany and other branches of the Natural History of the Himalaya Mountains and of the Flora of Cashmere*: Rhododendron; Gaultheria; Andromeda; Vaccinium; Thibaudia.

—— A. P. De Candolle, *Prodromus Systematis Naturalis Regni Vegetabilis*, vol. vii. sect. ii.: Agapetes, 16; Gaylussacia, 2; Vaccinium, 5; Diplycosia, 3; Gaultheria, 7; Pieris, 4; Cassiope, 1; Azalea, 5; Rhododendron, 20; Enkianthus, 3.

—— P. F. de Siebold, *Plantæ Japonicæ*; digessit J. G. Zuccarini, vol. i.: Rhododendron, 1.

1844. J. K. Hasskarl, *Catalogus Plantarum in Horto Bogorensi cultarum*: Rhododendron, 1.

1846. P. F. de Siebold et J. G. Zuccarini, *Floræ Japonicæ Familiæ Naturales*, Sect. altera, in Trans. of the Phys. and Mathem. Class. of the Royal Bavar. Acad. of Sciences, vol. iv.: Andromeda, 1; Pieris, 1; Meisteria, 1; Clethra, 2; Vaccinium, 1; Rhod. 4.

1846, 1847. R. Fortune, *Sketch of a visit to China in search of new plants*, in Journal of the Horticultural Society of London, vol. i. and ii.: Azalea, 3.

1846. J. Lindley, *New Plants, &c., from the Society's Garden*, in Journal of Hort. Soc. of London, vol. i.: Azalea, 2.

1847. R. Wight, *Icones Plantarum Indiæ Orientalis*, vol. iv. p. 1: Vaccinium 15; Gaultheria, 3; Andromeda, 3; Rhododendron, 3.

—— R. Fortune, *Three years' wanderings in the Northern Provinces of China*: Enkianthus, 1; Azalea, several.

1847, 1848. W. Griffith, *Posthumous Papers*: several Rhododendrons, Gaultheriæ, Thibaudiæ, Gaylussaciæ, Andromedæ, and Vaccinia.

1848. J. K. Hasskarl, *Plantæ Javanicæ Rariores*: Amphicalyx, 1; Agapetes, 2.

—— J. Lindley, *A notice of some species of Rhododendron inhabiting Borneo*, in Journal of the Hort. Society of London, vol. iii.: Rhododendron, 4.

1849. R. Wight, *Illustrations of Indian Botany*: Vaccinium, 9; Rhododendron, 1; Andromeda, 1; Gaultheria, 1.

—— C. G. C. Reinwardt, *The Vegetation of the Indian Archipelago*, in Journal of the Hort. Soc. of London, vol. iv.: simple allusion to Andromeda, Clethra, and Rhododendron.

1849, 1851. Joseph Hooker (the work at the head of this notice).

To the above list should have been added perhaps a variety of periodical publications like the Botanical Repository, Magazine, Register, Cabinet, Flower Garden, and others, both of this and foreign countries; but it was thought unnecessary.

In point of antiquity, the Ericeous family (in its extended sense) is among the oldest on record, inasmuch as there exist descriptions and figures of several of its members in ancient Chinese works; in which country, as well as in Japan, these plants have continued favourite objects of garden culture, almost from time immemorial. But we have to deal here with the botanical literature of European languages only; and we find accordingly, that our earliest account is from the pen of the incomparable Engelbrecht Kæmpfer, so far back as the year 1712. Since that period, and for more than a century, the additions, comparatively few, were made at distant intervals; so that, till the year 1819, all that we knew, including Kæmpfer's plants, was comprised in about sixteen species. Within the last thirty-two years the number has multiplied at more than a six-fold rate; thanks to the labours of Blumé, Griffith, Siebold, Hasskarl, Fortune, Wight, and Joseph Hooker. In De Candolle's 'Prodromus,' sixty-six species were recorded in 1839; more than one hundred are at present known. To Dr. Hooker alone we are indebted for the discovery of the amazing number of thirty-three new species of one genus alone, by far the noblest of the family, and, if viewed in the combined light of botanical and horticultural interest, undoubtedly the most magnificent and desirable woody plants in the whole vegetable kingdom, on account of their stature, foliage, and flowers. To those who have not had the advantage of seeing any of the Asiatic species of Rhododendrons in the wild luxuriance of their native growth, I may appeal, for a corroboration of my assertion, to the beautiful Rhododendron Exhibition, which took place a few years ago, at the Royal Botanical Society's Garden in Regent's Park. Bearing in mind, then, that each individual species of this gorgeous instalment forms a valuable object of cultivation; that many will be found perfectly well suited to the climate of this and other northern countries; that most of them are large shrubs or trees of surprising floral beauty, and that nearly all exist already in England in great vigour and profusion, having been raised, with many other things, from seeds sent home to the Royal Botanic Gardens at Kew by their discoverer; assuredly we may claim for Dr. Hooker a high rank among benefactors to the cause of gardening; no one having ever made anything like his important additions to our shrubberies and arboretums. The present monographic work takes up only one item among the multifarious objects which have engaged Dr. Hooker's attention, during the extensive Indian travels, from which he has only now returned, after an absence of three years and a half; but it may be considered as an exponent of his harvest of observations and collections, by which we may fairly regulate our expectations. The writer of this article is not permitted to indulge, as he otherwise should have done, by referring in detail to his distinguished friend's previous works, or the enormous treasures accumulated during his last travels; but this much he may say: they are entirely worthy of the name he bears, and of that high respect, as well as personal friendship, which have been accorded to him by a Humboldt and a Brown.

Very little remains to be added concerning the *Sikkim Rhododendrons*, except that both text and plates are excellent. The two parts hitherto published contain twenty highly finished representations of new species, made on the spot by the author himself, and not surpassed by any ever given to the botanical world, either as regards their truthfulness and elegance, their details, or the beauty of their colouring; in which respect, as well as in the lithography, they are worthy of that well-known and deserving artist, Mr. Fitch. A third part is announced as ready for immediate publication; it is to contain ten more plates, and will conclude this important work, not a slight recommendation of which is its moderate price. Prefixed to the first part is an historical account of the genus as respects India, by the author's father, and possessing the sterling value of everything emanating from that untiring pen. This is followed up by the author's own remarks on the physical and geographical distribution of the genus over the Sikkim mountains. In the second part we have a conspectus, or synopsis, of forty-three Indian Rhododendrons, of which thirty-six are natives of that range; thirty-three being, as has been mentioned before, Dr. Hooker's own discoveries. They are arranged under eight sections, dependent upon such floral and pericarpical distinctions as could be made out in a genus so purely natural as ours is. The largest-flowered species of the genus are *R. Dalhousiæ, Maddeni,* and *Aucklandii,* the first of which, with *pendulum,* are shrubby epiphytes. Others, such as *R. Falconeri, argenteum,* and *Hodgsonii,* exceed all the rest in the grandeur of their foliage. One of the most strikingly beautiful is *R. Thomsoni,* on account of its large, blood-red clusters of flowers, and broad, almost orbicular, leaves; named, in the words of the discoverer, after "Dr. Thomas Thomson, surgeon, H. E. I. C. S., late of the Thibetan Mission, son of the learned Professor of Chemistry of Glasgow University, my earliest friend and companion during my college life, and now my valued travelling companion in Eastern Hjmalaya." *R. pumilum* is the smallest and rarest of the Sikkim species, but exceedingly charming; "its elegant flowers are produced soon after the snow has melted: and then its pretty pink bells are seen peeping above the surrounding short heath-like vegetation, reminding the botanist of those of *Linnæa borealis.*"

I cannot take leave of this work without expressing a hearty wish that Dr. Hooker, and his zealous travelling companion, Dr. Thomson, having recently returned in safety from their travels, may obtain every encouragement required for speedily and extensively publishing the vast treasures of their harvest.

N. Wallich, M.D.

Tropical Scenery: Physiognomy of Tropical Vegetation; *drawn and lithographed by* M. De Berg.

Humboldt, in his 'Cosmos,' has observed, that "he who is endowed with susceptibility for the natural beauties of mountains, streams, and forest scenery; who has wandered through the countries of the torrid zone, and has seen the luxuriant vegetation, not only upon the cultivated shores, but in the vicinity of the snow-capped Andes, the Himalaya mountains, and the Neilgherry hills of the Mysore, or in the wide-spread forests between the Orinoco and the Amazon:—that man can alone understand what an immeasurable field for landscape-painting is open between the tropics of both continents, or in the islands of Sumatra, Borneo, and the Philippines, and how the most splendid and spirited works which man's genius has hitherto accomplished, cannot be compared with the vastness of the treasures of nature, of which Art may, at a future time, avail itself."— And again: "If in the frigid zones the bark of the trees is covered with discoloured spots, occasioned by the presence of Lichens or Mosses, in the regions of the feathery Palms *Cymbidium* and the aromatic *Vanilla* enliven the trunks of *Anacardium* and of gigantic Fig-trees. The fresh green of the *Dracontium*, and the deeply-cut leaves of Ferns, contrast with the many-coloured blossoms of Orchises; the twining *Bauhinia*, the Passion-flower, and the yellow-blossomed *Banisteria*, climb high into the air around the stems of the forest trees; delicate blossoms unfold themselves from among the roots of the *Theobroma*, as well as from the thick and rugged bark of *Crescentia* and *Gustavia*. In the multitude of flowers and leaves, in this luxuriant growth, and the confusion of climbing plants, it often becomes difficult to distinguish to which tree the blossoms, and to which the leaves, belong: indeed, a single tree, adorned with *Paullinias, Bignonias,* and *Dendrobium*, presents a multiplicity of plants, which, if separated one from another, would cover a considerable space." Influenced by these and other like passages in the writings of Humboldt, a talented Prussian artist has paid a visit of some months to New Grenada, penetrated to the snowy mountains of the Andes, where *Tolima* rears his gigantic head, and has brought home a rich collection of drawings of scenery and studies from nature, of a very high order of merit, especially made with a view to the grander forms of vegetation in the tropics, the enormous rooting trunks of the Fig-trees, the various large-leaved Arums, the Tree-ferns, but, above all, the lofty and most graceful Palms. These have received the approbation of His Prussian Majesty, of Baron Humboldt himself, and of Chevalier Bunsen. By the favour of the two latter gentlemen we have been permitted to inspect this collection, and we cannot but rejoice that a selection of them is to be made, to be lithographed by the author in Germany, and to be published in this country, by Colnaghi, early in next year. Besides their excellence, as works of art, they represent, faithfully, many of the most striking vegetable forms that exist, along with the attendant scenery, the rocky gorge, the sparkling water-fall, the still, sluggish river, or the snow-capped Andes. Descriptive letter-press will explain the plants, which the author takes great pains to have accurately named.

Second Report on MR. SPRUCE'S *Collections of Dried Plants from* NORTH BRAZIL; *by* GEORGE BENTHAM, Esq.

(*Continued from p.* 120.)

The last portion of this report was already in type when the remainder of Mr. Spruce's Santarem collection reached this country in excellent condition, containing, besides a considerable number of additional species, specimens in fruit or in flower of many others, which had previously been found only in one of these states. The chief additions to the few orders mentioned in the last number of this Journal, are two *Dilleniaceæ*, a *Doliocarpus* answering very well to Garcke's description of his *D. brevipedicellatus* (Linnæa, vol. xxii. p. 47), a Surinam species, found by Mr. Spruce on the shores of the Amazon, below Santarem, and the following new *Davila*, from the campos near Santarem.

Davila *pedicellaris*, sp. n.; scandens, glabra, foliis oblongis ellipticisve obtusis integerrimis rigide coriaceis reticulato-venosis scabris, racemis 1-4-floris, pedicellis elongatis, sepalis interioribus maximis induratis, ovariis geminis.

These specimens resemble so closely the figure of *D. flexuosa*, St. Hil. (Fl. Bras. Mer. p. 17, t. 2), that I should have considered them as belonging to that species, were it not that St. Hilaire describes his plant as erect (not twining), and the lateral veins of the leaves as "vix manifestis," whilst in our plant, which is certainly a twiner, the lateral veins, as well as the smaller reticulations, are remarkably conspicuous. The very hard inner lobes of the calyx, full half an inch in diameter when enclosing the fruit, the inflorescence, and the shape of the leaves, are as described in the above-quoted work. The flowers are, according to Mr. Spruce, mostly apetalous, rarely with five yellow obcordate petals.

Resuming the general Santarem collection with the Order *Polygaleæ*, we have six species of *Polygala*, all published ones, and most of them having a wide range in tropical America. Among them the delicate *P. subtilis*, H. B. K., whose minute flowers were carefully examined when fresh by Mr. Spruce, proves to be hexandrous, like *P. setacea*, Mich., not octandrous, as usually described with the great mass of species. To the *P. bryoides*, St. Hil., of which a broad-leaved variety has been distributed, should be reduced my *P. camporum* (Hook.

Journ. Bot., vol. iv. p. 100). The *P. spectabilis*, DC., has turned up again in sufficient quantity for general distribution.

The following new *Securidaca*, from the banks of the Amazon, near Santarem, is remarkable from the fruit having a second wing, arising from the back, giving the whole *samara* a very peculiar form, though not easy to describe.

Securidaca *bialata*, sp. n.; scandens, foliis ovalibus obtusis retusisve basi subcordatis coriaceis subtus ramulisque puberulis, alis ovato-orbicularibus, carina latiuscula apice brevissime triloba plicato-cristata, petalis superioribus anguste oblongis, alis fructus geminis fructu subbrevioribus, postica angusta antica latissima cristæformi.—*Frutex* super arbores alte scandens. *Folia* breviter petiolata, 2–3 poll. longa, 1–1¼ poll. lata. *Racemi* breves, multiflori. *Bracteas* caducas haud vidi. *Flores* majusculi. *Calycis* foliola 3, puberula, 1¼ lin. longa; alæ vix unguiculatæ, 5 lin. longæ, glabræ nec ciliatæ. *Carina* unguiculata, supra unguem 4 lin. longa; petala superiora 5 lin. longa, se invicem applicata, superne paulo latiora. *Stamina* 8, usque ad ⅔ in tubum incurvum basi latiorem connata; filamentis superne liberis. *Ovarium* subsessile, obovoideum, compressum, apice oblique truncatum, intus 1-loculare, 1-ovulatum. *Stylus* glaber, falcatus, a latere compressus, stamina æquans, apice oblique stigmatosus. *Fructus* absque alis oblongus, 8–9 lin. longus, reticulato-rugosus; ala superior ex suturæ posticæ parte superiore orta, oblonga, 5–6 lin. longa, 1¼–2 lin. lata, apice rotundata; ala major e tota fere sutura antica oriunda, in parte superiore 6–7 lin. longa, valde dilatata, plicato-rugosa, margine exteriore undulata v. sublobata.

There are likewise two species of *Catocoma*: *C. floribunda*, a species widely diffused over tropical Brazil, found by Mr. Spruce at Santarem, growing to the tops of the loftiest trees and covering them with a crown of white and yellow flowers spotted with purple; and *C. lucida*, Benth., a Guiana species, now gathered on the Rio Trombétas, and its affluent the Rio Caipurú. Klotzsch (Linnæa, vol. xxii. p. 51) refers this genus to *Bredemeyera* of Willdenow, but without stating on what grounds. If the identity has been established by inspection of original specimens, Willdenow's character will require considerable modification, for the "drupa ovata nuce biloculari" is at total variance with the essential character of *Catocoma*.

A small-flowered *Trigonia* (*T. parviflora*), from the neighbourhood of Santarem, resembles much in foliage what I take to be the *T. lævis*, Aubl., but the branches of the panicle are very much longer, and the flowers several together in clusters growing out into little racemes, and no larger than in *T. crotonoides*. Like most other species, it is a shrub with more or less twining branches. I have it also from Guiana, being n. 981 of Sir Robert Schomburgk's second journey, or n. 1251 of Richard Schomburgk, although apparently overlooked by Grisebach, in his account of the tropical American *Trigoniæ* of the Berlin herbarium, in the Linnæa, vol. xxii.

Trigonia *parviflora*, sp. n.; foliis ovali-ellipticis oblongisve adultis utrinque glabris nitidulis, paniculæ terminalis tomentosæ ramulis elongatis, floribus parvis numerosis, fasciculis demum racemiformibus, capsula obovoideo-triquetra.—*Frutex* biorgyalis, ramulis plus minus scandentibus, novellis tomentosis mox glabratis. *Folia* 2–3-pollicaria, pleræque breviter acuminata, obtusa v. acutiuscula, novella interdum lana parcissima hinc inde conspersa, et ad venas pilis paucis hirtella, adulta glabra, tenuiter coriacea, nitidula, penninervia et reticulato-venosa. *Petioli* 1–2 lin. longi, linea transversa connexi. *Stipulæ* parvæ, subulatæ, caducissimæ. *Panicula* terminalis, sæpe pedalis, basi foliata, ramis 3–6-pollicaribus, fere a basi floribundis. *Fasciculi* v. racemuli 3–10-flori, axi demum 2–4 lin. longa. *Bracteæ* acutæ, pedicello lineari breviores. *Flores* vix linea longiores, iis *T. crotonoidis* simillimi nisi minores. *Petalum* tamen anticum ad marginem sacci non ciliatum ut in icone (Fl. Bras. Mer. t. 105) depictum, et placentatio mihi videtur potius parietalis quam axilis dicenda. *Ovula* enim non angulo ipso centrali loculorum inserta, sed juxta angulum centralem ad latera placentarum (dissepimenta constituentium), in axi vix cohærentium. *Capsula* 4 lin. longa, extus tomentella, apice truncata et late triangulata, fere triloba.

The *Malvaceæ* are but few. Among them is a single specimen of a narrow-leaved very hairy variety of *Pavonia cancellata*, Cav., very different in aspect from the two or three specimens distributed from Pará, which are intermediate, as it were, between the commoner forms and the *P. modesta*, Mart.; and a few specimens of a variety of *Wissadula parviflora*, with golden yellow flowers. This is the *Abutilon parviflorum*, β *luteum*, A. St. Hil., and is closely allied to *W. hirsuta*, Presl (Bahia, Salzmann, Gardner, n. 867, Mart. Herb. Bras. n. 1002), only differing, indeed, in the absence of the long spreading hairs on the branches and panicle of that species or variety.

The *Sterculiaceæ* (including *Buettneriaceæ*) comprise—*Helicteres pentandra*, L., from Santarem, precisely similar to Hostmann's Surinam specimens; male specimens of *Sterculia striata*, St. Hil. et Naud., from Santarem; the *Cacao-rana* mentioned in Spruce's letters (vol. ii. p. 198), from the forests near Obidos, a slender tree of forty or fifty feet, bearing its flowers on the naked stem and leaves at the summit only, apparently the *Theobroma subincanum*, Mart.; *Buettneria divaricata*, Benth., from the borders of Lake Quiriquiry, and *B. scabra*, Aubl., from Santarem, both of them Guiana plants; two new species of *Buettneria*, described below; *Guazuma ulmifolia*, L.; a variety of the common *Waltheria Americana*, with the carpels usually more than one; *Melochia hirsuta*, Cav., from which *M. vestita*, Benth., and *M. clinopodifolia*, St. Hil. et Naud., are perhaps not really distinct; a large-leaved variety of *M. arenosa*, Benth., to which should apparently be referred the *M. cinerascens*, St. Hil. et Naud.; and a variety of *M. melissæfolia*, Benth., with smaller leaves, flowers, and fruits. In my description of the latter species (Hook. Journ. Bot. vol. iv. p. 130), by some error of the press or of copying, the seeds are said to be *hispid*, which has misled others, who have subsequently found the plant in other localities. The seeds are perfectly without hairs, and the manuscript notes being long since destroyed, I cannot now make out what was the word which has been so misread. Miquel's *Melochia concinna* is, however, distinct in other respects from *M. melissæfolia*, although it comes very near to the smoother forms of *M. hirsuta*.

The following are the new species :—

1. Buettneria *rhamnifolia*, sp. n.; glabriuscula, caule fruticoso aculeato (scandente?), foliis oblongis obtusiusculis basi rotundatis subtus ad axillas nervorum barbatis, pedunculis axillaribus umbelliferis tenuibus.—Affinis *C. tereticauli*, Lam., sed folia non acuminata, pedunculi longiores, flores majores, etc. *Ramuli* teretes, flexuosi, novelli minute puberuli. *Aculei* pauci, recurvi. *Folia* breviter petiolata, subtripollicaria, glabra, utrinque viridia, a basi penninervia, glandula costæ majuscula. *Pedunculi* in axillis foliorum novellorum solitarii v. ad basin ramulorum complures, tenues, glabri, infra umbellam semipollicares. *Pedicelli* capillares, pedicello æquilongi. *Calycis* laciniæ 2 lin. longæ, basi latæ, acuminato-acutæ, tomento minuto

extus canescentes. *Petalorum* lobus medius longe subulato-acuminatus. *Tubi staminei* laciniæ steriles, breves, latæ, obtusissimæ, glanduliformes. *Antheræ* intra sinus sessiles, extrorsæ, didymæ. Ovarium mox tuberculoso-echinatum. *Fructus* non visus.—From the south bank of the Amazon, opposite Monte Alegre.

2. Buettneria *discolor*, sp. n.; fruticosa, inermis, foliis ovali-ellipticis oblongisve supra glabris subtus tomento minuto albis, pedunculis brevissime umbellatis in glomerulos axillares confertis, petalorum lobo medio ungue multo breviore obtuso.—*Frutex* biorgyalis, ramosissimus, haud scandens, ramulis teretibus novellis petiolis venisque foliorum subtus ferrugineo-tomentellis. *Folia* majora 3–4 poll. longa, 2 poll. lata, breviter et obtuse acuminata, basi rotundata et trinervia, cæterum penninervia, transversim venulosa; floralia 1½–2-pollicaria, angusta. *Flores* ad axillas numerosissimi, parvi, pedunculis pedicellisve 1–2 lin. longis. *Calyces* extus tomentosi, laciniis latis haud acuminatis. *Petala* calyces vix superantia, linearia, apice leviter cucullata, lobis lateralibus dentiformibus reflexis, dorsali oblonga obtusa brevi. *Tubi staminei* lobi brevissimi; antheræ subdidymæ.

The *Tiliaceæ*, besides the common *Muntingia Calaburu*, L., only comprise two forms of *Triumfetta heterophylla*, Lam.; the one, gathered in May, is precisely similar to specimens from Jamaica and other West Indian Islands, and from Surinam; the latter of which (Hostmann, n. 499) are described by Miquel as new, under the name of *T. Hostmanni*, he having fancied he saw narrow linear petals in the very imperfect flowers of his specimen. I find, however, in several buds in very good condition which I have opened, no trace of petals, but ten stamens, whose somewhat dilated filaments may have been either mistaken for, or some of them accidentally converted into, petals. The second form, gathered in July, has almost the consistence of leaf and the down of *T. Lappula*, but no tendency to the peculiarly shaped lobes of the lower leaves of the latter species.

A parasitical Guttiferous shrub from Santarem has been distributed under the name of *Arrudea? bicolor*, which I refer to that genus on account of the great multiplication of the floral envelopes, notwithstanding a considerable difference between the stamens of our plant and St. Hilaire's original species. The flower, however, examined by Cambessèdes, was so far hermaphrodite, as to have a perfectly formed pistil, although without visible ovules, and a flower I possess of the same species is also hermaphrodite, whilst those of Mr. Spruce's specimens of *A. bicolor* are male, without any vestige of ovary; and we know how very different the stamens often are in the male and in the hermaphrodite flowers of the same species of *Clusia*. Our plant is specifically well distinguished from St. Hilaire's, by the foliage and petals. A third species from Surinam, *A. purpurea*, Splitg., is unknown to me: it is described as 5-petalous, like ours, but with purple flowers, and only nine sepals.

Arrudea? *bicolor*, sp. n.; foliis petiolatis obovali-cuneatis margine recurvis, venis vix conspicuis, corymbis (masculis) terminalibus paucifloris, sepalis ultra 20, petalis 5.—*Frutex* in arbores parasiticus, undique glaberrimus, ramis teretibus. *Folia* 3–5 poll. longa, supra medium 1½–2¼ poll. lata, apice rotundata, basi in petiolum semipollicarem longe angustata, costa media subtus prominente, venis lateralibus parallelis tenuibus v. sæpius omnino evanidis. *Flores* ad apices ramulorum 3–9, dispositi in cymas breves trichotomas breviter pedunculatas, multo minores quam in *A. clusioide*, vix enim 1¼ poll. diametro; in speciminibus suppetentibus omnes masculi. *Sepala* (v. bracteæ) circa 22, concava, coriacea, arcte appressa, exteriore brevissima et late orbiculata immarginata, interiora gradatim majora, intima late membranaceo-marginata et lacera. *Petala* 5, sepalis interioribus majora, alba, basi lutescentia, late obovali-orbiculata, apice emarginata, margine undulato-crispa, basi in unguem brevissimum latum angustata. *Stamina* in medio flore receptaculum totum obtegentia, numerosa (circa 80), arcte conferta, libera tamen: filamentis clavatis lineam longis; antheræ in summo vertice sessiles, biloculares, loculis brevibus divergentibus longitudinaliter dehiscentibus. *Ovarii* vestigium nullum vidi.

The two species of *Vismia* distributed—*V. dealbata*, H. B. K., and *V. Cayennensis*, Pers.—are both well known, though not so common in collections as the *V. Guianensis*. The *Malpighiaceæ* are interesting, but require further study, and will be referred to in a future report.

(*To be continued.*)

-Decades of Fungi; *by the* Rev. M. J. Berkeley, M.A., F.L.S.

Decade XXXV.

Sikkim-Himalayan Fungi collected by Dr. Hooker.

(*Continued from p.* 84.)

341. *Favolus tenerrimus*, Berk.; pileo pallido brevissime stipitato spathulato; poris 5–6-gonis dissepimentis tenerrimis; hymenio hinc exsiccatione tantum reticulato.

Hab. Darjeeling.

Stem extremely short, strictly lateral; pileus 1¼ inch long, very thin, forming a mere membrane when dry, spathulato-flabelliform, smooth, margin incurved; pores pentagonal or hexagonal, shallow, ¼ line broad; dissepiments extremely thin and delicate, so that in drying the hymenium appears to be simply reticulate.

A very beautiful species, of which only a single specimen was obtained. The dry specimen is brownish, but like *Pol. platyporus* it was evidently of a dirty ochraceous tinge when fresh. *Favolus tener*, Lév., of which I have a specimen, has far less delicate pores, neither do they shrink up in drying.

342. *F. intestinalis*, n. s.; pileo tenui molli subreniformi lobato; stipite obsoleto; poris hexagonis maximis.

Hab. Darjeeling.

White, turning olive-yellow and dirty ochraceous in drying. Pileus 3 inches or more broad, about half as much long, subreniform, elongated into a short or nearly obsolete stem, of a soft substance, becoming very thin and somewhat transparent in drying. Pores hexagonal, ¼ of an inch across. Spores white, broadly elliptic, with a small nucleus.

A very singular esculent species, looking like a piece of tripe. The pores resemble those of *F. cycloporus*, Mont., though very much larger. The substance dries up so completely, that they are visible from the upper side, as in some other species. Like several fungi of a similar texture, it seems when fresh to have a delicate farinaceous down on the pileus.

343. *Laschia subvelutina*, n. s.; pileo flabellato lobato fusco subtiliter velutino, lamellis paucis; interstitiis lamellato-reticulatis obscurioribus. Hook. fil., No. 133, cum ic.

HAB. On trunks of trees. September, October. Common. Darjeeling, 5–8,000 feet.

Pileus 1–2 inches broad, flabelliform, obtusely lobed, brown, paler and reddish towards the margin, rather convex, thinly clothed with short olive velvety down. Inodorous, not viscid. Substance subtremelloid. Gills few, darker than the pileus, their interstices beautifully lamellato-reticulate.

Laschia velutina, Lév. = *L. tremellosa*, Fr., has very irregular folds, and resembles strongly an *Exidia*. The present species is near to *Cantharellus congregatus*, Mont., which appears to be a *Laschia*, differing in its flabelliform, velvety, not cupulæform and resupinate smooth pilei.

* *Fistulina hepatica*, Fr. Ep. p. 504, *Fistulina buglossoides*, Bull. Champ. p. 314. Hook. fil., No. 71, 91.

HAB. Darjeeling, 4,000 feet, with *Polyporus flabelliformis*, Klotzsch. June. Abundant.

* *Hydnum coralloides*, Scop. Hook. fil., No. 97, cum ic.

HAB. On old trunks of trees, springing from the crevices. Darjeeling, 7,500 feet. June, July. Very abundant.

344. *H. gilvum*, n. s.; imbricatum tenue subcarnosum; pileo flabelliformi pallide gilvo postice virgato antice strigoso; aculeis tenuibus subulatis teretibus integris fuscescentibus.

HAB. On dead trunks. Darjeeling.

Imbricated. Pileus 2–3 inches long, flabelliform, sometimes laterally connate, thin but fleshy, pale reddish-grey, attenuated behind, strigose at the base, disc more or less virgate, rarely rough, margin strigoso-cirrhate, acute. Hymenium yellowish-brown, at length dark; aculei elongated, subulate, entire, margin generally sterile.

A very pretty species, allied to *Hydnum cirrhatum*, but thinner and differently coloured and shaped.

* *H. flabelliforme*, Berk., in Hook. Lond. Journ. Bot. vol. iv. p. 306.

HAB. On dead wood. Darjeeling, 7–8,000 feet.

Rather more lobed than the American specimens, but agreeing in every essential particular.

345. *Phlebia reflexa*, n. s.; e resupinato dimidiato-reflexa vinosa; pileo spongioso-tomentoso zonato; hymenio hic illic venoso-ruguloso.

HAB. On wood. Great Runjeet River, 2,000 feet. Tonglo, 10,000 feet.

At first resupinate and spreading widely, but easily detached, then broadly reflexed and dimidiate, 1½ inch long, 2 inches broad, sometimes laterally confluent, of a vinous brown, irregularly zoned, clothed with spongy down, which is sometimes collected into little tufts. Hymenium darker and of a purer brown, sometimes showing the impression of the zones of the upper surface; even towards the margin, but marked behind with little raised irregular lines.

Certainly the finest and most highly developed species of the genus. The spongy coat gives consistence to the pileus, in consequence of which it is strongly reflected.

346. *Stereum rimosum*, n. s.; umbonato-sessile coriaceum; pileo zonato subtiliter pubescente radiato ruguloso; hymenio pallido hic illic lutescente rimoso.

HAB. On vegetable soil, old trees, &c. Darjeeling, 7,500 feet.

Coriaceous, but probably when fresh of a more watery texture than others, in consequence of which it is minutely wrinkled longitudinally when dry; 1 inch or more long, 2–3 inches broad, umbonate, sessile, effused behind, wood-coloured, with brown fasciæ and numerous narrow zones minutely tomentose, wrinkled longitudinally. Hymenium whitish, here and there assuming a yellow tinge, much cracked, with the fissures silky within.

Undoubtedly allied to *S. Ostrea* and *lobatum*, but differing in its cracked, pale hymenium and other points. It is a very pretty species.

* *S. purpureum*, Fr. Ep. p. 548.

HAB. On dead wood. Darjeeling, 7–8,000 feet.

* *S. hirsutum*, Fr. Ep. p. 549.

HAB. On dead wood. Darjeeling, 7,000 feet.

* *S. spadiceum*, Fr. l. c. Hook. fil., No. 94, cum ic.

HAB. With the former, but more abundant.

It is not always easy to distinguish between dried specimens of these two species, as the main distinctive character is only visible in the fresh plant. The most certain point is the depressed pubescence and shining more closely zoned pileus of *S. spadiceum*, as distinguished from the hirsute dull surface of *S. hirsutum*. No certain dependence can be placed on the colour of the hymenium.

* *S. bicolor*, Fr. Ep. p. 549.

HAB. On dead wood. Darjeeling, 7,000 feet.

Exactly like specimens from North America, gathered by Schweinitz and others.

347. *Calocera sphærobasis*, n. s.; æruginosa, e tubere globoso cavo intus albo cartilagineo exoriens; clavulis simplicibus acutiusculis. Hook. fil., No. 106, cum ic.

HAB. On the ground, apparently springing from a twig. Darjeeling, 7,500 feet. August. Only a single specimen found.

Verdigris green. Clavulæ simple, $\frac{1}{3}$ of an inch high, scattered on a globose hollow body, which is white within, and of a cartilaginous substance, and about $\frac{3}{4}$ of an inch in diameter. The inner portion consists of closely packed threads, which form a reticulated mass towards the circumference.

Resembling, at first sight, an inverted *Rhizina*. I am not at all certain that this very curious production belongs to the genus *Calocera*, but I know not where else to place it, and the single specimen shows nothing of the fructification. The structure of the external coat is more like that of some lichen, but the habit is that of a fungus.

* *Tremella ferruginea*, Smith, Eng. Bot. Hook. fil., No. 21, cum ic.

HAB. In mossy and rocky wet places. Tonglo, 10,000 feet. May.

It might be suspected that, from the peculiar locality, the Himalaya species could not be the same as the British, but a comparison with a fine specimen sent to me from the Royal Botanic Garden, London, by Mr. J. D. C. Sowerby, shows a complete identity of structure, as also of colour, habit, and general appearance.

* *Exidia hispidula*, Berk. Ann. Nat. Hist. vol. iii. p. 396.

HAB. Attached to a specimen of *Pol. hirsutus*, No. 28. Darjeeling.

* *E. protracta*, Lév., Ann. d. Sc. Nat. Oct. 1844, p. 218. *E. Lesueurii*, Bory, MSS. Hook fil., No. 82, cum ic.

HAB. On trunks of living trees. Sikkim. Very abundant. June, July.

The underside is very obscurely velvety. The whole plant, when fresh, is soft, flabby, and transparent, with the hymenium lacunose. I refer the species to Léveillé's plant, on the authority of a specimen from Bory de St. Vincent's herbarium, which has been pronounced identical with that of Léveillé by Dr. Montagne.

348. *E. bursæformis*, n. s.; admodum gelatinosa aquosa bursæformis supra lacunosa, subtus subtilissime tomentosa. Hook. fil., No. 22, cum ic.

HAB. Abundant on moss and trunks of trees. Darjeeling, 7,000 - 10,000 feet. May, July.

Extremely gelatinous, tremelloid, very soft and mobile in the hand, as if alive, obconical, cup-shaped, sometimes spreading out for many feet, lacunose within, externally most minutely tomentose, of a dull umber; shrinking when dry to an extremely thin membrane, so as to be scarcely recognizable.

Unlike the other species, this will scarcely make a specimen for the herbarium, as it shrinks to a mere film. It is far more watery and gelatinous, though by no means viscid, than any other species of this difficult genus.

* *Lycoperdon cælatum*, Fr. Syst. Myc. vol. iii. p. 32. Hook. fil., No. 89, cum ic.

HAB. On the ground. Darjeeling, 7,500 feet. July. Rare.

There is a slight purple tinge at the top of the peridium when fresh, which is not, however, permanent in the dried specimens.

349. *L. sericellum*, n. s.; peridio obtuso apice dehiscente sericeo-corticato; strato sterili sericeo-spongioso cum capillitio continuo; sporis flavido-ochraceis. Hook. fil., No. 32, cum ic.

HAB. On the ground. Darjeeling, 7,000 feet. May, June. Rare.

Peridium 2–3 inches across, dark brown, obtuse, at length opening above, but not by a distinct orifice; outer coat silky, the down collected into little depressed fasciculi, cracking and separating from the inner yellowish coat. Stem various in length, obconical, sometimes paler than the peridium. Capillitium distinguished from the silky substance of the stem merely by a slightly different tinge, but not separated by any distinct membrane. Spores globose, dirty yellow.

Distinguished from *L. cælatum*, to which it is perhaps most nearly allied, by the confluence of the capillitium with the substance of the stem, which is silky, soft, and not rigidly lacunose, in which respect it agrees with *L. polymorphum*, Vitt. In form it comes very near to *L. saccatum*, but the spores are by no means fuliginous, and the stem is in that species "celluloso-spongiosus," which would seem to indicate a firmer texture. The Ceylon plant which I have referred to *L. saccatum*, is probably the same with the Darjeeling species, but unfortunately the specimens are not precisely in the same state of growth, and I have no specimen of *L. saccatum* to compare.

* *Lycoperdon gemmatum*, Fr. Syst. Myc. vol. iii. p. 36. Hook. fil., No. 17, 59, 105.

HAB. Abundant on the ground, paths, clay banks, &c., also on decayed timber. Darjeeling, Jillapahar, 7–8,000 feet. April, June.

There are two or three varieties; one approaching very near to *L. pyriforme*, and another very obtuse, and closely resembling *L. cælatum*.

* *L. pyriforme*, Schæff. Hook. fil., No. 126, cum ic.

HAB. On dead wood. Sikkim, 8,000 feet. October.

The specimens vary in length of the stem, but none are so strongly stipitate as the common form.

350. *L. microspermum*, n. s.; peridio toto flaccido persistente, ore rotundo dehiscente, cortice squamulis acutis demum exasperato; strato sterili parvo, capillitio uniformi; sporis globosis minimis. Hook. fil., No. 73.

HAB. On the ground. Darjeeling. June.

From ½–1 inch across, subglobose, obtuse, sending out one or more strong roots, from which proceed some white fibres; at first reddish-brown, obscurely rimose, rough with minute brown raised scales, at length paler, smooth below, opening by a small round aperture. Barren stratum small, scarcely distinguishable, except on accurate inspection, from the capillitium, which is uniform, without any columella. Spores very minute, globose, olive-green, quite smooth, rarely pedicellate.

This little species has all the characters of *L. pusillum*, but the diameter of the spores is not above half as large.

(*To be continued.*)

Characters of some GNAPHALIOID COMPOSITÆ *of the Division* ANGI-ANTHEÆ; *by* ASA GRAY.

(*Continued from p.* 158.)

BLENNOSPORA, nov. gen.

Capitula biflora, homogama, pauca, breviter pedicellata, in glomerulum ovatum subcompositum dense aggregata. *Involucrum generale* glomerulo brevius, pauciseriale; *squamis* ovatis scariosis viridi-carinatis extus laxe lanigeris. *Receptaculum generale* lineare, subramosum, ramis seu pedunculis inferioribus 2–3-cephalis paleis squamis involucr. gen. et partialibus consimilibus bracteatis. *Involucrum partiale* deciduum, floribus æquilongum, e squamis circ. 6 ovatis concavis scariosis exappendiculatis, nervo viridulo brevi apice dilatato dorso lanigero, constans. *Flores* hermaphroditi, conformes. *Corolla* tubo gracili (basi vix aut ne vix skirroso-tumido), limbo infundibulari 5-fido, lobis lanceolatis recurvis. *Antheræ* basi subcaudatæ. *Styli* rami apice truncato-capitellati. *Achænium* obovatum, glaberrimum, pellicula hyalina humectate valde tumescente gelatinosa tectum. *Pappus* corollam subæquans, caducus, e setis mollibus 8–10 laxe plumosis apice tenuissimis basi subdilatatis in annulum membranaceum concretis.—Herba annua, multicaulis, bi-triuncialis, laxe lanata, caulibus ramisve simplicibus foliosis glomerulo aphyllo terminatis, foliis alternis anguste linearibus apice subdilatatis mucrone apiculatis. Squamæ fuscæ.

B. *Drummondii.*

Swan River, *Drummond*.—Most allied, perhaps, to *Styloncerus*, but with a very different pappus, corolla, general receptacle, &c., the glomerule not foliose-bracteate.* The generic name alludes to the cellular pellicle of the achænium becoming gelatinous when moistened, as in *Cephalosorus phyllocephalus*, but even more strikingly. This pellicle consists of a close coating of linear or subclavate diaphanous cells, compactly arranged with their long diameter perpendicular to the smooth brown pericarp: on the application of water it promptly swells into a mass of transparent jelly, very much thicker than the enclosed achænium. The gorged mucous cells remain unbroken for a considerable time, and their extremely delicate walls show no markings, nor any contained coiled bands or fibres.

ANTHEIDOSORUS, nov. gen.

Capitula uniflora, super receptaculum generale planum arcte sessilia, in glomerulum hemisphæricum involucro generali pluriseriali radiante cinctum, capitulum simplex multiflorum prorsus mentiens, densissime aggregata. *Involucri generalis* squamæ scariosæ, oblongæ vel obovatæ, inferne appressæ, fere omnes appendice petaloidea aurea dilatata abrupte patente superatæ. *Involucrum partiale* e squamis paucis (3–5) oblongis vel linearibus, hyalinis, planiusculis, subpersistentibus, floribus brevioribus, paleas receptaculi generalis simu-

---

* The tube of the very short corolla of *Styloncerus humifusus* becomes tuberculate, thickened, and indurated, as in *Skirrophorus*.

lantibus. *Flores* omnes hermaphroditi, ii capitulorum marginalium fertiles, cæteri abortu steriles. *Corolla* tubulosa, limbo 5-partito, lobis oblongo-lanceolatis revolutis. *Antheræ* basi bicaudatæ, caudis piliferis. *Styli* rami apice capitellati. *Achænia* erostria, glabra; fertilia obovata, pellicula hyalina humectate creberrime gelatinosa induta; sterilia linearia, inania. *Pappus* florum fertilium parce plumosus, caducus,. corolla paulo brevior, e setis 5-6 tenuissimis basi annulatim coalitis, 1-2 paulo longioribus inferne nudiusculis et apice clavellato-plumosis; florum sterilium persistens, setis rigidioribus basi subdilatatis cæterum fere similibus.—Herba annua spithamea, ramosa, lanato-pubescens; foliis alternis anguste linearibus; caule ramisque gracilibus erectis; glomerulis capituliformibus parvis (2-3 lin. latis) aureis, ramulos graciles terminantibus, nutantibus. (Nomen ex ἄνθος, *flos*, εἶδος, *facies*, et σωρὸς, *acervus*, conflatum, quia glomerulum singulum florem compositum botan. antiquorum simulat.)

A. *gracilis*. (Ic. Pl. tab. ined.)

Swan River, *Drummond*.—This curious plant would probably on ordinary inspection be referred to the division *Cassinieæ*, and perhaps rightly; but the paleæ, although perfectly sessile on the general receptacle, are three or four for each flower, and surround it, so as to form a partial involucre, as is more readily discerned around the marginal fertile flowers. The structure of the glomerule, on this view, is much as in *Myriocephalus*, next to which genus I venture to place it.

MYRIOCEPHALUS, *Benth. Pl. Hugel.* p. 61.

Ad char. adde: Receptaculum generale latum, planum. Pappus aut nullus, aut 2-3-setosus, setis nudis.—Herbæ annuæ.

1. M. *appendiculatus*, Benth. l. c.—This is an annual, at least if specimens in Drummond's collection are the same as the original species. These are barely six inches high, the glomerule less than half an inch broad; the scales of the partial involucres are sparsely lanate, and the ovaries are not entirely glabrous.

2. M. *nudus* (n. sp.): involucro generali inconspicuo glomerulo multo breviore, squamis appendice minima fusca auctis; involucris partialibus 3-4-floris subglabris; achæniis calvis.

Swan River, *Drummond*.—Stems, foliage, &c., nearly as in the preceding. Glomerule from three to six lines broad, at length much depressed, appearing like a discoid capitulum, the appendiculate scales of the general involucre being so small, and the more inconspicuous because the tips are of the same yellowish-fuscous hue as the capitula.

3. *M. helichrysoides* (n. sp.) : involucro generali discum subsuperante, squamis glaberrimis appendice majuscula lactea radiantibus; involucris partialibus 3-floris; achæniis hirsutulis; pappo e setis 2-3 tenerrimis (sub lente vix denticulatis) corollam æquantibus constante; caule e basi decumbente seu repente erecto ramoso foliisque lineari-spathulatis elongatis glaberrimis.

Swan River, *Drummond.*—Accords with *Myriocephalus* in every respect, except in the pappus. The petaloid general involucre is longer in proportion, and the disc smaller than in *M. appendiculatus*.

CROSSOLEPIS, *Less.* ?

1. CROSSOLEPIS? *brevifolia* (n. sp.): foliis oblongis cum caule 1-2-pollicari mox glabratis, summis spathulatis vel obovatis glomerulum depressum lanatum laxe fulcrantibus; capitulis pedicellatis singulis bractea herbacea stipatis 10-15-floris.—*Caulis* e radice annua erectus, parce ramosus, glomerulo ratione plantæ majusculo 4-5 lin. lato terminatus. *Folia* caulina vix 2 lin. longa, ea glomerulum fulcrantia majora, cito glabra. *Capitula* tomento implexo dense lanata. *Involucrum* subduplex, nempe, squamis extimis, seu bracteolis, 2-4 spathulatis, herbaceis, laxis; cæteris circ. 10 imbricatis, scariosis, oblongis, inappendiculatis, margine fimbriato-laceris et lanigeris, flores subæquantibus. *Receptaculum* nudum. *Flores* hermaphroditi, consimiles. *Corollæ* graciles, tubo fere filiformi apice breviter infundibulari-ampliato, limbo 5-fido. *Antheræ* bicaudatæ. *Styli* rami apice truncati, subpenicillati. *Achænia* immatura obovata, glabra, calva.

S.W. Australia, *Drummond*, 1850.— This, with the two following evidently congeneric species, I doubtfully refer to Lessing's genus *Crossolepis*, of which nothing is known beyond the annexed most imperfect character, and that the author places it among his *Craspedieæ*, which nearly correspond with De Candolle's division *Angiantheæ*.* Here I

---

* "CROSSOLEPIS, nov. gen. Capitulum circ. 10-florum. Rachis ebracteolata. Pappus nullus.—Herba Novæ Hollandiæ annua, gracillima, lana mox evanida vestita; foliis angustissime linearibus; foliolis involucri vix biserialibus floribus parum brevioribus, scariosis, fimbriato-laceris.—*C. linifolia*, n. sp." (*Crossolepis? pusilla*, Benth., belongs to a very different genus, *Chrysocoryne*, Endl.)

should likewise refer these plants, although the glomerule and the subtending leaves are so lax, and the capitula so pedicellate from the axil of herbaceous bracts (although the bracts and pedicels are so densely clothed with wool, that the glomerule appears to be compact enough), that they might almost as well be placed among *Helichryseæ*. Having usually as many as ten flowers in each capitulum, a naked receptacle, no pappus, and scarious, fimbriate, involucral scales, they must needs be referred to *Crossolepis* until the plant so incompletely characterized by Lessing shall be further known.

2. C. ? *eriocephala* (n. sp.): foliis lineari-spathulatis linearibusve cum caulibus (2-3-pollic.) ramisque gracilibus diffusis mox glabratis, floralibus spathulatis capitulisque pedicellatis in glomerulum laxum confertis lanosissimis; involucri circ. 10-flori squamis oblongo-lanceolatis acutis intimis linearibus fimbriatis.

South-western Australia, *Drummond*, 1850.—The structure of the glomerule and capitula is the same as that of the preceding species, but they are densely clothed with long wool. The margins of the involucral scales, about ten in number, are beautifully fimbriate-dissected and lanigerous. Corolla sparsely glandular. Ovary obovate, glabrous.

3. C. ? *pygmæa* (n. sp.): caulibus vix semipollicaribus foliisque angustissime linearibus cito glabris; glomerulo globoso sublanato foliis paucis linearibus squamisque 5-6 ovatis latissime hyalinis involucrato; capitulis subsessilibus; involucri 5-flori squamis lineari-lanceolatis fimbriato-lanigeris; corollis 4-dentatis; achæniis anguste oblongis parce hirsutis.

South-western Australia, *Drummond*.—Stems numerous from the annual root, very short. Glomerule 2½ lines in diameter, much denser than in the preceding species, but exhibiting the same structure. Capitula subtended by one or two bracteate scarious scales: the proper involucral scales five or six, with a more rigid costa or axis than in the foregoing. Corolla very slender, as long as the involucre; the limb 4-toothed. Pappus none.

CHAMÆSPHÆRION, nov. gen.

Capitula 5-7-flora, homogama, in glomerulum sessile humi adpressum, foliis radicalibus involucrum generale simulantibus cinctum, conglobata. *Receptaculum* (partiale) angustum, planum, paleaceum; *paleis* scariosis, oblongis, acuminatis, basi linea viridula carinatis intusque excavatis, flores superantibus, persistentibus, squamis involucri referentibus. *Involucrum*, præter paleas exteriores basi tenuiter lanosas, nullum. *Flores* hermaphroditi, conformes. *Corollæ* longe tubulosæ, graciles, apice dilatatæ, 3-dentatæ. *Antheræ* 3, ecaudatæ. *Styli* rami apice capitellato-truncati. *Achænia* obovata, erostria, tenuissime 5-6-costata, glabra, pappo coroniformi lacerato caduco superata.—*Herbula* annua; radice exili; foliis rosulatis, lineari-subulatis basi dilatatis, glabellis, glomerulum globosum arcte involucrantibus.

C. ? *pygmæum*. (Ic. Pl. tab. ined.)

South-western Australia, *Drummond*.—The whole plant consists of a globular dense cluster of capitula, three or four lines in diameter, subtended by a rosulate tuft of narrow leaves (four or five lines long), which form a general involucre, and resting directly on the ground, to which it is affixed by a slender root. Capitula (2 lines long), cylindraceous, about twelve in the general glomerule, closely sessile, surrounded by no common involucre besides the rosulate leaves of the plant, the exterior, however, subtended by a subulate foliaceous bract. Proper scales of the involucre none; that is, there are no exterior scales destitute of a flower in their axil. Paleæ hyaline, whitish, all alike, except that the outermost are clothed with long woolly hairs at the base, otherwise all glabrous; the upper half of each more or less spreading; the lower appressed, broader and more concave, and with a narrow, greenish projecting keel, which is excavated within, and partly embraces the achænium; the base scarcely attenuated. Corolla rather longer than the appressed portion of the subtending palea. Pappus a delicate lacerately multifid crown, caducous.

This pigmy plant forms a genus (named from χαμαί, *on the ground*, and σφαιρίον, *a little sphere*), evidently related to *Chthonocephalus* of Steetz; from which it is at once distinguished by its few-flowered heads, flat receptacle, and coroniform pappus. The two would, perhaps strictly, belong, where Steetz placed the latter, to the *Gnaphalieæ Cassinieæ*; but I think they are more naturally placed, or at least *Chamæsphærion*, at the end of the *Angiantheæ*. The anthers are not at all caudate in our plant; nor are they perceptibly so in several *Angiantheæ*, and in a species of *Chthonocephalus* gathered by Drummond, which differs from *C. Pseudevax*, as described by Steetz, in the form of the paleæ and the want of any excavated and thickened portion at their base, as well as in the greater size of the plant. I add its characters:

CHTHONOCEPHALUS *Drummondii*, n. sp.: capitulis multis inter folia rosulata anguste spathulata glomeratis; paleis late ovalibus vel suborbiculatis obtusissimis undique hyalinis basi lata nec excavatis nec induratis.

Swan River, *Drummond*.—Plant, including the outstretched leaves, from one to two inches in diameter; the exterior leaves half an inch long, cinereous, tomentose. Heads forty or more, a dense depressed glomerule, sessile in the centre of the involucriform tuft of leaves, fuscous. The hyaline scales, all but two or three of the outermost, subtend flowers. They are much broader than described by Steetz, uniformly concave, thin, and hyaline throughout, and they are scarcely at all narrowed at the base.

(*To be continued.*)

Contributions to the Botany of WESTERN INDIA;
by N. A. DALZELL, Esq., M.A.

(*Continued from* p. 139.)

Nat. Ord. CONVOLVULACEÆ.

PHARBITIS.

P. *laciniata*, mihi; glabra, radice fibrosa, caule filiformi repente vel volubili angulato torto, foliis breve petiolatis subseptem-partitis, lobis linearibus angustissimis serrato-pinnatifidis dentibus inæqualibus mucronatis, pedunculis axillaribus solitariis angulatis clavatis 1–3-floris folio brevioribus, corolla (alba) *hypocrateriformi* 2¼ poll. longa, tubo gracili cylindrico intus purpureo, limbo plano patente, antheris exsertis, calycis foliolis oblongis æqualibus mucronatis crassis carnosis dorso 3-costatis, costis rugosis, stigmate 3-lobo, capsula sphærica 3-loculari loculis dispermis, seminibus fuscis sericeis.—Crescit in prov. Malwan; fl. Aug.

I have placed this plant in the genus *Pharbitis* entirely on account of the structure of its capsule, in accordance with Choisy's arrangement, although I believe that India does not possess a genuine indigenous *Pharbitis*; *P. Nil* being found only about villages. The present plant appears to me to be a true *Calonyction*, on account, partly, of the *form* of its corolla, and also (which is perhaps of greater value) of its habit of unfolding its flowers only at sunset.

IPOMÆA.

I. *rhyncorhiza*, mihi; radice tuberosa, tubere ovali compresso sublignoso *rostrato* mediocri, caule filiformi volubili glabro, foliis longiuscule petiolatis subseptem-partitis supra et in nervis subtus pilis fulvis hispidulis, lobis inæqualiter pinnatifidis acuminatis, pedunculis axillaribus solitariis uni- v. bifloris filiformibus glabris folio longioribus, floribus mediocribus flavis, calycis glabri foliolis inæqualibus interioribus longioribus lanceolatis 7-lin. exterioribus oblongis acutis 5-lin.

Corolla 1¼ poll. longa, folia diametro 2-pollicaria, pedunculi 3-pollicares. —Crescit rarissime in montibus Syhadree, prope Tulkut-ghât; fl. Aug. et Sept.; fructum non vidi.

The plant is greedily sought after by the Ghaut people, who eat the tubers, and also make use of the leaves as *Bajee* (greens); hence its rarity. This interesting plant, as well as the former, has been discovered for the first time during the present year (1850).

Nat. Ord. URTICACEÆ.

ELATOSTEMMA.

E. *oppositifolium*; herbacea pedalis, caule simplici glabro basi nudo, foliis longiuscule petiolatis oppositis lanceolatis acuminatis grosse dentato-serratis trinerviis lineolatis supra parce pilosis subtus glaberrimis supremis majoribus, capitulis in axillis alternis solitariis pedunculatis, pedunculis petiolo brevioribus, receptaculo communi plano discoideo simplici, floribus masculis et fœmineis mixtis.

Folia majora cum petiolo semipollicari 5 poll. longa, 1¼ poll. lata, tenerrima, basi inæquilatera. *Pedunculus* filiformis, 5–6 lin. longus. *Capitulum* diametro 4–5 lin.—Crescit in montibus Syhadree; fl. Sept.

This plant has also been discovered for the first time during the present year (1850).

Nat. Ord. AURANTIACEÆ.

CLAUSENA.

C. *simplicifolia*; arborea, foliis simplicibus ovali-oblongis basi cuneatim attenuatis nigro-punctatis glabris, floribus ex axillis supremis cymoso-trichotomo-paniculatis folia superantibus, alabastris lineari-oblongis, sepalis parvis rotundatis, petalis linearibus obtusis reflexis basi intus sericeo-tomentosis, filamentis stylum æquantibus basi dilatatis ibique pilis fulvis sericeis vestitis, stylo gracili, stigmate brevissime 4-fido, stylo haud crassiore, ovario in toro elevato sulcato insidente tomentoso 4-loculari, loculis 2-spermis, fructu demum subglabro pisi magnitudine.

Folia cum pet. ½-poll. 4–4¼ poll. longa, 2 poll. lata.—Floret Aug. et Sept.; fr. Oct. Crescit in montibus Syhadree, prope Tulkut-ghât.

(*To be continued.*)

A new species of ARNEBIA, *detected by* Dr. J. E. STOCKS *in Beloochistan*.

TAB. VI.

ARNEBIA FIMBRIOPETALA, *J. E. Stocks*.

Annua parva simplex v. ramosa hispido-hirsuta, foliis copiosis erectis strictis anguste linearibus (infimis subspathulatis), calyce corollæ tubo breviore post anthesin elongato pubescente basi longe setoso laciniis foliis conformibus, cor. limbi laciniis ovalibus pulcherrime fimbriatis sinubus plica dentiformi fimbriata, stigmate integro, nuculis trigonis acutis lævibus. (TAB. VI.)

Arnebia fimbriopetala, *J. E. Stocks, MSS. in Herb. nostr.* (n. 977.)

Hab. Not uncommon on the hills of Upper Beloochistan, *Dr. J. E. Stocks*.

Among a valuable assortment of Scinde plants sent to me by my very valued correspondent Dr. Stocks, the present one is marked as "the gem of the collection; the expanded flower being perfectly lovely." Its nearest affinity is with *A. linearifolia*, De Candolle, Aucher-Eloy, Pl. Exs. n. 2368 (in Herb. nostr.) from the desert of Sinai: but the corolla in this is very different, and the leaves are broader: in both, the segments of the calyx very much resemble the leaves. How far the genus *Arnebia* is really distinct from *Lithospermum* must be left for decision to those who are familiar with the species in a living state.

Neither here nor in *Arnebia echioides*, DC. (Bot. Mag. t. 4409), do we find the style bifid at the apex.

TAB. VI. ARNEBIA FIMBRIOPETALA.—Fig. 1, flower; fig. 2, stamens; fig. 3, pistil; fig. 4, calyx with fruit; fig. 5, fruit :—*magnified*.

BOTANICAL INFORMATION.

The Botanic Gardens of MADRID *and* VALENCIA. By Dr. MORITZ WILLKOMM. (Translated from the *Regensburg Flora of March* 7, 1851, *p.* 129 *seq.*; by N. WALLICH, M.D., V.P.L.S.)

The Garden at Madrid, founded in the second half of the past century by Charles III., a king who was fond of the arts and sciences, is the principal botanical institution in Spain: it has again revived, after being almost entirely neglected during a long series of years, and promises to become, in time, a garden of importance. This state of things is not much due to the Government, which does almost nothing for the garden, though belonging to the Royal domains, nor yet to its direction, but is owing to Professor Vincente Cutanda's indefatigable zeal in restoring the establishment, without being himself a professional botanist (for he was formerly a barrister); and he would be still better enabled to effect his object, if he had the entire direction of it. But this last is, unfortunately, not the case: neither he nor the two other professors have any share in the direction, which belongs exclusively to the Gefe local del Museo nacional de Ciencias. The garden, with its botanical museum, forms part of the national museum just named, whose chief director is the celebrated zoologist, Professor Don Mariano de la Paz Graëlle, a Catalonian; and under whom an English gardener (Jardinero mayor) is placed. This person, who is said to know very little of his profession, enjoys nevertheless a much larger salary than any of the three professors attached to the garden. One of these is the above-mentioned Don Vincente Cutanda, professor of organology and physiology, and director of the botanical museum; another is Don Pascual Asensio, professor of agriculture and inspector of the agronomical branch of the botanical museum; and the third is Don Jose Alonso y Quintanilla, professor of descriptive botany, who conducts also botanical excursions, as well as exercises in determining plants. Of these three gentlemen, the first is a tolerably good botanist, well acquainted with the progress and literature of the science, and, although past forty years of age, is still full of youthful ardour and attachment to botany, and devoted to it from his youth, from inclination.

The botanical museum is placed under the immediate direction of Cutanda. It comprises—besides the agronomical branch already alluded to, consisting of a library and a collection of models, woods, cerealia, and fruits—the botanical library, the herbariums, and the store of seeds. The library, which is well arranged, is seemingly complete as regards the older works; but it is poor in more recent publications. The Seed-store is arranged according to the Linnæan system, and has an especial seed-collector (semillero), who gathers the seeds in the garden, and distributes them among other gardens. He stands under Cutanda, who is the director of the garden of Madrid, only as regards corresponding with other gardens, which are connected with it by exchanging seeds, superintending generally the garden cultivation, and enriching it with new species; but he has nothing to do with the cultivation itself. The herbariums constitute the most important portion of the botanical museum; those of Cavanilles, Rodriguez, Née, Clemente, part of the collections of Lagasca, Pourret, and others, being kept there; likewise many plants of Boissier and Reuter, some gathered by the writer of this notice, and by several of the pupils of the botanical institution. All these collections were lying in the greatest confusion in Rodriguez's time, so that it was utterly impossible to compare any plant, or examine any particular original specimen. Cutanda has made it a point of primary importance to introduce some order into this chaos, after four years of constant exertion, aided by the semillero, Don Francisco Alea, a young, zealous, and clever botanist. All the said collections form now one general herbarium, of about 30,000 species, arranged according to De Candolle's method. The specimens of each species, in the several herbariums, are placed separately in sheets of paper, having a printed label with the name of the herbarium attached; and a detailed catalogue renders the search after any particular species very easy. Cutanda is now engaged in determining all the species in this general herbarium, from first to last, because there are many plants in it, either not at all, or wrongly named.

It is likely to obtain soon a very considerable addition in Lagasca's herbarium, which the Government intends purchasing. One portion of this latter, consisting of some hundreds of packets, is in the natural-history building; the other, in about twenty cases, lies at the Custom-house in Malaga, where both have continued many years, because Lagasca's heirs, who are uneducated people, caring nothing about botany, have declined to defray the expenses of warehousing the collections. Lagasca, it appears, had left the cases at the Custom-house in Malaga, on his return from England, not being able, probably, to pay the charges for duty, &c. It is to be feared that the contents are in a very bad condition. Piolongo, who has seen the cases, states, that they are kept in a wet vault, that the bottom parts are quite decayed, and mice have attacked them. The packets at the Natural History Museum at Madrid, where they were deposited on the death of Lagasca, to save them from destruction, are now offered for sale by his creditors and heirs; but the good specimens cannot be very numerous, as it is reported, that the worms have caused much mischief among them. It is probable that the noble collections of the late Don Mutis, of Santa Fé de Bogotá, are in an equally deplorable state of decay. They are kept in a separate room of the botanical museum, inscribed "Direccion del Museo botanico de la Nueva Granada." Only a small part of the collections, made in New Granada by Mutis and others, has been unpacked; the rest, about sixty cases, has remained unopened these fifty years. The small portion which was unpacked by Rodriguez, and has been loosely arranged according to their genera, is so much injured from insects, as to be scarcely distinguishable, *Cyperaceæ* and *Gramineæ* only excepted. What must not be the havoc among the unopened cases! These immense collections came to Madrid before the war with Napoleon had commenced, and were to have been published at the expense of Government, together with the fine drawings, made partly by Mutis himself, and partly by experienced artists, under his own direction. The professors at the garden were charged with putting the entire herbarium into proper order; and for this purpose a very large and superior kind of paper was expressly manufactured at the public cost, which is still kept, in large quantities, in an adjoining closet. In short, everything was prepared for producing a noble herbarium and flora of New Granada, when the disastrous war broke out, succeeded by years of distress and tyranny, and at length by a civil war; and thus was that grand scheme annihilated. In fact, nothing remains at present of those noble collections, except the above-mentioned drawings—the finest I have ever seen. They are in large royal folio; the size of the figures, that of life; and generally there are triplicates of each drawing: one being an outline only; another, shaded, is executed in Indian ink; and the third is coloured after nature. As regards style and execution nearly all these drawings are perfect; they amount to many hundreds. The herbarium of the Flora Peruviana, formed from Ruiz and Pavon's collections, is in a better condition, and preserved in a separate room in the Museum.

The number of cultivated plants in the Madrid garden very little exceeds 5,000 species. The catalogue,* published in 1849 by the three professors, at their own expense, comprises 3,780 species,—that is, such only as they were able to determine since the death of Rodriguez, which took place in the summer of 1847. He had—it is impossible to guess for what reason—removed all the labels of the plants! Cutanda takes much pains to increase the number of plants, and is particularly anxious that the Madrid garden should cultivate all the plants of the peninsula. As a member of the Comicion de la Carta geologica de Espanna (which chart, at present merely an accurate geognostic-botanical one of the province of Madrid, is to be published at the charge of the Government) Cutanda is obliged to undertake annual journeys in order to study the vegetation of the country; on which occasion, he is always accompanied by the semillero, who collects seeds and plants for the garden. If this honest, zealous, and disinterested young man is long spared, the Madrid garden may be expected gradually to recover the rank it held in Cavanilles's time. Last year the Government built a hot-house, which was hitherto entirely wanting. It is still more to be wished that a better supply of water could be obtained; at present it is scarcely adequate for watering one-half of the very considerable area of the garden, especially in summer.

What is hoped for in regard to the Madrid garden has partly been accomplished in that of Valencia. When the author visited it for the first time in 1844, it was only nominally a botanical garden, in which little more was cultivated than oranges, limes, roses, and common ornamental plants; whereas it is at present in tolerable order,

* Católogo de las plantas del Jardin botánico de Madrid en el anno 1849.

and contains more than 6,000 species. There is a pretty large glasshouse, one-half being a caldarium, the other a tepidarium: in the former are cultivated nearly 130 species of *Orchideæ*, and 50 of Palms; in the latter, among others, a considerable number of tropical and subtropical Ferns. A second house is to be erected in the course of the present year. A number of *Crassulaceæ* and *Cacteæ*, and similar plants of New Holland and the Cape, grow in the open air. The general number is constantly augmenting, and everything is done to cultivate plants of colder climates than the Valencian, by means of watering, artificial rocks, shrubberies, &c. This sudden and advantageous change in the state of things is almost exclusively due to the then Rector of the University of Valencia, Don Francisco Carbonell. This learned, energetic, and wealthy gentleman, was political chief of Valencia in 1844, and was much dreaded throughout the kingdom, on account of his inflexible and rather despotic procedure; but he made it a point, it seems, to restore, at any cost, the university garden. Though a diplomatist, and not a botanist, he interests himself actively in natural history, especially zoology and botany. The hitherto very insignificant zoological museum of the university was considerably enlarged during his rectorship; for instance, the indigenous birds of Valencia, especially the numerous water-birds of the Albufera Sea, have been added, and form a very interesting collection. The director is Professor Don Ignacio Vidal, who is said to be a good zoologist. But Carbonell's real hobby is the botanic garden. He has removed, somewhat arbitrarily, the old *personnel*, with the exception of D. José Pizcueta, Professor of Botany, who was garden-director in 1844, and continues so still, though, of course, only nominally; and he has attached to it a clever, scientific French gardener, M. Jean Robillard, a zealous young man; and as the public funds were too insignificant to restore and support the garden, he has contributed large sums out of his own means. M. Robillard has placed himself in communication with the leading gardens in Europe, and will be able, under the powerful patronage of Carbonell, to double and treble the number of plants in a short time. If we take into account the excellence of the climate of Valencia—in which New Holland and Cape plants, as well as many plants of tropical countries, thrive in the open ground,—the superiority of the soil, the abundant supply of water, the continually moist and never too hot air,—it must be admitted that we have here a combination of all the conditions required for a grand botanic establishment; and such the Valencia garden will become, if Carbonell's life is spared and his rectorship continued. I will, in conclusion, specify some of the rarities in the garden; rarities, at least, as concern the individual specimens. The large water basin is filled with tropical aquatics, such as several plants of *Nelumbium speciosum* in full bloom at the time I speak of (August), and remarkable on account of the great size of the flowers and leaves. In the open air grow small trees of *Gleditschia caspica*, the stem of which is armed with compound spines a span long; *Parkinsonia aculeata*; *Araucaria excelsa* and *imbricata*; and a splendid specimen of *Yucca filamentosa*, with a stem eight feet high and nearly one foot thick. The *Parkinsonia* is a layer from an old large tree, which was ignorantly cut down by the Canon Carrascosa, formerly director of the agronomical garden, now united with the botanic garden. The *Chamærops humilis*, which so much astonished me in 1844, is fortunately still in existence, and it measures nearly twenty feet in height. The proper "botanical school" remains still a Linnæan arrangement, but it is intended to put it in order according to the natural system. May the Valencia garden continue its progress towards perfection, and serve as a praiseworthy pattern of imitation for all the other botanical establishments in Spain!

Observations upon the elevated temperature of the male inflorescence of CYCADEOUS PLANTS; *communicated by* Dr. W. H. DE VRIESE, *Professor of Botany and Director of the Royal Garden of the University of Leyden.*

All living bodies have a temperature peculiar to themselves; that is to say, they have a temperature different from, and independent of, those that surround them. This temperature is intimately connected with their nature, and is modified according to the different conditions in which they may be. This necessary consequence of the successive changes which organic matter undergoes during life, is in its turn one of the causes which preserve organized bodies, and by which animal and vegetable life are protected from destruction or dissolution, which external circumstances would not be long in producing. It is this peculiar temperature which permits animals to inhabit regions of the globe that on account of their cold would be uninhabitable; which allows the development of aquatic vegetables in frozen water; which defends trees against winter, and which, in tropical regions, causes vegetables to withstand a temperature often too high for their organization. The observations upon the elevated temperature in the flowers of Aroideous plants in general, have shown that this phenomenon takes place in a high degree, and originates in a sort of combustion, that is to say, an absorption of oxygen and emission of carbonic acid.

Very recently a high degree of temperature has been observed in a plant belonging to a family in which that phenomenon has not been noticed before. Mr. Teysman, chief gardener at Burtenzorg (in Java) in 1845, has informed me that he has observed an elevated temperature, and at the same time a very strong smell, in the male cone of *Cycas circinalis*. I received from him, in October 1849 and November 1850, seven series of observations, made in the aforesaid garden, upon male flowers of this plant. What is most remarkable in these observations is connected with the following facts. The elevation of the temperature always takes place between 6-10 o'clock in the evening. Messrs. Bory (at the Isle of France) and Hasscarl (at Java) have observed the maximum at 6 o'clock in the morning. De Saussure observed it in the *Arum Italicum* between 4-7 in the evening; and the *Colocasia odora* in the gardens of Paris, Amsterdam, and Leyden has always attained its maximum at noon. This periodical production of heat, differing in different climates and in flowers of different families, has not yet been accounted for. It appears from the inspection of the tables of several hundreds of observations, that the maximum has varied between 9-14° C., and the difference has been 3·75-4·50°.

It is acknowledged that in general the coloured parts among the appendicular organs in vegetables have an absorption and exhalation contrary to those of green parts. The oxygen is absorbed, carbonic acid is exhaled. Both take place in organs where the elevated temperature is shown in a high degree. It is proved that this phenomenon is constantly preceded and accompanied by rapid growth in the flower. Nothing prevents us from admitting that the same action actually takes place in the male cone of *Cycas*, where the rapid development of pollen, or the formation of cells which compose it, should surpass all that has been observed in this respect in the vegetable kingdom.

We shall endeavour to prove it by the following calculation. The male cone, of which I have given the description elsewhere, is (in metres) 0·450 long, and 0·200 broad. The sum of the external surface is difficult to estimate on account of the irregular form of the organ, but it cannot be considerable. In calculating the number of scales at 3500, and the surface of each of them at four square centimetres, the whole sum of the organs which compose the cone should be equal to 14,000 square centimetres. The surface of the scales at the underside is covered with unilocular anthers almost contiguous, and the number of these anthers may be calculated at 400. Thus the total number of these anthers might be calculated at 1,400,000. Each anther contains several thousands of granules of pollen, which in a very short space of time undergo, in their cavities, all the necessary organic, physical, and chemical changes. It is easy to admit that the alternate absorption and emission of gas, in so rapid a process, must have an important part. The whole leads us to believe that, in which there is so great an analogy in the functions (as in the flowers of Aroideous and Cycadeous plants), the same agents should regulate and preside over the phenomena of life, of which, all that modern science has been able to discover as to its mode of action, belongs to physics and chemistry.

Sale of the extensive HERBARIUM *and of the* BOOKS *of the late* GEORGE GARDNER, ESQ., F.R.S., *Director of the Royal Botanic Garden, Peradenia, Ceylon.*

In consequence of the lamented death of Mr. Gardner, instructions have been given to the executors to sell, without reserve, the entire of the above-mentioned collections of this gentleman, which have recently been received in London for that purpose. The whole Gardnerian Herbarium, that is, the collection arranged by himself for his own use, it is wished should be disposed of separately and by private contract. It is admirably arranged, and as fully and correctly named as probably any of like extent; all the specimens are fastened upon the best stout white demy folio paper, measuring sixteen inches long, by ten and a half inches broad. Every genus is included in one or more envelopes of the same paper in folded sheets, and marked on the outside with the name of the genus, that of the natural family, and numbered according to the numbers and arrangement in Endlicher's 'Genera Plantarum.' The specimens are invariably in excellent condition, no trace of insects having been seen among them, and we have reason to believe they are all poisoned. From as accurate a calculation as can be yet made, there are about 14,000 papers containing specimens, and we think we are within bounds when we say that there are 12,000 species of Phænogamous plants and Ferns. The collection is, as may be anticipated, extremely rich in Brazilian and Ceylon plants, gathered mostly by Mr. Gardner during his five and a half years' travels in the former country (and they are the authority for his many *published* species) and during his four years' sojourn in Ceylon. It further includes numerous plants prepared by himself in Mauritius, and a still more extended assortment from the Nielgherries; a rich collection of Malacca plants from the late Mr. Griffiths; Hong-Kong plants from Capt. Champion; South European plants from Mr. Bentham; and others from various quarters of the globe; the whole forming an extensive and well-authenticated Herbarium, such as is seldom offered for sale to the botanical world.

Mr. Samuel Stevens, 24, Bloomsbury Street, London, is charged with the disposal of this, and further particulars may be obtained on inquiry of him.

The books, almost exclusively botanical, and a few unarranged bundles of duplicate plants, will be sold by public sale at Mr. Stevens's Auction Room, King Street, Covent Garden, and full particulars will be announced previous to the sale.

Papyrus of SICILY.

Our valued friend, Professor Parlatore, writes to us as follows, from Florence:—" Je vous dirai, à propos de *Papyrus*, que j'ai découvert, il y a déjà quelques mois, que le *Papyrus* de Sicile, que tout le monde a cru la même chose du Papyrus d'Egypte, est une espèce bien distincte. Je viens de recevoir tous les détails que j'espérais de l'Egypte, et même un dessein de l'espèce égyptienne. J'en possédais déjà un exemplaire dans l'herbier qui m'a mis à même de connaître la différence des deux espèces : j'ai tracé l'histoire de deux papyrus dans ma 'Flore d'Italie,' au commencement du second volume, qui paraîtra plus tard : l'espèce, qui se trouve ordinairement dans les jardins d'Europe, c'est cela de Sicile, que j'ai nommé *Papyrus Sicula*."

Death of Professor Kunze.

It is with regret we have to announce the recent death of Professor Gustav Kunze, of Leipzig, of apoplexy; this event took place on the 30th of April. "Thus," a friend writes to us, "besides this distinguished writer on Ferns, has botany to deplore recently the loss, in rapid succession, of Stephen Endlicher, of Vienna; Koch, in Erlangen; Kunth and Link, in Berlin; Hornschuch, in Greifswald; Wahlenberg, in Upsala; Dr. Anton Sprengel, in Jena (son of Curt. Sprengel); and Professor Fried. Gottl. Dietrich, in Eisenach."

Lindheimer's and Fendler's *American Plants*.

Mr. Samuel Stevens, 24, Bloomsbury-street, London, is charged with the disposal of—

1. A set of plants of *Texas*, collected by Lindheimer, containing 315 species: price, 3*l.* 15*s.*
2. Several sets of Fendler's Plants of *New Mexico*, the largest containing 80 species, the smallest 48: price, 30*s.* per 100.
3. Several small sets of Fendler's *Chagres* plants for sale, at 2*l.* per 100.

Welwitzsch's Plants of Portugal.

Some few sets of the plants collected in Portugal by Dr. Fried. Welwitzsch remain in the hands of his London agent, Mr. Pamplin. The Phænogamous plants contain about 800 species, including many of the rarest species of the Lusitanian Flora; the whole are named and localized: the price, 25*s.* per 100: if the Cryptogamous plants be taken separately, they are charged at the rate of 30*s.* per 100. Early application is recommended. 45, Frith-street, Soho, London, May 1851.

Second Report on MR. SPRUCE'S *Collections of Dried Plants from* NORTH BRAZIL; *by* GEORGE BENTHAM, Esq.

(*Continued from p.* 166.)

The *Sapindaceæ* are more numerous, several of them new, and, many being both in fruit and flower, I shall in describing them add a few notes on some allied species.

The *Cardiospermum*, distributed as *C. microcarpum*, H.B.K., appears to accompany the *C. Halicacabum* over the whole of the tropical world, and to be at least equally common; and as it seems impossible to distinguish the two by any other character than that of the fruit, it becomes doubtful whether they may not be varieties of one species. I have them both from western and eastern tropical America, from the warmer regions of North America, east and west tropical Africa, South Africa, various parts of East India, and the Pacific Islands.

1. Serjania *nitidula*, sp. n.; glaberrima, ramis teretiusculis, foliolis ternis ovato-oblongis obtusis paucidentatis, intermedio in petiolulum contracto, paniculis racemiformibus subsimplicibus, sepalis 5, interioribus exteriores paullo superantibus, fructu basi late cordato glaberrimo nitidulo reticulato, alis dimidio longitudinis angustioribus.— *Rami* tenues, juniores leviter angulati. *Stipulæ* lineam longæ, late triangulares, obtusæ. *Foliorum* petiolus communis 1–3-pollicaris; foliolum intermedium 2–3-pollicare, apice obtuso dentibusque paucis callosis, basi rotundatum v. cuneatum et in petiolulum 2–3 lin. longum abrupte contractum, lateralia subsessilia quam terminale nunc paullo nunc dimidio minora, omnia integra v. intermedium raro subtrilobum, penninervia et reticulato-venosa, chartacea v. demum subcoriacea, supra nitidula, subtus pariter viridia sed pallidiora. *Thyrsi* (v. paniculæ racemiformes) axillares, breviter pedunculati v. ad apices ramorum terni v. subpaniculati, 3–5-pollicares, glaberrimi, basi cirrhosi; ramuli per anthesin breves, recurvi, demum sæpe semipollicares, 5–10-flori. *Pedicelli* 2–3 lin. longi, infra medium articulati. *Sepala* exteriora late ovata, concava, 1½ lin. longa, interiora angustiora et longiora. *Petala* calycem vix superantia, anguste oblonga, in unguem longe contracta, appendice basilari cucullata petalo breviore villoso, apice glanduliformi glabra. *Ovarium* glabrum. *Fructus* 10 lin. longus, apice emarginatus, undique reticulatus et nitidulus, alis basi rotundatis singulis 4 lin. latis.

Semen et *embryo* omnino *S. velutinæ*, Camb.—Santarem, distributed as "*Serjania*, sp. n., *S. sinuatæ* aff."

Six Brazilian trifoliate *Serjaniæ* are enumerated by Schlechtendahl (Linnæa, vol. xviii. p. 55). Amongst them the *S. Guaruminea*, Mart., is surely the same as *S. cuspidata*, St. Hil., a common Rio Janeiro species since described by Walpers under the name of *Paullinia Meyeniana*. The *S. lanceolata* is unknown to me, unless a specimen of Pohl's, with leaves much larger than those described by Cambessèdes, belongs to it. *S. Mansiana*, Mart., occurs in Pohl's as well as in Martius's collection. *S. monogyna*, with "alæ capsulæ apice dilatatæ," must be a *Paullinia*, as suggested by Schlechtendahl. *S. Regnelii* is evidently a distinct species unknown to me. No. 1245 of Martius's 'Herbarium Floræ Brasiliensis' referred to by Schlechtendahl appears to be the *Urvillæa glabra*. The two following are new:—

2. Serjania *platycarpa*, sp. n.; cinereo-pubescens, ramis obtusangulis, foliolis ternis ovatis serratis subtus præsertim molliter pubescentibus, paniculis compositis subcorymbosis longe pedunculatis, sepalis 5, intimis petalisque orbiculatis, fructu maximo glaberrimo glauco basi profunde cordato, alis dimidio longitudinis latioribus.—*Sepala* extus cano-tomentosa, interiora 3 lin. longa. *Petala* glabra, unguiculata, orbiculata, calyce dimidio longiora. *Glandulæ* receptaculi oblongæ. *Fructus* 2 poll. longus et latus.

A fine species, remarkable for its large flowers, broad petals, and large, broadly-winged fruits, found in the province of Goyaz by Pohl and by Gardner (n. 3629).

3. Serjania *hebecarpa*, sp. n.; ramis teretibus glabriusculis, foliolis ternis ovatis acute acuminatis grosse paucidentatis supra scabriusculis subtus junioribus pubescentibus, paniculis compositis tomentosis, sepalis 5 subæquilongis, petalis obovali-oblongis, fructu undique pubescente apice villoso basi rotundato.

Agrees in some respects with the character of *S. Regnelii*, but the leaves are less hairy, and the structure and proportions of the calyx and corolla are different. Gathered by Gardner in the provinces of Ceará (n. 1498) and Minas Geraes (n. 4479).

Besides the above, I have another Brazilian plant which, if a *Serjania*, would belong to the same groupe, and be nearly allied to the West Indian *S. sinuata*, but which I refrain from describing, as in the absence of fruit it is impossible to say with certainty whether it be a *Serjania* or a *Paullinia*.

Among the numerous *Serjaniæ* with biternate leaves, Mr. Spruce has but one, distributed as a species allied to *S. paucidentata*, DC., which on a closer comparison with Guiana specimens (British Guiana, *Robt. Schomburgk*, 2nd coll., n. 992, *Richard Schomburgk*, n. 1710; Cayenne, *Martin*, and Surinam, *Hostmann*, n. 998) does not appear to be really distinct from that species.

There are five *Paulliniæ*; three new species described below, a single specimen of *P. riparia*, H.B.K., from the Rio Aripecurù, and another from the same locality, of *P. pinnata*, Linn., with rather narrower wings to the petiole than usual. To the various synonyms of this widely-diffused species must be added the *P. diversiflora*, Miq., and *P. Hostmanni*, Steud.

1. Paullinia *spicata*, sp. n.; caule obtusangulo foliisque pinnatis glabris, petiolo aptero, foliolis 5 amplis ovali-oblongis breviter acuminatis obtuse paucidentatis viridibus nitidulis, racemis subsessilibus spicæformibus dense multifloris, capsula elongato-pyriformi obtusa glabra.—*Ramuli* crassi, angulis 3-4 elevatis obtusis sulcatisve. *Stipulæ* caducissimæ. *Foliorum* petiolus communis 4-7-pollicaris, teres, leviter sulcatus. *Foliola* 4-5-pollicaria v. terminale 6-7 poll. longum, 2-3 poll. lata, margine more *P. pinnatæ* grosse paucidentata, basi rotundata v. terminale cuneatum, utrinque viridia et glaberrima exceptis barbis minutis ad axillas venarum paginæ inferioris, supra nitidula, petiolulis 1-3 lin. longis. *Racemi* axillares, 2-5-pollicares, tomento tenui canescentes, fere a basi densiflori, fasciculis florum approximatis sessilibus. *Bracteæ* lineares, caducæ. *Flores* albi, brevissime pedicellati. *Sepala* 5, ovato-orbicularia, concava, extus cano-tomentella, exteriora fere 1 lin., interiora $1\frac{1}{2}$ lin. longa. *Petala* 4, oblonga, calyce paullo longiora, squamis inferiorum appendicula inflexa auctis. *Ovarium* breviter stipitatum, fere glabrum, stylo usque ad medium trifido, lobis intus stigmatosis. *Capsula* fere *P. pinnatæ*, sed longior et longius stipitata.—On the Lake Quiriquiry.

I should have taken the above plant, which I have also from Martin's Cayenne collection, for the *Guaranà* or *Paullinia sorbilis*, Mart. (Reise, vol. ii. p. 1098), but that its stem is certainly climbing and the capsule not at all rostrate. As Martius's species has not been taken up either by Walpers or by Steudel, and as his interesting notes on the use and properties of the Guaranà appear to have been misunderstood (for it is used either medicinally or as furnishing a refreshing drink, not as a bread or article of food) I here extract the greater portion of the passage from that distinguished traveller's relation.—

"From the *Mauhés* (Indians of the Rio Mauhé) the Brazilians, as well as the civilized Indians of the same race, receive cloves, sarsaparilla, cacao, and especially the *Guaranà*, a drug whose preparation most especially occupies the Mauhés. The Guaranà is a very hard paste, of a chocolate-brown colour, and with little odour. For use, it is reduced to a fine powder and mixed with sugar and water as a cooling stomachic draught, to be taken like lemonade merely for its flavour, or, medicinally, against diarrhœa. Its use is so widely diffused that it is sent from Topinambarana through the whole empire and even beyond the limits of Brazil to the provinces of Mochos and Chiquitos. A good-natured Indian, of the race of the Mauhés, presented me with several sticks of Guaranà which he had himself prepared, and allowed me to witness the operation, of which I now give an account, together with some other data relating to this remarkable product." (Reise, vol. ii. p. 1061.)

"The *Guaranà* (which must not be confounded with the *Caranna gum*) was originally prepared by the Mauhés alone. But since its use has been so widely spread as to become an object of no inconsiderable commerce, it is made also by other settlers, particularly at Villa Boa, and here and there on the Rio Tapajoz. The genuine article is distinguished from the adulterations by its greater hardness and density, and in that, when powdered, it does not assume a white colour, but a greyish-red tint. . The preparation showed me by the Indian in Topinambarana was as follows:—The Guaranà plant (*Paullinia sorbilis*, Mart., glabra, caule erecto angulato, foliis pinnatis bijugis, foliolis oblongis remote sinuato-obtuse-dentatis, lateralibus basi rotundatis, extimo basi cuneato, petiolo nudo angulato, racemis pubescentibus erectis, capsulis pyriformibus apteris rostratis, valvulis intus villosis) ripens its seeds in the months of October and November. These are taken out of their capsules and exposed to the sun. When they are sufficiently dried to allow of the white arillus, in which they are half enveloped, being rubbed off with the fingers, they are emptied into a stone mortar, or deep dish of hard sandstone, which is heated over a charcoal fire, and they are then ground to a fine powder, which, mixed with a little water, or exposed to the night dew, is kneaded into a dough. In this dough are mixed a few seeds, either whole or broken into two or three pieces, and the whole is made up into sticks or cakes of the requisite form, usually cylindrical or spindle-shaped, containing from twelve to fifteen ounces of the paste, and from five to eight inches long; it is more rarely compounded into balls. These sticks are then dried in the sun, or in the smoke of the huts, or by the fire, till they become so hard and tough that it requires an axe to break them. They are now packed among the broad leaves of *Scitamineæ* into baskets or sacks, and when not exposed to much damp, will keep uninjured for several years. In the province of Parà, the Guaranà is grated for use on the jawbone of the Piracurù fish, which is covered with bony asperities, and which, kept in a basket made of *Uarumà* stalks (*Maranta Tonchat*, Aubl.), is a common article of household furniture. An inferior paste is prepared by mixing powdered cocoa or mandiocca flour with the true Guaranà: this is less hard and tough, and is whitish when broken." (Mart., ibid. p. 1098, in a note which concludes with some medical and chemical observations, the result of the experiments of Dr. Theodore von Martius, the traveller's brother, and with the expression of his belief that it would make a valuable addition to our Materia Medica.)

2. Paullinia *interrupta*, sp. n.; ramulis petiolisque ferrugineo-puberulis glabratisve, foliis pinnatis, petiolo non alato, foliolis 5 ovatis oblongisve acuminatis integerrimis v. obsolete sinuato-dentatis utrinque viridibus glabris v. ad venas marginesque pilosulis, racemis sessilibus v. breviter pedicellatis, floribus interrupte fasciculatis, sepalis 5 exterioribus duplo minoribus, capsula breviter stipitata globoso-triquetra ferrugineo-tomentosa.—*Frutex* scandens; rami teretiusculi, juniores leviter angulato-striati, nunc uti petioli dense ferrugineo-tomentosi, nunc fere glabri. *Stipulæ* oblongo-lineares, caducæ. *Petioli* communes 3–6 poll. longi. *Foliola* majora (3 superiora foliorum majorum) 5–6 poll. longa, 2–3 poll. lata, inferiora cujusve folii et omnia foliorum superiorum minora, et hæc sæpius proportione angustiora, terminale basi cuneatum et ima basi in petiolulum contractum, lateralia basi rotundata et brevissime petiolulata, omnia rigide chartacea v. subcoriacea, penninervia, rete venarum subtus prominula supra vix conspicua, præter pilos nunc rarissimos nunc copiosores ad venas marginesque glabra, utrinque punctis minutis

plus minus copiosis quasi irrorata, nec nitentia. *Racemi* 3–8-pollicares, rhachi ferruginea, fasciculis florum subsessilibus. *Bracteæ* minutæ, acutissimæ. *Flores* albi, breviter pedicellati. *Sepala* lato-ovata, concava, extus minute tomentella v. fere glabra, interiora 1¼ lin. longa. *Petala* obovata, calyces vix superantia; squamæ inferiorum latæ, apice biglandulosæ, intus appendice lata reflexa auctæ. *Stamina* brevia. *Ovarium* globosum, hirsutum. *Styli* 3 distincti, extus puberuli, intus a basi papillosi. *Capsulas* 2 tantum vidi, alteram immaturam dense ferrugineo-villosam, alteram vetustam 5 lin. diametro coriaceo-lignosam apice brevissime obtuseque acuminatam.—Near Santarem.

3. Paullinia *pachycarpa*, sp. n.; glabra, foliis pinnatis, pinnis trijugis cum impari infimis trifoliatis, petiolo subalato, foliolis oblongis obtuse acuminatis grosse paucidentatis, racemis laxis paniculatis tomentosis, sepalis 4 magnis canescenti-tomentosis, capsulis longe stipitatis globosis tomentosis, valvulis crassis sublignosis.—*Frutex* alte scandens. *Foliorum* petiolus communis semipedalis, undique anguste alatus v. ala versus basin evanescente. *Foliola* glaberrima, chartacea, 3–5 poll. longa, omnia basi angustata et subsessilia. *Panicula* ampla, ferrugineo-tomentosa, e racemis 3–6-pollicaribus composita. *Ramuli* racemorum 2–4 lin. longi, pedicelli breves. *Bracteæ* oblongæ, caducissimæ. *Flores* in genere majusculi, extus cano-tomentelli. *Sepala* 4, intimum orbiculare, 3 lin. diametro, interdum breviter bifidum, exteriora minora. *Petala* 3 lin. longa, elliptica. *Squamæ* interiores appendice inflexa barbata aucta. *Filamenta* villis albis dense hirsuta. *Styli* 3, breves, distincti, intus a basi stigmatosi. *Capsulæ* 9 lin. diametro stipite semipollicari, subglobosæ, costis sex percursis quarum 3 magis prominent, extus tomento brevissimo vestitæ, in vivo ex cl. Spruce roseæ, siccitate corrugatæ. *Semina* magna, irregulariter hemisphærica, canescentia, opaca excepto hilo basilari subdorsali orbiculari nitido. *Arillam* non vidi, in specimine omnes a vermibus jam destructæ. *Testa* crustacea. *Cotyledones* crasso-carnosæ, subæquales. *Radicula* brevissima.—On the shores of the Tapajoz, near Santarem.

1. Schmidelia *leptostachya*, sp. n.; subglabra, foliolis ternis ellipticis v. obovali-oblongis obtusis remote serratis subtus pallidis ad axillas venarum barbatis, racemis gracillimis folio longioribus, floribus minimis numerosissimis.—Primo intuitu *S. lævi*, St. Hil., affinis, sed

racemis elongatis et floribus multo minoribus facile distincta. *Frutex* est biorgyalis. *Foliorum* petiolus communis semipollicaris, pilis paucis puberulus. *Foliola*, exceptis barbis ad axillas venarum, glabra, terminale 2–4-pollicare, lateralia vulgo minora, omnia basi angustata, margine dentibus paucis brevibus subcallosis notata, supra siccitate nigricantia, subtus pallide virentia. *Racemi* vulgo semipedales vel longiores, fere a basi fasciculato-floriferi. *Flores* albi, aperti vix ¾ lin. diametro. *Sepala* 2 exteriora interioribus dimidio breviora, omnia orbiculata, concava. *Petala* sepalis breviora, cuneato-oblonga. *Fructus* sessiles, subglobosi, magnitudine pisi.—Near Santarem.

There are fine specimens in flower and fruit of *Sapindus inæqualis*, DC., a tree of twenty to forty feet, common about Santarem. As this has been ascertained to be the same as *S. divaricatus* of Willdenow's herbarium (Schlecht. Linnæa, vol. vi. p. 419), it is probably also the *S. divaricatus* of Camb. and St. Hil., and has a wide range in Brazil. I have it from Utinga, prov. Bahia (Blanchet, n. 2755), and from Pohl's collection. It is nearly allied to the West Indian *S. marginatus*, and differs from all other known American *Sapindi* in its small flowers. The following handsome large-flowered species appears to be undescribed, although probably near to *S. Surinamensis*, Poir., a species unknown to me. That is, however, said to have thin membranous leaflets rounded at the base and blunt at the apex, whilst those of our *S. cerasinus* are stiff and narrowed at both ends. Mr. Spruce found it in gravelly situations about Santarem, where it forms a shrub of eight or ten feet, with yellowish flowers, and a fruit resembling a cherry in appearance. It is called *Pitomba* by the Brazilians, who eat its thin pulp.

1. Sapindus *cerasinus*, sp. n.; foliolis 3–8-jugis oblongis sublanceolatisve acuminatis basi inæqualiter angustatis coriaceis utrinque glaberrimis nitidis, panicula ampla floribunda, cymulis laxis calycibusque puberulis, petalis glabris calyce triplo longioribus squamas ligulatas brevissime bilobas paullo superantibus.—*Foliorum* petiolus communis semipedalis ad pedalis v. etiam longior. *Foliola* foliorum majorum 6–8 poll. longa, 2–2½ poll. lata, utrinque viridia, venulis crebris. *Panicula* pyramidata, floribunda, 1–1½-pedalis. *Sepala* vix lineam longa, orbicularia. *Petala* 3 lin. longa, oblonga, crassiuscula, glaberrima, squama intus dense et longe barbata. *Ovarium* sessile, breviter pubescens, stylo simplici ovario æquilongo.

I refer also the following plant, a single specimen from Santarem, to *Sapindus*, on account of the similarity of its flowers to those of the last species, although in the absence of fruit the genus cannot be determined with certainty. There are, however, no American *Cupaniæ* known with flowers at all like it.

2. Sapindus *oblongus*, sp. n. ; glaber, foliolis trijugis oblongis vix acuminatis utrinque viridibus nitidulis basi cuneatis, thyrsis simplicibus ramosisve folio subbrevioribus, cymulis pedunculatis, petalis calyce duplo longioribus squamam ligulatam subintegram paullo superantibus.—*Arbor* parva. *Petioli* communes 3-5-pollicares. *Foliola* ultima 3 poll. longa, inferiora minora, consistentia chartacea v. fere laurina. *Inflorescentiæ* in axillis supremis ¦3-5-pollicares, laxæ, cymulis breviter pedunculatis paucifloris. *Sepala* interiora linea paullo longiora, exteriora breviora, ovato-orbiculata, concava, crassiuscula, margine ciliolata, dorso puberula. *Petala* flavicantia, oblonga, 3 lin. longa, glabra, patentia, squama erecta intus villosissima. *Discus* cupuliformis, crassus, ovarii basin cingens. *Stamina* pistillo adpressa et eo dimidio breviora : filamenta brevia ; antheræ lineares, glabræ. *Ovarium* pubescens, in stylum desinens petalis paullo breviorem simplicem, summo apice simpliciter stigmatosum. *Ovula* e basi loculorum erecta, solitaria.

The plant described (vol. ii. p. 212) as *Talisia laxiflora* has been again gathered in abundance with a few specimens in fruit, which show that it cannot in fact be generically separated from *Cupania*, to which should most probably be referred Aublet's original *Talisia*. The name of our species must therefore be changed to that of *Cupania laxiflora*, and the following particulars added to the description :—*Capsula* sessilis, depresso-globosa, obtuse subtriquetra, extus tomentella et siccitate corrugata, 6 lin. diametro, trilocularis, loculicide trivalvis, valvulis crasso-coriaceis intus dense villosis. *Semina* solitaria, erecta, ovoideocompressa, nigra, nitida ; arillus tenuis, semine dimidio brevior ; testa crustacea ; cotyledones crasso-carnosæ ; radicula brevissima. There are, besides, three other *Cupaniæ* ; one appears to be the *C. geminata*, Poir., of which I subjoin a somewhat fuller description than that given in the Encyclopédie, and two new species :—

1. Cupania *geminata*, Poir. Dict. Suppl. vol. ii. p. 419.—*Frutex* biorgyalis, ramulis subteretibus striatis rufo-tomentosis. *Petioli* communes $\frac{1}{2}$-1-pollicares. *Foliola* gemina, breviter petiolulata, obovata,

ovalia v. elliptica, 3–8-pollicaria, obtusissima v. acutiuscula, integerrima v. obscure sinuata, margine recurva, basi sinuata, vetustiora bullato-rugosa, supra glabra v. ad venas impressas rufo-tomentella, subtus pube brevi rufescentia, costa nervis primariis intra marginem anastomosantibus reteque venularum prominentibus. *Racemi* axillares, foliis multo breviores, simplices v. parce ramosi, ferrugineo-pubescentes. *Pedicelli* breves. *Calyces* extus pubescentes, profunde 5-fidi, lobis ovatis obtusis ¾ lin. longis. *Petala* 5 (v. interdum nulla?), calyce longiora, oblonga v. ovata, squama lineari-cuneata apice villosula petalo breviore. *Discus* 8–10-lobus. *Stamina* 9 (v. 10?), petalis duplo longiora, filamentis villosis, v. in floribus subfœmineis abbreviata. *Ovarium* ovatum, hirsutissimum. *Styli* 3, breves, recurvi, intus a basi stigmatosi. *Capsula* qualis a Poiretio descripta.—From the vicinity of Santarem.

2. Cupania *Spruceana*, sp. n.; foliolis 6–10 oblongis sublanceolatisve basi acutis concoloribus ramisque glabris, racemis ramealibus simplicibus fasciculatis v. subramosis folio multo brevioribus, floribus parvis, calycis laciniis petala subæquantibus, squamis bipartitis hirsutissimis petala superantibus, staminibus longe exsertis, capsulis stipitatis glabris.—*Arbor* procera, a basi racemosa, ramulis teretibus, novellis minute tomentellis mox glabratis. *Petioli* communes 3–6-pollicares, glabri. *Foliola* 2–4-pollicaria, vix acuminata, integerrima, basi subæqualia acuta et breviter petiolulata, consistentia chartacea v. sublaurina, utrinque reticulato-venosa, supra nitidula. *Racemi* ad nodos vetustos fasciculati, v. in axillis foliorum pauciores, 2–3-pollicares, simplices v. ramis paucis divaricatis instructi, rhachide tenui ferrugineo-tomentella. *Flores* parvi, solitarii v. subgemini, pedicello fere lineam longo, omnes quos vidi abortu masculi. *Calyces* semilineam longi, extus puberuli, in partes 5 triangulares acutiusculas fere ad basin divisi. *Petala* 5, rhombea. *Ovarii* rudimentum parvum. *Capsulæ* juniores stipitatæ, stylo breviter rostratæ, mox obtusatæ stylo evanido; maturas tamen non vidi.—Near Santarem.

3. Cupania *frondosa*, sp. n.; foliolis 6–8 ovali- v. oblongo-ellipticis vix acuminatis paucicrenatis basi acutis glabris v. subtus petiolisque puberulis, panicula terminali tomentoso-canescente, floribus secus ramos fasciculatis, sepalis squamisque bifidis petala æquantibus, capsula triangulari ferrugineo-tomentosa siccitate transverse rugosa.

—*Arbor* 20–30-pedalis, frondosa, ramulis teretibus apice rubiginoso-tomentellis mox glabratis. *Foliorum* petiolus communis subteres, 4–6-pollicaris. *Foliola* alterna, 3–5-pollicaria, obtusa v. breviter acuminata, crenis paucis sæpe obsoletis, basi cuneata v. acuta et breviter petiolulata, consistentia laurina, utrinque reticulato-venosa, nervis primariis parallelis subtus prominentibus vix intra marginem anastomosantibus. *Panicula* semipedalis, late pyramidata, floribunda. *Flores* albi, mediocres, secus ramos dense fasciculati, breviter pedicellati, omnes quos vidi abortu masculi. *Sepala* fere lineam longa, ovata, concava, extus puberula. *Petala* rhombea, breviter unguiculata. *Discus* crasso-carnosus, subinteger. *Stamina* flore dimidio longiora. *Capsulæ* obtusæ, triquetræ, fere trilobæ, 6–7 lin. latæ, supra planæ, basi in stipitem brevem attenuatæ, siccitate insigniter transverse rugosæ, in vivo tamen teste Spruceo rugæ haud apparent; juniores stylo trifido coronatæ; valvulæ intus villosæ.—Near Santarem.

(*To be continued.*)

DECADES OF FUNGI; *by the* Rev. M. J. BERKELEY, M.A., F.L.S.

Decade XXXVI.

Sikkim-Himalayan Fungi collected by Dr. Hooker.

(*Continued from p.* 172.)

351. *Trichocoma paradoxum*, Jungh., Præm. in Fl. Javæ, p. 9, tab. 2, fig. 7. Hook. fil., No. 63, cum ic.

HAB. On dry dead wood. Sikkim, 7–9,000 feet. October.[*]

I have little to add to the long description given by Junghuhn, except that the upright fascicles of threads which proceed from the base give off, during their course, more or less horizontal branched threads, on which the irregular echinulate spores are produced. Some of these threads are thicker and slightly nodulose, as in *Arcyria*. There are two distinct peridia, the interior springing not from the edge of the other, but from the base, and separated from the outer and shorter by a few flocci, which are mostly bent towards the base, as though the outer peridium had at first grown more rapidly, and the elongation of the inner peridium has taken place principally above.

[*] It has been found also on the Santee River, South Carolina, by Mr. Ravenel.

The genus appears clearly allied to *Geastrum* and *Broomeia*. Unfortunately I have seen no specimen in an early stage of growth, but the whole structure is that of *Trichosperma* rather than *Myxogastres*, especially the filamentous nature of the yellow upright septa, which at first sight call to mind those of *Æ. thalium*, though they do not appear, as far as we have seen, to constitute spurious cells, but rather to represent a multitude of columellæ.

* *Scleroderma Bovista*, Fr. Syst. Myc. vol. iii. p. 48.

HAB. On the ground. Sikkim.

There is but a single specimen which resembles externally stipitate forms of *S. vulgare*, but the flocci are yellow.

852. *Mitremyces viridis*, n. s.; peridio amplo stipiteque lacunoso-costato cartilagineo viridibus; squamis oris margine coccineo-granulatis. Hook. fil., No. 19, cum ic.

HAB. On the ground and on dead timber. Tonglo and Sinchul, 7–9,000 feet. May, June. Rare.

Inodorous. Stem consisting of numerous anastomosing olive-green cartilaginous threads, 1¼ inch high, 1 inch thick. Peridium of the same colour as the stem, inflated, sprinkled with small, flat, granular scales; margin of the scales of the orifice rough with little scarlet warts. Spores globose, strongly granulated.

Distinguished at once by its green tint from *M. lutescens*, which it resembles in form. Its spores, instead of being broadly elliptic as in that species, *M. luridus*, and *M. australis*, are globose and rough, as in *M. Junghuhnii*. Ic. Pl. ined.

* *Lycogala epidendrum*, Fr. Syst. Myc. vol. iii. p. 80. Hook. fil., No. 83, cum ic.

HAB. On the ground. Darjeeling. June. Rare.

853. *Reticularia entoxantha*, n. s.; peridio nigro granulato, hypothallo distincto albo-marginato; sporis flavis; floccis parcis subcylindricis concoloribus.

HAB. On dead wood. Sikkim-Himalaya.

About ¾–1¼ inch broad; hypothallus distinct, extending beyond the peridium. Peridium depressed, very dark brown, minutely but distinctly granulated. Spores bright yellow, mixed with a few subcylindrical coarse flocci, principally attached to the upper portion of the peridium.

A beautiful and most distinct species, of which only two specimens were gathered, and those scarcely mature, the spores being still closely compacted. The peridium looks very like a specimen of *Sphæria nummularia*, its granulations reminding one of *Ret. maxima*. In the colour of its spores it approaches *R. olivacea*, but recedes in its other characters, as, for instance, in its dark granulated, not plicate and hyaline peridium.

354. *Ustilago Emodensis*, n. s.; sporis ellipticis ovatisve lævibus minimis saturate lilacinis filamentis radiantibus furcatis immixtis.

HAB. On some *Polygonum*. Tonglo, 10,000 feet.

Forming a lobed tubercle, apparently exterior to the ochrea, but probably formed by a short deformed spike bursting through the sheath, every triple nodule of which corresponds with the original site of a flower-bud, no trace, however, being left of floral envelopes or stamens. Spores ovate or elliptic, deep lilac, smooth, very minute, traversed by radiating forked threads.

Of this remarkable production I have seen a single specimen only, from which, in the absence of all information as to the normal mode of inflorescence, I can form no judgment as to the nature of the change produced by the parasite. It is certainly closely allied to *U. Candollei*, but differs in the presence of the forked threads and the different mode of growth. The spores, too, are far redder and have not a third of the diameter of those of that species. I do not suppose that the threads have any real connection with the fungus, but they differ very greatly from the columella, which exists in the described forms of the species just mentioned.

* *Leotia lubrica*, Pers. Syn. p. 613. Hook. fil., No. 131, cum ic.

HAB. On clay banks. Sinchul, 8,600 feet. October. Rare.

In habit and fructification exactly like the well-known species. The colours are brighter than usual, the stem being orange, and the head of a metallic blue-black.

355. *Peziza Darjeelensis*, n. s.; cupula expansa subcochleata umbrina, extus pallida aleuriata; sporidiis minoribus scabro-punctatis, endosporio simplici.

HAB. On the ground. Darjeeling.

Cup two inches or more broad, at first subcochleæform, split on one side, at length expanded, umber-brown; externally paler, mealy. Asci slender; sporidia small, elliptic, beset with minute scabrous points.

Resembling at first sight *P. repanda*, but the sporidia are much smaller and remarkable for their minutely scabrous surface. Those of *P. repanda* are quite smooth. I have a *Peziza* from Bristol with larger rough spores, which is, I believe, on comparison with French specimens, *P. umbrina*, P., a species referred by Fries to *P. cochleata*, and it should seem correctly, for in undoubted *P. cochleata* the spores are rough. They are, however, larger than in the Himalaya species, more coarsely granulated, and contain two large nuclei.

356. *P. macrotis*, n. s.; cupulis elongatis obliquis auriformibus basi connato-ramosis coriaceo-subcartilagineis hepaticis, extus glabris; hymenio purpurascente. Hook. fil., No. 87, cum ic.

HAB. On rotten wood. Darjeeling, 7,500 feet. June, July. Abundant.

Inodorous, dry, firm, leathery, subcartilaginous, varying in size, sometimes five inches long, erect, tufted, connate below and thence branched. Cups elongated, oblique, auriform, of a bright liver-colour, smooth externally, margin subinvolute. Hymenium even, purplish. Sporidia oblongo-elliptic, with one side in general more convex. Nucleus single in the dry specimens.

This splendid species is evidently closely allied to *P. onotica* and *P. leporina*, but it differs not only in colour and its very elongated narrow cups, but in its firmer, tougher substance, and the mostly connate base. It is one of the finest species of the genus, being exceeded in size only by *P. Cacabus*.

* *P. aurantia*, Pers. Hook. fil., No. 34, cum ic.

HAB. On clay banks. Darjeeling, 7,000 feet. Very abundant from June to August, when it dies away.

"So conspicuous that every one asks whether you have seen the scarlet fungus." It is precisely the European species, agreeing in the verrucose sporidia as in all other points. It is sometimes six inches across.

357. *P. geneospora*, n. s. media; cupula expansa concaviuscula aurantio-coccinea extus pilis badiis vestita; ascis amplis; sporidiis magnis ellipticis verrucosis. Hook. fil., No. 132, cum ic.

HAB. On rotten wood. Sinchul, 8,000 feet. October.

Cup one inch or more across, convex below, expanded, slightly hollowed out above, scarlet, clothed and fringed with bay-brown septate bristles. Asci rather thick, often torulose beneath the sporidia, which are large, elliptic, verrucose, colourless, containing a single nucleus. Paraphyses slender, linear.

This species, like *P. trechispora*, Berk. and Br., resembles very closely *P. scutellata*, but differs from the former in its elliptic not globose sporidia, from the latter in these organs being much larger and verrucose instead of smooth. It attains a larger size than either.

The sporidia are properly eight in each ascus, but the number occasionally varies, being frequently six. A portion of the inner membrane of the ascus often separates with the spores, giving them the appearance of having two transparent, smooth, apiculate tips, with a large elliptic verrucose nucleus, and this separation often takes place before they are discharged, while in other asci the sporidia have the usual elliptic form.

* *P. clandestina*, Bull., p. 251.

In small quantities scattered over the upper side of the leaves of a species of *Pyrus*. Tonglo.

* *P. fructigena*, Bull., t. 228. Hook. fil., No. 128, cum ic.

HAB. On stems of dead *Umbelliferæ*. Sikkim, 8–9,000 feet. October.

I can see no difference between these specimens and the European species. The sporidia in both are oblong, with one end rather thicker, and have a single obscure colourless septum, which is not, however, always distinctly visible.

358. *P. turbinella*, n. s.; minuta turbinato-clavata stipitata; stipite glabro cum cupula sursum dilatata confluente; disco demum convexo; sporidiis filiformibus.

HAB. On the underside of the leaves of a *Pyrus*. Tonglo.

Minute, not ¼ of a line high, at first pale yellow, at length opake white; stem smooth, confluent with the turbinate or cyathiform cup; hymenium at length quite plane or convex. Asci clavate; sporidia filiform, of various lengths.

Closely resembling *P. clavata*, Fr., but a smaller species, and remarkable for the perfectly plane or convex hymenium. When moistened, it has very much the form of *Craterium leucocephalum*.

359. *P. stilboidea*, n. s.; minuta fulvo-lutea glabra; stipite subelongato, cupula hemisphærica concava margine incurvo; sporidiis filiformibus.

HAB. On the main nerve on the underside of the leaves of a *Pyrus*. Tonglo.

Not half a line high, tawny-yellow; stem equal, slender, somewhat elongated; cup hemispherical, concave, margin incurved. Asci clavate; sporidia filiform.

Near allied to *P. clavata*, Fr., and *P. Ciborium*, but distinguished by its regular cup, which is not much broader than the stem. It is even more minute than *P. clavata*, which is itself a far less species than *P. Ciborium*.

* *P. citrina*, Pers. Hook. fil., No. 125, cum ic.

HAB. On rotten wood. Sikkim, 8–9,000 feet. October.

Exactly the European species.

* *Bulgaria inquinans*, Fr., Syst. Myc. vol. ii. p. 167.

Var. *chalybea*, Hook. fil., No. 52, cum ic.

HAB. On trunks of trees. Jillapahar, 7,500 feet. May and June. Rare.

The disc, in the Darjeeling specimens, has a steel-blue tint, and they contract more in drying, in consequence of which the under surface appears more rugose and spiculato-verrucose than in British and North American specimens. The sporidia in both are brown and subcymbiform, and the inner substance marbled. The Sikkim plant presents a distinct variety, but is not, I think, entitled to be separated as species.

* *Hypoxylon tabacinum*, Kickx in Bull. de l'Ac. Roy. de Bruxelles, vol. viii. n. 8. *Sph. involuta*, Klotzsch. Hook. fil., No. 95, cum ic.

HAB. On dead wood. Darjeeling, 7,500 feet. June, July.

The specimens are all young and therefore pale. The surface, when fresh, is of a chalky ochre, dotted with the black ostiola. Most of the specimens are hollow, and many deeply grooved at the sides, as if composed of two confluent individuals. I have not been able to compare the sporidia.

* *Hypoxylon polymorphum*, Ehr.

HAB. Sinchul, 8,000 feet.

* *Hypoxylon vulgare* (*Sphæria Hypoxylon*, Ehr.).

HAB. On old wood. Darjeeling, 7,500 feet.

A flagelliform variety, the upper divisions being very long and slender. It resembles in habit Montagne's *Hypoxylon ianthino-velutinum*.

* *Hypocrea peltata*, Berk. (= *Sphæria peltata*, Jungh.) in Hook. Lond. Journ. vol. i. p. 156. t. vii. f. 7.

HAB. On dead bark. Darjeeling, 7,500 feet. June, July.

Inodorous, dry, tough, coriaceous, even, smooth, opake, hollow towards the centre; plicate below. Perithecia elliptic, with a short ostiolum. Asci linear, containing a row of thirteen or more globose sporidia. Substance consisting of branched threads, which penetrate between the perithecia and form a close network, the perithecia themselves being closely cellular.

I have not seen mature specimens of *Sphæria peltata*, Jungh. These are plicate or almost lamellose beneath; the Philippine Island specimens, on the contrary, are even. The habit and general appearance, however, is so exactly the same, that in the absence of perfect specimens of Junghuhn's plant, I cannot but consider them as mere forms of the same species. The folds, it should be observed, are stronger in the specimens than in Dr. Hooker's figure, and probably arise partly from contraction. The fungus *in situ* reminds one of some crawling *Natica*, the centre being swollen and the margin thin and free.

360. *H. grossa*, n. s.; receptaculo erecto crasso sursum breviter diviso miniato opaco; lobis obtusis; intus pallide stramineo; contextu lento radiato; peritheciis irregularibus confluentibus. Hook. fil., No. 99, cum ic.

HAB. On rotten wood. Darjeeling, 7–8,000 feet. July. Very rare.

Inodorous. Receptacle 2 inches high, ¾ inch thick below, springing from a tuberous base, erect, clavate, divided above into short obtuse lobes, externally of an opake vermilion which fades in the dry specimens to a pale reddish-brown. Substance pale yellow, dry, and somewhat coriaceous. Perithecia irregular, confluent.

Unfortunately the sporidia are imperfect, but so far as could be ascertained they appear to be minute, colourless, and elliptic.

One of the most curious species in the collection.

(*To be continued.*)

Contributions to the Botany of WESTERN INDIA;
by N. A. DALZELL, Esq., M.A.
(*Continued from p.* 180.)
Nat. Ord. ANONACEÆ.
GUATTERIA.

G. *fragrans*, n. sp.; foliis breve petiolatis ellipticis acuminatis basi obtusiusculis glabris membranaceis nitidis, pedunculis axillaribus brevibus ramosis velutinis 5-12-floris, floribus pedicellatis fasciculatis ternis, pedicellis pollicaribus infra medium bracteola cucullata obtusa caduca instructis.

Folia 7-9 poll. longa, venæ costales numerosæ, parallelæ, subtus prominulæ. *Pedicelli* pedunculum ramosum æquantes. *Calyx* 3-partitus; segmenta brevia, obtusa, apice (etiam in alabastro) recurva, caduca. *Petala* omnia æquilonga, anguste linearia, acuta, lutea, sesquipollicaria. *Ovaria* numerosa, dense tomentosa; *ovulum* unicum e basi loculi oriens. *Fructus* ovalis, pollicaris.—Crescit ad pedem jugi Syhadrensis, prope Sivapore; fl. Nov. Flores fragrantes.

Inflorescence an abbreviated, rigid, divaricating, leafless panicle, one to two inches long, from the axils of the fallen leaves, and resembling that of *G. longifolia* only.

UNONA.

U. *pannosa*; arborea, foliis parvis breve petiolatis lanceolatis acuminatis utrinque glabris nitidis crebre *pellucido-punctatis*, floribus pannosis majusculis axillaribus solitariis breve pedunculatis, pedunculis basi bibracteolatis petiolum brevem æquantibus.

Folia cum petiolo 2-lineari 2¼ poll. longa, 9-12 lin. lata. *Calycis* tomentosi *foliola* ovato-lanceolata, acuta, 3 lin. longa. *Corollæ petala* 6, biseriata, lineari-lanceolata, basi late unguiculata, utrinque fusco-velutina; interiora pollicaria, *basi callositate nuda* carnosa rugosa instructa, exteriora paulo majora. *Ovaria* 10-12, densissime strigosa, biovulata; *ovula* in angulo centrali superposita. *Stigma* capitatum. *Fructus* ignotus.—Crescit in jugo Syhadrensi, prope Tullawarree; fl. Oct.

SAGERÆA, genus novum. (Tribe BOCAGEÆ.)

GEN. CHAR. *Calyx* triphyllus. *Corollæ petala* 6, hypogyna, libera, biseriata, subæqualia, crassa, carnosa, orbicularia, concava, *æstivatione imbricata*. *Stamina* 12, biseriata: *filamenta* nulla; *antheræ* biloculares, loculis linearibus connectivo crasso carnoso truncato squamæformi extus adnatis, longitudinaliter dehiscentibus. *Ovaria* 3-5-linearia, oblonga, in tori convexi apice sessilia, unilocularia; *ovula* 10, biseriata, horizontalia, angulo ventrali inserta. *Stigmata* sessilia, obtusa, emarginata. *Bacca* globosa, glabra, hexasperma, cerasi magnitudine.

S. *laurina*; foliis alternis lineari-oblongis integerrimis coriaceis glabris breve petiolatis supra nitidis, floribus pedunculatis in axillis foliorum fasciculatis ternis v. quinis, pedunculis basi squamis obtusis imbricatis suffultis, floribus mediocribus albis.

Arbor mediocris.—Crescit in Concano australiore; fl. Oct. et Nov.

A beautiful laurel-looking tree, producing a valuable reddish timber. The native name is *Sageree*, hence the generic name here given.

Nat. Ord. LEGUMINOSÆ.

SMITHIA.

1. S. *capitata* (non Desv.); caule glabro ramoso, foliolis 9-15-jugis lineari-oblongis obtusis ciliatis subtusque secus costam setuloso-strigosis, petiolo communi hispido, stipulis adnatis ovato-lanceolatis seta terminatis deorsum longe- productis, capitulis sphæricis multifloris solitariis terminalibus, pedunculis glabris folio brevioribus, bracteis obovato-lanceolatis calycem æquantibus, calycis *reticulati* glabri labiis rotundatis indivisis dentatis, dentibus longe setaceo-acuminatis, leguminis articulis 6-7 glabris lævibus.

Herba sesquipedalis, tota foliis exceptis glabra.—Crescit in jugo Syhadrensi, prope Parwar-ghât; fl. Oct.

This species is well distinguished by its spherical and terminal heads of flowers.

2. S. *setulosa*; 3-4-pedalis, caule dichotome ramoso setuloso-hispido, foliolis 5-7-jugis lineari-oblongis obtusis margine ciliatis utrinque glaberrimis, petiolo communi hispido foliolisque seta **terminatis**, stipulis longe setaceo-acuminatis paulo infra medium adnatis glabris, floribus in paniculam terminalem aphyllam dispositis, racemis in dichotomiæ ramulis terminalibus secundis, calycis *striati* strigosi labiis *integerrimis* minute ciliatis valde inæqualibus superiore orbiculari majore inferiore oblongo acuto, bracteis ovalibus obtusis calyce dimidio brevioribus, floribus flavis, carina apice fusca, legumine 10-12-spermis, articulis prominulo-reticulatis.

Folia 4 poll. longa; *foliola* fere bipollicaria, subtus glauca. *Calyx* 4 lin. longus. *Corolla* 8 lin.—Crescit cum præcedente; fl. eod. temp.

The whole plant is covered with bright yellow bristles; it seems allied to *S. paniculata* (Arnott), but differs in not being glandular, in having leaves of a different shape, calyx segments quite entire, and legumes with many seeds.

3. S. *bigemina*; pedalis e basi ramosa, ramis filiformibus pilis patulis

bulbosis hirsutis, foliolis 2-jugis obovato-cuneatis margine ciliatis subtusque parce strigosis seta terminatis, racemis axillaribus pauci- (2–8)-floris pedunculatis, pedunculis glabris filiformibus folio multo longioribus, calycis labiis margine setaceo-ciliatis superiore cuneato emarginato mucronato inferiore 3-lobo lobis lateralibus obtusis intermedio longiore acuminato, bracteis calyce dimidio brevioribus lanceolatis setaceo-acuminatis, floribus flavis, legumine 7-spermo, articulis grosse tuberculatis.

Foliola 5–6 lin. longa. *Calyx* fere 2 lin. *Corolla* 4 lin.—*S. racemosæ* affinis.—Crescit cum præcedente; fl. eod. temp.

This species has the calyx scariose, neither striated nor reticulate, but having many dichotomous veins proceeding from the base to the apex; it is almost glabrous and not in the least glandular, I therefore conclude that this is not *S. racemosa*, of which I know nothing except from the too short description in Wight and Arnott's Prodromus. The present is the smallest and most delicate species which I have met with, and makes the fifth new species recently discovered in this Presidency, the first being *S. purpurea* of the Bot. Mag. t. 4283, discovered by Mr. Law.

GALACTIA.

G. simplicifolia, n. sp.; caule reptante filiformi pilis brunneis retrorsum hispido, foliis *simplicibus* petiolatis ovatis, stipulis infra medium adnatis nervosis acutis, stipellis ad apicem petioli setaceis, floribus axillaribus terminalibusque fasciculato-racemosis, racemis folio brevioribus.

Calyx 3 lin. longus, brunneo-pilosus, bibracteolatus, ultra medium 4-fidus, laciniis acuminatis, superiore latiore bifida, infima lateralibus longiore. *Flores* parvi, cærulei, breve pedicellati, fasciculati, fasciculis bracteis adnatis, pedicellisque bracteolis oblongis acutis suffultis. *Vexillum* latum, obovatum, emarginatum, ecallosum, alis carinaque longius. *Stamina* diadelpha, omnia fertilia. *Ovarium* dense strigosum, pluriovulatum. *Stylus* incurvus; *stigma* capitatum. *Legumen* brunneo-pilosum, lineare, apicem versus latius, parum compressum, multiloculare, $2\frac{1}{2}$–3 poll. longum, $3\frac{1}{4}$ lin. latum. *Semina* orbicularia, compressa, testa brunnea nitida.—Crescit in jugo Syhadrensi, prope Tullawarree; fl. Oct.

PHASEOLUS.

P. *pauciflorus*; volubilis, radice fibrosa, caule striato filiformi pilis albis minutis retrorsum hispidulo, foliolis membranaceis rhombeo-ovatis acuminatis petioli longitudine pilis raris minutis albis utrinque conspersis, stipulis lanceolatis acutis infra medium adnatis, pedunculis petiolo brevioribus, floribus racemoso-capitatis paucis (2-3), pedicellis calycis longitudine geminis e glandularum basi ortis, calycis glabri campanulati labio superiore emarginato inferiore 3-dentato, dentibus lateralibus triangularibus obtusis intermedio longiore acuminato, bracteolis lineari-subulatis calyce subduplo longioribus, legumine tereti stricto bipollicari *glaberrimo* 9-10-spermo, seminibus cylindricis truncatis glabris.

Flores lutei.—Crescit ubique; fl. Sept.

Much smaller in every way than *P. trinervius*, with a stem creeping and rooting when no support is near. There is sometimes a tendency in the leaflets to be trilobate by the formation of a tooth on the lateral margin. Keel with a long, acute, ascending horn. Most nearly allied to *P. setulosus*, mihi.

CROTALARIA.

C. *epunctata*; suffruticosa diffusa e basi ramosa, ramis teretibus filiformibus basi nudis apice ramosis cum racemis foliisque subtus adpresse strigoso-pubescentibus, stipulis minutis patentibus vel interdum nullis, foliis epunctatis lineari-oblongis supra glabris vel junioribus sparse strigosis, racemis terminalibus 4-10-floris, bracteis linearibus patentibus pedicellos breves æquantibus, bracteolis minutis, calyce corolla dimidio breviore strigoso, labio superiore profunde bifido inferiore 3-fido, segmentis subulatis, legumine cernuo glabro transverse reticulato oblongo apice latiore calyce dimidio longiore, poly-(20-)spermo.

C. *vimineæ* (Graham) valde affinis; differt foliis epunctatis, calyce subæqualiter 5-fido. Crescit ubique in Concano australiòre.

GLYCINE.

G. *Warreensis*; staminibus diadelphis, foliolis ovato-oblongis submembranaceis supra *glabris* subtus strigosis pallidis foliolis lateralibus inæquilateris, racemis compositis folio 2-3-plo longioribus multifloris, floribus *approximatis*, legumine transverse venoso pilis albis adpressis strigoso 6-spermo, seminibus distantibus.

Calyx nervoso-striatus, basi bracteolis 2 ovatis acuminatis suffultus, tubularis, corollam subæquans, vix ad medium 4-*fidus*, segmentis *omnibus integris* acuminatis, supremo majore.—Crescit in regno *Warreensi*; fl. temp. frigido.

I might have supposed this species to be *G. mollis* of W. and A., but it would appear from the observations of these authors that they had never seen a *Glycine* with diadelphous stamens, the entire upper lip also of my plant distinguishes it from *G. mollis*. I have also another species from Ceylon, with diadelphous stamens.

Alyssicarpus.

A. *parviflorus*; herbaceus erectus, caule ramoso basi glabro apice patenti-piloso, foliis simplicibus et trifoliolatis oblongo-ellipticis mucronatis basi subcordatis supra glabris subtus adpresse strigosis, stipulis petiolis brevioribus, calyce fere quinquefido, segmentis subulatis ciliatis obscure nervosis, legumine semimoniliformi reticulato subglabro 5–6-spermo calycem duplo excedente.

Folia simplicia et foliola terminalia cum petiolo semipollicari 2 poll. longa, 7–9 lin. lata; foliola lateralia multo minora. *Calyx* 2 lin. longus. *Bracteæ* ovatæ, longe cuspidato-acuminatæ. *Pedicelli* patenti-pilosi. *Stylus* longus, pubescens, infra apicem incrassatus, ibique glaber.—Crescit in jugo Syhadrensi, ad Phonda-ghât; fl. et fr. Nov.

This is the second trifoliolate species discovered in this Presidency since the publication of W. and A.'s Prodromus, in which all the species described have simple leaves only. The first trifoliolate species is the *A. Belgaumensis*, Wight, Ic. 92, discovered by my friend, Mr. J. S. Law, from which the present is readily distinguished by its smaller size and more slender habit. The stem of *A. Belgaumensis* is covered with white adpressed hairs, this with spreading fulvous hairs; the calyx in the former is quadrifid, the segments being lanceolate, acuminate, and very strongly nerved, and, along with the flowers and legumes, are four times larger than in the present species.

Nat. Ord. GENTIANEÆ.
Ophelia.

O. *pauciflora*, n. sp.; caule erecto tetraptero glabro apice tantum ramoso, ramis fastigiatis, foliis sessilibus lanceolatis acuminatis 3-nerviis, cymis fastigiatis *paucifloris*, calycis segmentis subulatis *corollam æquantibus*, corollæ 4-partitæ *albæ* segmentis obovato-ellipticis mucronulatis, foveis *magnis* orbiculatis solitariis squamula apice fimbriata tectis fimbriarumque brevium serie cinctis, filamentis linearibus.

Corolla 4 lin. longa.   *Foliorum* basis *lata*, rotundato-truncata.—Crescit in jugo Syhadrensi; fl. Sept.

This species seems most nearly allied to *O. Grisebachiana* of Wight's 'Icones,' n. 1330, but differs from that species in having few flowers; in the shape of the leaves, which in that species are attenuated towards the base; also in the very different shape of the segments of the corolla, which in *O. Grisebachiana* are acuminated.

Nat. Ord. UMBELLIFERÆ.
PIMPINELLA.

P. *monoica*, n. sp.; caule elato tereti glabro lævi stricto parte inferiore simplici apice ramoso, ramis ramulisque alternatim bifariamque dispositis, foliis inferioribus longe petiolatis pinnatim trifoliolatis, foliolis longe petiolulatis cordato-lanceolatis minute cartilagineo-serratis basi 5-nerviis pubescentibus, foliis superioribus multifidis laciniis filiformibus vel ad vaginas redactis, petiolis vaginantibus margine ciliatis, involucro nullo v. monophyllo, involucellis oligo-(1–4)-phyllis, umbellis ramos terminantibus fœmineis fructiferis, lateralibus (scilicet in ramulis ultimis) masculis, fructu juniore granulis pellucidis obsito.—Crescit in jugo Syhadrensi; fl. Nov.

This is the tallest of the three species found in this Presidency, being seven to eight feet high, with a stem as thick as a swan's quill. It is exclusively confined to the higher ranges of the ghauts. It is evidently closely allied to *P. Candolleana* and *Javana*, from which the trifoliolate leaves sufficiently distinguish it. The terminal umbels are destitute of stamens, while the lateral ones are equally so of styles. Our other species are *P. involucrata*, W. and A., and *P. adscendens* (mihi), confined to the banks of rivers in the low country.

(*To be continued.*)

Letter from Dr. ANDREW SINCLAIR *on the Vegetation, &c., of the neighbourhood of Auckland, New Zealand.*[*]

Auckland, New Zealand, Dec. 16, 1850.

As I have sent you specimens of the plants I have collected near

[*] The valuable information contained in this letter will be perused with the more interest at the present time, when the first number of Dr. Hooker's 'Flora of New Zealand' is on the point of appearing.—ED.

Auckland, and the scenery is therefore interesting, and access to it may not be impracticable to any one prepared with tents for following up the observations.

The isthmus of rice fields extends across from the Manukau forest, on the western shore of the harbour, to the easterly direction fourteen miles to the Tamaki river, and is in breadth from five to ten miles. It is crossed in two parts by Manukau basin, and in the north of the Waitemata and Shouraki Gulf. [illegible] ...ses in general pretty abrupt from the waters, but from two hundred feet, and the coast line is broken up, particularly on northern side, by gullies and green bays, which at the [illegible] ... mer produce a water communication merely across the [illegible] ... in such place a portage of less than a mile in breadth.

The isthmus may be considered as that only of a portion of it the south extends away to the Manukau to the west of the Waikato River, and on the north nearly as far as the containing much land generally clear of wood or fern. This great part is at a much lower level than the [illegible] north of it, and consequently towards it [illegible] and the Waikato come from the south [illegible] flow from the north.

The surface is varied by gentle [illegible] by several conical hills from 300 [illegible] craters of extinct volcanoes. [illegible] size, and the rain which falls [illegible] which generally terminate [illegible] part of New Zealand, has [illegible] subject to long continued [illegible] eruptions take place now and [illegible] quake is felt. In some places [illegible] ral springs, abundance of [illegible] canic productions, and [illegible] met with in the clay and [illegible] place of their growth [illegible] conclusion that [illegible] elevation.

The geological formation [illegible] volcanic, overlaid [illegible] sandstone. The rocks of [illegible] columnar and always more or less cellular. In the ravines and on the coast, where the loose friable soil crumbles away rapidly, the strata are generally horizontal, but sometimes they are very tortuous. Such deviations from a straight line would at first sight appear to have been caused by volcanic action; but the strata being uninterrupted and seldom attenuated at the flexure, which must have been the case had they been bent to their present position, I think there can be little doubt that this phenomenon has been occasioned by the beds having been deposited on an uneven surface in still water, such as may be supposed to be now going on in the muddy creeks along the coast.

The ranges of hills in the Manukan forest, which bound the plain on the west, and the headlands, and in the islands on the opposite coast, at the entrance of the Thames, the composition of the rocks is nearly the same, being chiefly conglomerate with greenstone, and other primitive rocks in a few places. Several metallic veins have been discovered, and two copper mines have been opened; one on the Great Barrier Island, and the other in the island of Kuwau, where mining operations are carried on at present on a large scale.

At the time the site of Auckland was fixed on for the capital of the colony, there were almost no inhabitants in the neighbourhood. Independent of what is known, however, of the history of the locality, the remains of pahs, and marks of cultivation chiefly on the scoria land, and the heaps of pepe-shells everywhere, show that the country was, at no distant period, highly cultivated and thickly peopled. It is now twenty years since Shungee, on his return from Europe, armed his followers with the fire-arms he brought with him and could procure from whalers, and laid waste the country, destroying the inhabitants wherever he went. The escarpments on the slopes of the volcanic hills, and the ditches and ramparts that enclosed the pahs on the headlands along the shores, still remain as proofs of these places having been fortified, and the great quantities of human bones found in the caves and among the rocks attest the fate of the defenders.

The general aspect of the country, covered with dingy fern (*Pteris esculenta*), the tea-tree (*Leptospermum scoparium*), relieved only here and there by flax bushes (*Phormium tenax*), and the tufts of foliage on the stems of the cabbage-tree (*Cordyline stricta*), is not very encouraging to a botanist. In going over the country, however, he will often come suddenly on the edge of a steep ravine, and be agreeably surprised to find its sides covered with most luxuriant vegetation, and

[illegible]

The most [illegible] … [illegible] … beyond [illegible] must have grown [illegible] … quantity of Kauri timber [illegible] … harbour [illegible] the heads of the [illegible] … and a great portion of [illegible] … on the banks of the [illegible] … Royal Navy can be found. Though the [illegible] … gone on since the establishment of the colony [illegible] been made on the forest, and in places where surveyors have been at work for years, their labours have rarely extended a gunshot from their houses. Besides the Kauri, the other trees felled are chiefly the Pohutukana and Rata (*Metrosideros tomentosa* and *robusta*), the Puriri (*Vitex littoralis*), for ship-timbers and other purposes requiring great durability and strength. The number of Kauri-trees must be diminishing, for in many places, where the felling of timber has been carried on, there are no young trees rising up to supply the place of others decayed from age, or cut for removal; but that is not the case in the part of the Manukau forest nearest Auckland, where the young trees of all sizes are very numerous.

Though the Kauri does not grow to such a large size in the Manukau forest as in others farther north, vegetation is exceedingly vigorous, and it presents an inexhaustible field of interest to the botanist. The trunks of the old trees are clothed and festooned with *Astelias*, climbing *Metrosideros*, *Orchidaceæ*, *Ferns*, *Mosses*, and *Jungermanniæ*, in the greatest profusion. The deep hollows within the forest are penetrated with difficulty, from the interlaced stems of the *Ripogonum* and other under-shrubs. In the deep sheltered parts of the forest, some plants are found of extraordinary size, and amongst them I have measured the *Areca sapida* thirty-six feet high, and the *Cyathea dealbata* attaining a height of fifty-four feet. It is along the margin of the forest, however, and up the abrupt winding ravines, and at the sawing stations, where the falling of lofty trees brings down masses of vegetation generally beyond reach, that botanizing is pursued with most success.

With respect to the progress of the colony and the prospects of the settlers who have made it their home, I think the information I can give you will be satisfactory. Ten years have passed since the Government was established, and the European population at the present time must be not less than 25,000. Near one-third of this number are in Auckland, or living in villages or places almost within sight of it. In all parts of the colony, I believe the frugal, sober, and industrious settlers have done exceedingly well. Privations, which have been so generally felt in the early years of the settlement of other colonies, have been unknown here. The number of those who came hither without or with little means, and have already realized a competency, is very great. The large capitalists, of which there have been few, have not done comparatively so well, although some even of them have made small fortunes. It has been objected to the colony that there is no staple article of export; but it takes some time for a colony to overflow in production, and it is not easy to say in what direction that will take place. When British enterprise and capital are engaged, and with such active settlers as have come to this country, there is no reason to fear that it will be behind other colonies in due time. We are going on increasing our exports rapidly, and decreasing the importation of articles producible in the country. The native flax is gradually rising, as was expected long ago, into importance. In Auckland, we have the largest rope-walk on this side of the tropics, from which, on several occasions, above twenty tons of rope have been exported at a time. The raw material is now prepared of a superior quality and in greater quantity than before, and the demand for both it and the manufactured form is greater than can be supplied.

The settlers have gone on as their means would allow in the importation of cattle and horses, till they have obtained a pretty good stock. Sheep are now being imported in large number, and excellent artificial pastures are spreading over the country, which will make them a profitable investment, and furnish, in the wool, another article of export.

The experience we have had of the climate proves its excellence. There is no malaria in the country, and contagious diseases are almost unknown. Pectoral complaints, as shown by the military returns, are much fewer and less fatal than at any other station we know of, either at home or abroad.

I must now close this hurried sketch of the country. I have collected the plants I have sent you lately and send you now. There are still many which I have not been able yet to obtain, as I am obliged to pursue botany by snatches only. As I find more, I will transmit them; and I shall try to send you both seeds and living plants, which I have been disappointed in not having done long before the present time.

BOTANICAL INFORMATION.

A letter from Dr. DE VRIESE *to* ROBERT BROWN, Esq., *on a new species of* RAFFLESIA *in the Island of Java, discovered by* MM. J. E. Teysman *and* S. Binnendijk.

Leyden, May 16, 1851.

The recollection of our botanical discussions during your last year's visit to Leyden, induces me to send the following particulars respecting the discovery of a new *Rafflesia*, which grows in the Dutch East Indian Colonies. We owe the detection of this interesting plant to MM. Teysman, the head gardener, and Binnendijk, the second gardener, at Bintenzorg.

The new species, *Rafflesia Rochussenii*, so called after the late Governor-General of Java, J. V. Rochussen, is, like its congeners, a parasite. It grows on a *Cissus*, the *C. serrulatus* of Roxburgh, and was found at the foot of the Manellawangi, a mountain-chain, extending from the Panzzerango to Salak, and above the line of coffee-plantations named Pondok Tjatting. The exact locality is to the west of the tea-plantations which belong to Count Van den Bosch, and which form part of his estate, called Tjawi.

Several specimens were dug up, and transplanted, together with the *Cissus* to which they were attached, into the Royal Gardens, where the *Rafflesia* developed itself and produced its flowers. The following account will show the difference between this new species and those already described. And first comes

Rafflesia Arnoldi, Br. This species is the most beautiful and largest of all. It was detected in 1848 by Dr. Arnold, in Sumatra, and living plants of it are shortly expected in the Botanical Gardens of Bintenzorg. The central column is furnished with from forty to sixty irregular processes, which are divided at the points (see Transactions of the Linnean Society, vol. xiii. t. 20) and covered with hairs (vol. xix. t. 22). The interior surface of the perianth is moreover beset with thick glandular hairs. We may judge of the difference of size, for the bud of *R. Arnoldi* measures about twelve inches and a half in diameter, and that of *R. Patma*, Bl., thirteen inches and a half.*

The *R. Patma*, Bl., is finely figured in Professor Blume's 'Flora Javæ.' Living plants of it may be seen in the Gardens of Bintenzorg, where they have twice flowered. The first blossom expanded in March, and the second in October, 1850. The colour of these flowers differed considerably from those represented in the 'Flora Javæ;' for the processes were pale, nearly flesh-coloured, while the annulus and lobes, on the first day of expansion, were of a light rose-colour: the warts, moreover, which exist on these parts, are not so regular in size. The margin of the perianth appears, in the delineation, to be elevated above

---

* The buds of *R. Patma*, as figured in Blume's 'Flora Javæ,' Tab. I., measure seven inches and a half in diameter.—ED.

the lobes of the flower; but in the living specimen it springs from the place where the lobes of the flower are united. In the growing plant, the processes of the central column were thirty-six in number, viz., four in the middle, ten in the second series, and twenty-two in the outermost row. The elevated margin of the disc of the central column hardly attained the height of half an inch. This species is immediately distinguishable from the one now described as new, and inhabits a different species of *Cissus*.

No description exists of *R. Horsfieldi*, Br., which is admitted to be a species of *Rafflesia*; the only notice of it is in the Linn. Trans. vol. xix. p. 242, " stylis indefinite numerosis."

Again, *R. Cumingi*, Br., bears on the surface of its central column ten to thirteen processes, one of which is central, while the rest are regularly ranged around it. The interior perianth is covered with differently-formed warts.

Of this novel species of *Rafflesia* (*R. Rochussenii*) the following description is given by MM. Teysman and Binnendijk :—

"Dioica; antheris 15-19 serie simplici disci inferum marginem cingentibus; columna 15-16 sulcis descendentibus antheris oppositis; disco subpatellæformi glabro stellato, vel processibus 1-2 tecto, polline rotundato hyalino. *Hab.* Manellawangi, costam Pondok Tjatting dictam, in sylvis umbrosis."

Diameter of expanded flower	0·145*
Circumference ,, ,,	0.430
Breadth of lobes	0·08 to 0·09
Length ,,	0·06
Breadth of exterior annulus perianthii	0·34
,, interior ,, ,,	0·155
Diameter of neck	0·05¼
Disc of central column from annulus perianthii	0·02
Breadth of column	0·70

The bud, about to expand, was far smaller than that of *R. Patma*, and of a somewhat different hue; but the surprise of the discoverers was much excited by finding that, when the flower opened, the disc was seen destitute of processes. In the centre alone rose a small needle-like point, an eighth of an inch long. The most beautiful

---

* These measurements are possibly in millimetres, and are given as in the original; the proportions of the different parts of the flower are unintelligible to us.—ED.

part of the disc was a red five-pointed star. The ground of the disc was dirty white, and the divisions of the star (0·025 long) were directed towards the fissures of the perianth. The margin of the disc, which was somewhat elevated, was darker in colour than the star. The inner surface of the perianth presented a striking dissimilarity to that of *R. Patma*; for whereas in the latter species it is smooth and even, in this it is entirely covered with long stalks, bearing warts from one to six lines long and one line thick, cylindrical, the longest of them being situated at the base of the perianth, and the shortest in the upper part. The hue of the expanded flower was dark red, with the warts of the same colour, and much smaller than in *R. Patma*. It departs from *R. Cumingi* in the processes, and other peculiarities already indicated. The flower seen by these gentlemen was a male, and no trace of a pericarp was discernible.

I trust, ere long, to send you a full description of this interesting discovery, for I am in expectation of receiving from Java a specimen, growing on the *Cissus*, which M. Teysman has kindly promised to transmit to the Royal Garden of Leyden.

Extract from a Letter of the Rev. W. COLENSO, *relating to a second species of* "NEW ZEALAND FLAX," PHORMIUM.

I have been both gratified and amused with the remarks of M. Auguste Le Jolis, in the 7th volume of your 'London Journal of Botany,' p. 533, &c., "on a new kind of *Phormium*." His is a very plain and true statement, as far as the description, &c., of his *Phormium* goes; but, without doubt, it is the very identical plant which I first mentioned to you in my letter of July 20, 1841 (an extract from which you published in the Lond. Journ. Bot., vol. i. p. 305), and which I subsequently showed, growing and flowering in my garden, to your son, Dr. Joseph Hooker, in the spring of the same year, and which I brought thither, a small plant, from the east coast some considerable time previous. In my long journal-like letter to you, dated Sept. 1, 1842 (and which you published in the Lond. Journ. Bot. vol. iii.), I spoke of this "new species" as "*P. Fosterianum*" (vide p. 8); and, in a subsequent letter to you, dated Dec. 1, 1842, I again refer to it. And, in the more elaborate account of that ramble (subsequently published in the 'Tasmanian Journal,' vol. ii. p. 219), I also allude to it by its then published name, "*P. Fosterianum*," adding (in a note), "I intend, at some future day, giving a descriptive account of this very elegant and useful, and very distinct species of *Phormium*." True, I have not yet done this; but, inasmuch as its name has been long ago published in the two hemispheres, and that, too, with no lack of living botanical testimony to its identity, I think enough has been done to secure its *first* published specific name. There is, however, a slight error; and that is in *your* letter to M. Le Jolis (as printed in the Lond. Journ. Bot. vol. vii. p. 535), in which you speak of a "small *red*-flowered kind." Now, this species under consideration (*P. Fosterianum*, alias *Colensoi*, alias *Cookianum*) is not wholly *red*-flowered, but much as M. Le Jolis describes it, reddish without and green within. And it should not be forgotten, that the flowers of this plant darken considerably in drying, especially the lighter-coloured parts of the perianth; and your having only imperfect and dried specimens to examine, while M. Le Jolis had living ones, sufficiently accounts for your calling it a "*red*-flowered species." Further; the fact is that both the *Phormiums* are highly sportive in the *colour* of their perianths, in the *size* of every part, and in the *shape* and size of their pods, scarcely any two of the latter being alike. Some are sharp-, others obtuse-angled; some very acute at the apex, others quite as obtuse; some *twisted*, others plain, and *that on the same plant*. Again, if a pod is allowed to remain until it is fully ripe—*to become dry*—before it be gathered, it will generally be membranaceous; if, however, it is plucked before that period of dryness, it will be more or less thick and tough. And further, the whole flower (especially its perianth and styles) changes its colour considerably after impregnation, always becoming darker; so that, on a large full-flowered scape, flowers may be found of various hues.

There is, however, a *large* variety of the Common (or first-known) *Phormium* (for both species are equally common), which almost invariably bears *red* flowers. This is the *largest* kind in New Zealand, and is, on account of its large size and little use, universally rejected by the natives, who state its fibre to be less in quantity, as well as inferior in quality, to that of the other kinds; while its bulky *parenchyma* and superabundant gumminess make its manipulation a heavy work. This variety is found on the brink of rich alluvial ever-wet swamps, in deep rich soil on the margins of rivers, and, not unfrequently, near native villages, and on the edges of woods; and, also, on dry and barren sand-hills near the sea, as well as on certain sandy shores a little above high-water mark. I once thought that this variety might prove to be a distinct species—and it is, it must be confessed, in certain spots a constant variety,—but I have seen it gradually losing its bulk and length, and red flowers, and descending (*gradatim*) to the common yellowish-flowered state of that species; so that I have long ago set it down as being merely a variety. This variety is very common all about the Bay of Islands, and, in fact, to be everywhere met with throughout the whole of this North Island. At present there are several varieties known to me, but only two species. One of those varieties is a very curious one, every leaf being highly and differently variegated in alternate stripes of white and green, running from the apex to the base of the leaf. This kind grew well in my garden at Paihia, Bay of Islands (as it does in my garden here); but, although I have met with several plants in different parts of the island, I have never yet found one in flower. Nor have I seen a single plant growing quite wild: they always occur where they have been, at some time or other, planted by the natives.

The principal points of difference,—points, I mean, which may be obvious to the most unscientific eye at first sight,—between the two present known species, are, the reflexed, subacute, and mostly greenish perianth, and the narrower, and strictly tapering, almost pungent leaf of the second species :—this latter character never fails; the smallest plant at all seasons may be correctly distinguished by it. Whereas the leaves of the first species (*P. tenax*) are *never* strictly tapering, *never* acuminate; they (their apices) always present, so to speak, a segment of a circle, a kind of pointed arch. For, the carinated back of the leaf abruptly inclining towards its edges at about two inches from the apex, gives the top of the *young unopened* leaf seen in profile the appearance of the bow of a canoe. Hence it is that a mature leaf of *P. tenax*, unbroken, unslit at the apex, I have never yet seen; on the contrary, fullgrown leaves of the second species, with unsevered apices, may be obtained from every plant.

I could offer further remarks upon this peculiarly New Zealand genus; but time presses. I will merely add, that I have often attempted to secure good specimens of both species for you, which some years ago I promised to do, and have as often failed. What with the size of the plant, its stout woody scapes, perianths overflowing with honey, and harsh and gummy leaves, which invariably roll up while drying, it is a most unpleasing subject to handle, and as ugly when dry. But I will have another trial yet.

Characters of a New Genus of COMPOSITÆ-EUPATORIACEÆ, *with remarks on some other Genera of the same Tribe;* by ASA GRAY.

(TAB. V.)

DISSOTHRIX, nov. gen.

Capitulum 6–9-florum, homogamum. *Involucrum* biseriale, floribus æquilongum; squamis laxis mox patentibus, lanceolatis, cuspidato-acuminatis, membranaceis, bicostatis, margine scariosis, exterioribus brevioribus. *Receptaculum* parvum, nudum. *Corolla* anguste cylindrica, apice breviter 5-dentata, dentibus erectis. *Achænium* breve, acute pentagonum, angulis hispidulis, callo basilari majusculo. *Pappus* plurisetosus, biformis, nempe, setis 5 rigidis ad angulos achænii respondentibus corollam superantibus, cæteris (30–40) capillaribus brevioribus valde inæquilongis, omnibus sursum hirtello-scaberrimis. —Herba Brasiliana annua, glabriuscula; caule erecto striato superne paniculato laxe polycephalo; foliis membranaceis ovato-lanceolatis petiolatis dentatis, inferioribus oppositis, summis alternis. Corollæ apice purpurascentes.

D. *Gardneri* (Tab. V.).—*Stevia imbricata*, Gardner, in Lond. Journ. Bot. vol. v. p. 458 (n. 1744 and n. 2211).

This plant is certainly to be distinguished from *Stevia*, as well by the truly setose pappus, of which even the five stronger rays can scarcely be termed aristæ, as by the minutely five-toothed and connivent limb of the corolla, and the biserial, imbricated, and lax or spreading involucre. The thin and smooth scales of the latter are strongly two-nerved (in the manner of *Brickellia*, &c.), and have also three very slender intermediate nerves, often scarcely visible except by transmitted light. When the achænia fall the scales become reflexed.

Judging evidently from the description alone, Gardner compared this plant with the *Stevia calycina* of De Candolle, which is doubtless the *Eupatorium?* *calyculatum*, Hook. and Arn. Comp. Bot. Mag. vol. ii. p. 242, and therefore the *Disynaphia Montevidensis*, DC. Prodr. vol. vii. p. 267, to which *Dissothrix* bears no particular resemblance.

Respecting *Disynaphia*, I may remark that the setæ of its pappus are rather *barbellulate* than scabrous. Also that true congeners of *D. Montevidensis* are, *Eupatorium spathulatum*, Hook. and Arn. Comp. Bot. Mag. l. c., which is the same as *E. halimifolium*, DC. (named *E. passerinioides* by De Candolle in the Parisian herbarium),

E. ligulæfolium, Hook. and Arn. l. c., *E. gnidioides*, DC., and *E. ericoides*, DC. It is not probable, however, that *Disynaphia* can be admitted except as a section of *Eupatorium*, some genuine species of which have the pappus nearly or quite as biserial; and in many others it is barbellulate or barbellate-denticulate.

Among the species with a barbellate and strictly uniserial pappus which have been referred to *Eupatorium*, are the ambiguous *Eupatorium decipiens* and *E. paradoxum*, Hook. and Arn. l. c.; two closely allied species, although the former has nine- to twelve-flowered, the latter five-flowered capitula. The latter is *Nothites baccharidea* of De Candolle (who was not acquainted with any of the original species of this genus); the former is *Eupatorium foliosum*, DC., and also doubtless, as Mr. Bentham has suggested to me, the *Ophryosporus triangularis* of Meyen.

Another groupe of extra-tropical South American *Eupatoria* with a barbellate pappus, the setæ of which more or less evidently occupy two series, and are about as rigid as in *Disynaphia*, comprises *Eupatorium elongatum*, *E. ceratophyllum*, *E. tanacetifolium* (*E. subplumosum*, Don, ined.), and *E. lanigerum*, of Hooker and Arnott, l. c., with an apparently unpublished species allied to the last (of which all, except the first, have more or less dissected or lobed leaves). These make so near an approach to *Trichogonia*, Gard. (*Kuhnia § Trichogonia*, DC.), that there is only the difference between a barbellate and a plumose pappus to separate them.

On the other hand, *Trichogonia* is a perfectly good genus as distinguished from *Kuhnia*, which has a terete and multistriate achænium. It must needs include Gardner's n. 1723, referred by him to *Eupatorium adenanthum*, DC., I think incorrectly, as the corollas are not glanduliferous (and only rarely sprinkled with a few resinous atoms), the leaves are cordate, and the pappus is so manifestly plumose, that De Candolle could not have overlooked it.*

---

* TRICHOGONIA *Gardneri* (n. sp.): herbacea, glanduloso-pubescens; caule erecto ramoso; foliis cordatis acuminatis dentatis membranaceis triplinerviis in petiolum longiusculum alatum contractis; capitulis laxe subcorymbosis multifloris; involucri squamis acuminatis; corollis glabris; achæniis glabris ad angulos tenuiter hispidulis; pappi setis longiuscule plumosis.—The heads and flowers are twice as large as those of *T. salviæfolia* and *T. menthæfolia*, and the pappus plumose with (rather sparse) longer hairs. It will be noticed that Gardner has cited the same number (4839) to *Trichogonia salviæfolia* and to *Isocarpha eupatorioides*. Planchon has remarked, in Herb. Hook., that the two are states of the same plant, in some specimens of which either all the flowers or some of them are destitute of pappus!

Brickellia, Ell. (including *Bulbostylis*, DC.), is in like manner distinguished by its terete and striate or 10-costate achænia: in no other way can the line be drawn between it and *Eupatorium*. This excludes I believe all of Gardner's Brazilian *Bulbostyles*; also De Candolle's *B. glabra, B. spinaciæfolia, B.? pauciflora,* and *B. triangularis* (the latter, perhaps, a *Carphephorus*). *B. annua*, Nutt. Pl. Gambell., is not even an Eupatoriaceous plant. I know, however, about twenty-six genuine, chiefly Mexican and New Mexican, species, several of which are yet unpublished;* besides three or four species of *Clavigera*, DC., which must evidently be reduced to a section of *Brickellia*; the more or less barbellate pappus not affording an adequate or sufficiently well-marked generic character.†

TAB. V. *Dissothrix Gardneri* :—*natural size*. Fig. 1. A capitulum enlarged. 2. An involucral scale. 3. A flower. 4. Corolla laid open. 5. A stamen. 6. Achænium and pappus. 7. One of the larger and one of the smaller setæ of the pappus :—*all the dissections more or less magnified.*

Contributions to the Botany of WESTERN INDIA ;
by N. A. DALZELL, Esq., M.A.

(*Continued from p.* 212.)

Nat. Ord. CYRTANDRACEÆ.

DIDYMOCARPUS.

D. *cristata*; caule herbaceo 8–9-pollicari simplici erecto tereti carnoso, foliis oppositis petiolatis amplis late cordato-ovatis obtusis utrinque parce pilosis, inflorescentia in axillis oppositis insidente et cum petiolis concreta cristata pilosa folio breviore, sc. pedicellis numerosis seriatim sursum spectantibus basi inter se in pedunculum brevem

---

* The following are in Wright's Texano-New-Mexican collection, viz., no. 249, *B.* (*Clavigera*, Gray, Pl. Fendl.) *brachyphylla*; n. 250, *B. cylindracea*, Gray, Pl. Lindh.; n. 251, *B. reniformis*, n. sp.; n. 252, *B. baccharidea*, n. sp.; n. 253, *B. laciniata*, n. sp.—N. 2104, coll. Galeotti=*B. Galeottii*, n. sp.; n. 293, Coulter, Calif. coll.=*B. Coulteri*, n. sp.; n. 598, Pl. Hartweg.=*B. Hartwegi* (*Eupatorium rigidum*, Benth.).

† *Leucopodum campestre*, Gardner, in Lond. Journ. Bot. vol. iv. p. 125 (n. 5788) is *Chevreuilia acuminata*, Less., from which *C. filiformis*, Hook. and Arn., does not appear to be distinct.

crassum connatis, calyce 5-partito, laciniis linearibus acutis inæqualibus.

Folia profunde cordata, subrotundo-ovata, lobis basilaribus rotundatis, 4–5 poll. longa, 4 poll. lata, obscure penninervia. *Calyx* 2–3 lin. longus. *Corolla* alba, 5–6 lin. longa. *Stamina* 2. *Ovarium* dense strigoso-pilosum. *Capsula* siliquæformis, 1½–2 poll. longa, sesquilineam lata, curvata, pubescens. *Semina* nuda, oblonga, utrinque acuta, longitudinaliter 5-angulata.—Crescit in rupibus, prope Parwarghât; fl. Sept. et Oct.

The crested disposition of the flowers, seated on the petiole, which this species has in common with *D. crinita*, Jack, renders it a remarkable plant. The leaves are comparatively very large, and have the texture of those of a cabbage.

Nat. Ord. COMMELYNEÆ.

CYANOTIS.

C. *vivipara*, n. sp.; epiphyta acaulis tota pilis rufis patulis conspersa, foliis radicalibus lineari-ensiformibus planis crassis carnosis fasciculatis, scapis ex radice orientibus filiformibus radicantibus viviparis, pedunculis e scapi nodis solitariis alternis umbellam bi-trifloram bractea duplici suffultam gerentibus, bracteis ad basin pedunculi foliaceis parvis oblongis acutis vaginantibus, capsulæ valvis post dehiscentiam valde recurvis, loculis 2-spermis, seminibus cylindricis.— Crescit in arboribus in jugo Syhadrensi; fl. Aug. et Sept.

This is a plant of a very peculiar habit, probably resembling *C. tuberosa*. The tuft of radical leaves, which are two to four inches long, are liliaceous in appearance, and from below these spring several long slender scapes, each bearing a young plant on its apex. The ordinary leaves observable in the genus, *i. e.*, with tubular sheaths, are here reduced to the size of bracts, subtending each umbelliferous peduncle. Flower and fruit as in the genus.

Nat. Ord. ACANTHACEÆ.

LEPIDAGATHIS.

L. *mitis*; caule diffuso ramoso, ramis trichotomis glabris purpureis subtetragonis, foliis lineari-oblongis acutis sessilibus basin versus angustatis utrinque glabris margine setulis minutis ciliatis, spicis circa radicem in globum diametro bi-tripollicarem glomeratis, bracteis e basi lata acuminatis, bracteolis calycisque laciniis linearibus acutis omnibus *muticis* pellucidis subcartilagineis basi glabris apicem versus sericeo-villosis.

Folia 18-20 lin. longa, 3 lin. lata. *Bracteæ* dense imbricatæ, bracteolæque semipollicares. *Calycis laciniæ* majores 5¼ lin. longæ. *Corolla* extus pubescens, alba, intus roseo luteoque punctata, 6-7 lin. longa.—Crescit in rupibus in jugo Syhadrensi, prope Phonda-ghât; fl. Nov.

The habit of this plant is entirely similar to that of *L. cristata*, Willd., *L. rupestris*, N. ab E., and *L. lutea*, mihi, but is readily distinguished from all three by the bracts and calyces being unfurnished with a pungent point, and being quite harmless to the touch.

BARLERIA.

B. *elata*, n. sp.; fruticosa, caule tereti strigoso ad nodos tumido, foliis herbaceis inæqualibus longe petiolatis ellipticis acuminatis subito in petiolum attenuatis utrinque pubescentibus, spicis terminalibus et in axillis supremis solitariis brevibus patentibus robustis sesquipollicaribus bi-trifloris, floribus secundis brevissime pedicellatis, pedicellis bracteis lanceolatis foliaceis calycem æquantibus suffultis.

Calycis laciniæ majores (altera inferior sesquipollicaris, altera 1¼ poll. longa) basi subcuneatæ, apice longe acuminatæ, extus densissime strigosæ, obscure palmatim 17-nerviæ. *Corolla* 3-3¾ poll. longa : limbus læte cæruleus; tubus rubro-purpureus; laciniæ obtusæ, superior cæteris brevior. *Stamina* antheris perfectis 4, quorum duo multo breviora, cum rudimento quinti. *Flores* sursum spectantes. *Folia* cum petiolo sesquipollicari 6-8 poll. longa, 2-3 poll. lata.— Crescit in jugo Syhadrensi, prope Phonda-ghât; fl. Nov.—Quoad inflorescentiam *B. dichotomæ*, Roxb. proxima.

Next to *B. grandiflora*, mihi, this is the largest-flowered species which I have met with, in a district rich in *Barlerias* : it rises to the height of six feet, and when in flower has a very showy appearance. *B. dichotoma*, I may remark, is not herbaceous, as stated by N. ab E.: it is a true shrub, at least when cultivated—the only state in which it is to be seen on this side of India.

Nat. Ord. COMBRETACEÆ.

TERMINALIA.

T. *Gella*; foliis sparsis amplis late ovalibus utrinque puberulis, junioribus pilis fulvis lanato-tomentosis, spicis axillaribus solitariis tomentosis folio brevioribus, floribus inferioribus fertilibus breve pedicellatis, superioribus masculis sessilibus, omnibus intus dense lanatis, calycis segmentis triangularibus acutis sub anthesi revolutis, drupa sphærica sericea diametro 9–10-lin.—Crescit raro in Concano australiore; fl. Apr.; fr. Jan.

A large and handsome tree, with the leaves nine inches in length and five in breadth, finely reticulated, and the petioles destitute of glands. The fruit, when procurable in any quantity, is an article of commerce, and has the same qualities as that of *T. Bellirica*, being used in dyeing cloth and also medicinally by the natives. The flowers have the same disagreeable smell as those of most species of this genus. In Graham's 'Catalogue of Bombay Plants' this tree is entered as *T. nitida* (Roxb. in Herb. Lamb.), but it cannot be that species, as the leaves do not taper to both ends, nor is their apex acuminated.

Nat. Ord. EUPHORBIACEÆ.

Tribe PHYLLANTHEÆ.

ANOMOSPERMUM, genus novum.

GEN. CHAR. *Flores* monoici. *Calyx* 5-phyllus. *Corollæ petala* 5 calyce multoties minora, glandulis minutis dentatis utrinque. *Stamina* 5: *filamentis* inferne in ovarii stipitem disco plano orbiculato impositum connatis, superne liberis patentibus, *antheris* demum extrorsis. FŒM. *Ovarium* disco orbiculato impositum, triloculare, loculis biovulatis. *Styli* 3, semibifidi, breves, obtusi. *Fructus capsularis*, depresse globosus, tricoccus, coccis dispermis. *Semina exalbuminosa.*—*Arbores* in India orientali indigenæ, foliis alternis stipulatis integerrimis coriaceis glabris lævibus, *venis costalibus inter se arcus anastomosantes formantibus*, floribus axillaribus fasciculatis, masculis et fœmineis mixtis.

A. *excelsum*; foliis ellipticis utrinque acutis nitidis coriaceis glabris cum petiolo subpollicari 6–7 poll. longis 2½–3 poll. latis, stipulis e basi lata acuminatis squamæformibus glabris, floribus longiuscule pedicellatis, pedicellis fructiferis cernuis suprapollicaribus, petalis obovatis integris, sepalis rotundatis, capsula glabra diametro pollicari.—Crescit in jugo Syhadrensi; fl. Aug.; fr. Nov.

That the species of *Cluytia*, figured in Roxb. Cor. pl. 169–173, are the representatives of two very distinct genera, there cannot be the least doubt; this, too, must have been Willdenow's belief, when he separated three of them under the name of *Bridelia*; and as *Cluytia* is exclusively an African genus, the rest actually to this day remain without a name. Endlicher has included all Roxburgh's *Cluytias* under *Bridelia*, which is certainly an error. The true *Bridelias* have baccate, indehiscent, two-celled fruit, seeds with an osseous testa, and an embryo within albumen, with flat foliaceous cotyledons. The texture and venation of the leaves is also peculiar, the costal veins being numerous, prominent, parallel, and reaching the margin of the leaf undivided. The tree under consideration, which is closely allied to *Cluytia collina* and *patula* of Roxb., has leaves like a *Croton*, the veins hidden within the substance of the leaf, and forming anastomosing arcs at a good distance from the margin; besides, the capsular, trilocular, dehiscent fruit, seeds without albumen (singular to relate), and folded fleshy cotyledons, remove it far from *Bridelia*. *Cluytia collina* and *patula*, Roxb., have not been discovered on this side of India, and therefore I have no opportunity of examining the seeds of those species; but it is highly probable that they are also exalbuminous, and, if so, they will be referred to the genus now described, as well as the other three *Cluytias* described in the 'Flora Indica.'

ROTTLERA.

R. *urandra*; ramulis glabris, foliis anguste oblongis obtuse acuminatis, basi in petiolum subpollicarem cuneatim attenuatis calloso-serrulatis coriaceis nitidis, floribus masculis racemosis glabris, racemis axillaribus solitariis folio brevioribus, staminibus in receptaculo nudo liberis connectivo ultra antheram *caudatim producto*, floribus fœmineis axillaribus solitariis longissime pedunculatis, pedunculis 3-pollicaribus nudis glabris, stylis 2 profunde bifidis ramis longis filiformibus velutinis, capsulis bilocularibus glabris.

Folia cum petiolo subpollicari 5–6 poll. longa, $1\frac{1}{2}$–2 lata. *Racemi* graciles, simplices, 3–$3\frac{1}{4}$ poll. longi; *pedicelli* 2–3-ni, bractea squamæformi suffulti.—Crescit in jugo Syhadrensi, prope Phondaghât; fl. et fr. Nov.

This is well distinguished by the unusually long peduncles of the solitary female flowers. In habit allied to *R. aureo-punctata*, mihi, the leaves being also somewhat like those of that species.

EUPHORBIA.

E. *strobilifera*; floribus in ramis ramulisque terminalibus racemosis bracteis imbricatis cordato-ovatis obliquis mucronulatis scariosis palmatinerviis reticulatis suffultis, bracteis geminis oppositis flore longioribus, racemis 1-2-pollicaribus, capsula pubescente.

Herba 2-3-pedalis, glabra. *Caulis* erectus, teres, basi nudus, apicem versus dichotome ramosus, ramulis apice floriferis. *Folia* nunquam vidi; an aphylla? *Squamæ* 4 petaloideæ basi glandula præditæ, obliquæ, subcuneatæ, apice eroso-dentatæ. *Bractearum* reticulatio singularis, ramulis ultimis abrupte desinentibus liberis.—Crescit in rupibus nudis in regno Warreensi; fl. Feb.

Nat. Ord. BEGONIACEÆ.
BEGONIA.

1. B. *integrifolia*; rhizomate tuberoso, caule herbaceo, foliis oblique ovatis obtusis integris basi cordatis utrinque pilis albis hispidis subtus sanguineis, pedunculis dichotomis, floribus parvis, capsulis glabris trialatis, alarum una cæteris multo majore.

Tuber pisi majoris magnitudine, caule vix crassior. *Caulis* 6-8 poll. altus. *Folia* radicalia nulla, caulina alterna, breve (1-poll.) petiolata, majora 7 poll. longa, 4 poll. lata, valde obliqua, inæquilatera, atro-virentia cum zona concentrica pallida. *Flores* albi, diametro 5-lin.—Crescit ad pedem jugi Syhadrensis, in rupibus; fl. temp. pluviali.

2. B. *trichocarpa*; rhizomate tuberoso, caule herbaceo sesquipedali, foliis amplis subinæqualiter cordatis acuminatis sinuato-dentatis rugosis margine minute laceratis supra et in nervis subtus hispidis utrinque viridibus, floribus terminalibus umbellatis, pedicellis hispidis pollicaribus, capsula hispida alis 3 æqualibus obtusis.

Tuber parvum, sphæricum. *Folium* radicale unicum, longe petiolatum, 7 poll. longum, 7 poll. latum, *petiolo* crasso, carnoso, colorato, 12-14 poll. longo; folia cætera caulina minora. *Caulis* basi nudus. *Flores* albi, 4-7-ni, diametro bipollicares; perigonium fœmineum pentaphyllum; foliola exteriora 2, rotundata, palmatinervia, margine basin versus setaceo-lacerata, dorso hispida.—Crescit in rupibus, cum præcedente.

Nat. Ord. BALSAMINEÆ.
IMPATIENS.

I. *ramosissima*; herbacea erecta dichotomo-ramosa, ramis plurimis glabris acute quadrangularibus, foliis oppositis sessilibus ellipticis pollicaribus supra hispidulo-scabridis subtus glabris pallidis minute et distanter spinuloso-serratis, pedicellis in axillis oppositis solitariis unifloris villosis, floriferis folio brevioribus, fructiferis deflexis folium æquantibus, sepalis lateralibus linearibus villosis anterioris longitudine, posteriore orbiculari apice acutiusculo dorso carinato villoso, anteriore longo naviculari extus villoso, calcare brevissimo obtuso incurvo, petalorum lobo posteriore minimo, anteriore cuneato longe unguiculato, capsula oblonga glabra utrinque attenuata oligosperma, seminibus nigris nitidis.—Crescit in montibus Syhadree, prope Phonda-ghât; fl. Sept.

This species seems nearly allied to *I. tomentosa* and *rufescens*. From the former it is readily distinguished by the very different form of the anterior sepal, which in the present species is cymbiform, being four times longer than broad; and from the latter by the presence of a spur. The flowers are of the size of *I. Kleinii*.

Nat. Ord. COMPOSITÆ.

ADENOSTEMMA.

A. *rivale*; caule erecto simplici *tereti* glabro bipedali, foliis *lineari-lanceolatis* basi longe attenuatis serrato-dentatis glabris, panicula corymbosa puberula laxa oligocephala, involucri squamis linearibus vel lineari-spathulatis obtusis herbaceis subæqualibus subglabris, achæniis glanduloso-tuberculatis.—Crescit in rivorum marginibus umbrosis utriusque Concani.

This is the *Ageratum aquaticum* of Roxb., which I have inserted here under its proper genus, as it is not to be found in De Candolle's 'Prodromus.' The achænia are exactly like those of *A. leiocarpum*. Steudel is in error in his synonym of Roxburgh's plant, such as it is described in his 'Flora Indica,' where he says, "stem *round*" (as I find it); there must have been some mistake about the plant found in Roxburgh's herbarium, under the name of *A. aquaticum*, as recorded by De Candolle.

DECANEURUM (Gymnanthemum).

D. *microcephalum*; caule pubescenti-scabro ramoso, foliis petiolatis ellipticis acuminatis basi in petiolum longe attenuatis remote aristato-serratis supra pubescentibus subtus dense albo-tomentosis, capitulis parvis in apice ramorum solitariis, involucri ovati squamis squarrosis dorso albo-tomentosis, exterioribus lanceolatis acuminatis aristato-cuspidatis margine ciliatis, *achænio* basi vix attenuato pallido glabro nitido subcylindrico *ecostato*.

Flores 12–14. *Pappus* valde caducus.—Crescit ad Parwar-ghât; fl. Nov.

This plant forms a remarkable exception to all the published species of this well-marked unmistakeable genus, by its unribbed seeds; but although it belies its generic name of *Decaneurum*, it would be doing violence to nature to remove, on this account, a plant agreeing so completely in every other respect with the character and habit of the genus. When fresh gathered, it has a strong odour of *Chamomile*, which it loses by drying.

Nat. Ord. ARTOCARPEÆ.

ANTIARIS.

A. *saccidora*, Dalz.—*Lepurandra saccidora*, Nimmo, Plants of Bombay, p. 193.

Crescit in regno Warreensi, ad pedem jugi Syhadrensis; fructum maturum habet Januario.

This is a gigantic tree, with a trunk eighteen feet in circumference at the base. On wounding the fruit, a milky viscid fluid exudes in considerable quantity, which shortly hardens into the appearance and consistence of bees' wax, but eventually becomes black and shining. The inner bark of the tree is composed of very strong tenacious fibres, and seems excellently adapted for the manufacture of cordage and mattings. Mr. Nimmo speaks of this tree as being discovered by Dr. Lush at Kandalla in 1837, where it grows in deep ravines, and he adds, "It is common in the jungles near Coorg, where the people manufacture very curious sacks, and by a most simple process, which will be hardly credited in Europe. A branch is cut, corresponding to the length and breadth of the sack wanted. It is soaked a little and then beaten with clubs until the liber separates from the wood. This done, the sack formed of the bark is turned inside out until the wood is sawed off, with the exception of a small piece left at the bottom of the sack, and which is carefully left untouched. These sacks are in general use among the villagers for carrying rice, and are sold for about six annas each. Some of them have been sent to England as curiosities[*]

---

[*] Two of these curious sacks, sent by Mr. Nimmo, are deposited in the Museum of the Royal Botanic Gardens of Kew.—ED.

by Mr. P. Ewart. The nuts are intensely bitter, and contain an azotized principle, which may prove an active medical agent."—The natives call the tree "*Juzoogry*," others "*Kurwut*," but this last name is applied to many kinds of trees with scabrous leaves.

(*To be continued.*)

Sketch of the VEGETATION *of the Isthmus of* PANAMA; by M. BER-THOLD SEEMANN, Naturalist of H. M. S. Herald.

That part of the Republic of New Granada which connects the two great Continents of America, and is called the Isthmus of Panamà or Darien, lies between the fourth and tenth parallels of north latitude, and the seventy-seventh and eighty-third of west longitude. It comprises the provinces of Veraguas and Panamà, and the territories of Darien and Bocas del Toro; it is bounded on the N. and N.E. by the Atlantic, on the S. and S.W. by the Pacific Ocean, on the E. by the rivers Atrato and San Juan, and on the W. by Mosquitia and the State of Costarica; and it presents a surface, including the adjacent islands, of 34,700 square miles, an extent of territory nearly equal to that of Portugal.

Its coasts are fringed with islands. The largest, on the Atlantic side, are the Escudo de Veraguas and those situated in the Lagoon of Chiriqui: while various others, of smaller size, generally known to voyagers by the name of Cayos, are scattered along the shores, and form, as is the case with the Sambaloes, regular chains: they are, however, but thinly inhabited, and little frequented. Of greater importance, and more populous, are the islands of the Pacific. Several groupes, viz., Secos, Paredez, Ladrones, and Contreras, lie off the coast of Veraguas; another cluster, of which Coyba, Gobernadora, and Cebaco are the largest, exists near the Bay of Montijo, while a little archipelago, the Pearl Islands, also known by their synonyms of Islas del Rey, Islas del Istmo, and Islas de Colombia, valuable from the number of pearls which are annually collected there, is situated at the entrance of the Bay of Panamà. Smaller, but scarcely less important, is the groupe in the vicinity of Panamà, consisting of Taboga, Taboguilla, Otoque, Flaminco, and Perico, islands which are highly cultivated, and whose beauty has designated them emphatically the Garden of Panamà.

The Isthmus is not distinguished for the height of its mountains. The mighty chain of the Andes, after traversing the whole continent of

South America, gradually decreases when approaching this narrow neck of land, and in the province of Panamà is hardly recognizable in a ridge of low hills, seldom exceeding 1,000 feet in height. A new series of mountains seems to commence at Punta de Chame, which visibly attains a greater elevation on entering the province of Veraguas, and in the volcano of Chiriqui presents the most elevated part of the Isthmus, a peak 7,000 feet in height. This ridge is thickly covered with forests, and chiefly confined to the centre and northern parts of the country; the districts on the coast of the Pacific, especially the cantons of Natà, Santiago, and Alanje, abound in grassy plains, or savanas, of great extent, which, by affording pasture to numerous herds of cattle, constitute the principal riches of the country.

Volcanoes, all now extinct, rise in different parts. The highest is the volcano of Chiriqui, already mentioned: another, of considerable elevation, about 3,000 feet, called the Jananò, is seen at Cape Corrientes in Darien; and several others are reported to exist in Veraguas: even "the island of Taboga," says Mr. E. Hopkins, "appears to have been a portion of a former volcanic crater." But, although destitute of active volcanoes, the Isthmus by no means therefore enjoys an immunity from earthquakes. Some pretty severe shocks are now and then experienced, especially during the dry season, from January until May. They consist of undulating movements, coming from the west, and apparently having their origin in Central America; but they do not seem to exercise any baneful influence on the vegetation, as is frequently the case in Peru, where, after a severe shock, whole fields have been known to wither.

The geographical position of the Isthmus, the almost entire absence of elevated mountains, and the vast extent of the forests and other uncultivated parts, tend to produce a hot and rainy climate, which, nevertheless, is, with the exception of a few localities, healthy, and more favourable to the constitution of the Caucasian race than that of most tropical countries. The seasons are regularly distributed into wet and dry. The rains are expected with the new moon in April, and they last eight months, till the end of December: they are prolonged, however, in the south of Darien, and in some localities on the Atlantic Ocean, to ten, and even to eleven months. In the beginning, the rains are very slight, but they gradually increase, and are fully established towards the end of May, when they fall in torrents sometimes for days in succession, and are accompanied by thunder and lightning of the most terrific description. The air is loaded with moisture and fogs; calms and light variable winds prevail. The temperature does not vary more than from 75° to 87° Fahr., but still, the perspiration being impeded, the atmosphere feels exceedingly hot and close. Towards the end of December, the violent rains diminish in frequency, the clouds begin to disperse, and with the commencement of the new year the N.W. wind sets in. An immediate change follows. The air becomes pure and refreshing, the sun brilliant, the sky blue and serene, hardly a cloud is to be seen, and the climate exhibits all its tropical beauties. The heat, although much greater, ranging between 75° and 94° Fahr., is less felt, because the atmosphere is almost free from moisture. The rays of the sun, however, bear very great power: and the rising of the thermometer to 124° Fahr., when it is at noon exposed to their influence, is no uncommon phenomenon. These observations, however, only refer to the lower regions; on the higher mountains the climate is, of course, subject to various modifications, and better adapted, on account of its lower temperature, to an European constitution.

A country so much visited by heavy rains, naturally abounds in rivers. Not counting the smaller streams and ravine torrents, their number cannot fall short of 200. Of those emptying themselves into the Pacific, the San Juan, Churchunque, Bayano, Rio grande de Natà, Santamaria, Tavasarà, and Chiriqui are the largest: among those flowing into the Atlantic Ocean are the Chagres, Belen, Veraguas, and the nine-mouthed Atrato. Most of them have one or more deltas, which in many instances assume the appearance of islands. Their vegetation is a curious mixture of littoral and inland plants, and it often presents species of the higher mountains, by which the remote sources of the river may be traced.

As the Isthmus connects the continents, so does its vegetation combine the floras of tropical North and South America: the virgin forests of Guayana, the Vegetable Soory groves of the Magdalena, and the oak-woods of the Mexican highlands, are all equally represented. It is, therefore, not to be expected that the Panamian flora should exhibit any very striking character, or be distinguished by the presence of strongly delineated forms, like Mexico by its *Cactuses*, Australia by its *Epacrideæ*, capsular *Myrtaceæ*, and phyllodineous *Acacias*, or the Cape of Good Hope by its *Heaths, succulent Aloes, Stapelias*, and *Mesem-*

bryanthemums.. The want of such forms is so obvious, that a superficial observer would be induced to declare the flora identical with those of the bordering states: a person, however, who investigates more closely, cannot fail to notice the prevailing clothing of the leaves with hair and tomentum, the abundance of greenish, yellow, and white flowers, and the numerical superiority of the Natural Orders *Leguminosæ, Melastomaceæ, Compositæ, Rubiaceæ, Orchideæ,* and *Ferns*,—features, which, although less prominent than those alluded to, still exercise a decided influence on the physiognomy of the vegetation.*

But it must not be supposed that the flora is without certain peculiarities of its own; indeed, it has peculiarities which distinguish it from that of all other countries, and which are calculated to show many a genus and many a natural order in an entirely new light. The most important, perhaps, that might be adduced, is the *Balboa odorata*, Seem., whose discovery has established the union of *Passifloreæ* and *Turneraceæ*, embracing, as it does, the chief characteristics of these two families (n. 1922). The genus *Pentagonia* is equally curious, on account of its being the only *Rubiacea* which has yet been found with pinnatifid leaves; it belongs to the subdivision of *Gardenieæ*, and thus forms a clear transition to the order of *Lonicereæ*, in which a pinnatifid foliage and a baccate fruit are not uncommon features. Remarkable are two species of *Begonia*, *B. oppositifolia*, Seem. (n. 1099), and *B. centradenioides*, Seem. (n. 561), both with leaves which are opposite and of unequal size, as is the case with *Centradenia rosea*, Lindl., *Clidemia cyanocarpa*, Bth., *C. fenestrata*, Bth., *C. barbinervis*, Bth., and numerous other *Melastomaceæ*. Their similarity in habit to some of the *Melastomaceæ* is really very striking, and will give additional weight to the arguments of those who favour the relationship between the two orders. The *Carludovica palmata*, R. et Pav., is another production of the Isthmus, though not exclusively confined to it, which deserves notice. It has large fan-shaped leaves, and resembles many of the Palms so closely that it must always be con-

* The vegetation of Guayana is, probably, most like that of the Isthmus; for not only do the common plants of that country occur, but also a great number of those more recently discovered by Sir R. Schomburgk. However, at present, this must be mere conjecture: after the materials, collected by me, shall have been more carefully examined, it will be a task of little difficulty to show to what flora the Panamian approximates the nearest; and in the meantime to furnish a more defined general character than that which I am at present able to give.—*B. S.*

sidered as constituting an important link between *Pandaneæ* and *Palms*. No less surprising is the occurrence of the genera *Macleania*, *Sphyrospermum*, and *Cypripedium* in the low coast region, in a temperature far exceeding that in which any of their species have hitherto been known to exist.

The aspect of the flora is much more diversified than the uniformity of the climate and surface of the country would lead us to expect. The sea-coast and those parts influenced by the tides and the immediate evaporation of the ocean, produce a quite peculiar vegetation, among which the *Mangroves*, chiefly composed of different *Rhizophoras* and *Avicennias*, are the most remarkable features. Growing in all muddy places, down to the very verge of the ocean, they form impenetrable thickets, the abode of alligators, sand-flies, and mosquitoes, and the places, from whose exhalation of putrid miasmata, sickness spreads over the adjacent districts. To destroy these dreaded swamps is almost impossible: the *Avicennias*, with their asparagus-like rhizomata, send up innumerable young shoots whenever the main stem is disturbed; the *Rhizophoras* extend in all directions their long aerial roots, which soon reach the ground, and preserve the tree from falling, after its terrestrial roots have lifted it high above its original level. At Panamá, where the tide rises to the height of twenty-two feet, these trees are frequently under water, apparently without being in the least injured or checked in their growth. Rivers, as far as subjected to the influence of the ebb and flow, are full of Mangroves; and the highest *Rhizophoras*, which, growing always on that side where there is the deepest water, assist the natives in conducting their canoes through the mud and sand-banks. Among many other littoral plants will be recognized the *Prosopsis horrida*, Kunth, *Crescentia obovata*, Bth., *Cereus Pitajaya*, De Cand., *Hippomane Mancinella*, Linn., *Acrostichum aureum*, Linn., *Ipomœa pes-capræ*, Swartz, *Hibiscus arboreus*, Desv., *Cocos nucifera*, Linn., *Pithecolobium macrostachyum*, Bth., *Guilandina Bonduc*, Ait., and several species of *Jacquinia*, *Plumieria*, and *Ruyschia*.

Far different is the vegetation of the savanas. The ground, being level or slightly undulated, is clothed during the greater part of the year with a turf of brilliant green; groupes of trees rise here and there; silvery streams, herds of cattle, and the isolated huts of the natives enliven a scene, over which the absence of *Palms* and *Tree-ferns* throws almost an European character, giving the whole more the appearance of an English park than that of a tract of land in tropical America. The turf contains, besides numerous species of Grasses, many elegant *Papilionaceæ*, *Polygaleæ*, and *Gentianeæ*: the tender sensitive-plant (*Mimosa pudica*, Linn.) prevails in many localities, shutting up its irritable leaves even upon the approach of a heavy footstep. The clumps of trees and shrubs, over which the Garumos (*Cecropia peltata*, Linn.) and *Pava-trees* (*Panax*, sp. nov.) wave their large foliage, are composed of *Myrtaceæ*, *Melastomaceæ*, *Chrysobalaneæ*, *Papilionaceæ*, *Compositæ*, *Anacardiaceæ*, *Malpighiaceæ*, *Hederaceæ*, *Anonàceæ*, *Dilleniaceæ*, and *Acanthaceæ*, and often overspread by masses of *Convolvulaceæ*, *Aristolochieæ*, *Apocyneæ*, and other creeping or winding plants. *Orchideæ* are very plentiful in the vicinity of the rivers and rivulets, where the trees are literally loaded with them. The *Vainilla* (*Vanilla*, sp.) is seen climbing in great abundance up the stems of young trees, and increasing in many instances so much in weight that it causes the downfall of its supporter. The *Chumicales*, or groves of *Sand-paper trees* (*Curatella Americana*, Linn.), form occasionally curious features in the landscape. They extend over whole districts, and their presence always indicates a soil strongly impregnated with iron. The trees are about forty feet high, with crooked branches, an approximation to the winding habit of their tribe, and the paper-like leaves occasion a rattling noise if stirred by the wind, which strongly recalls the European autumn, when northerly breezes strip the trees of their foliage.

Dense forests of immense extent cover at least two-thirds of the whole territory. From the height of their trees, and the numerous creepers overhanging them, the rays of the sun are almost interrupted; and, in all forests of this description, little underwood exists. The principal forms belong to *Sterculiaceæ*, *Tiliaceæ*, *Mimoseæ*, *Papilionaceæ*, *Euphorbiaceæ*, *Anacardieæ*, *Acanthaceæ*, *Piperaceæ*, *Rubiaceæ*, *Myrtaceæ*, and *Melastomaceæ*; these, and the prevalence of *Tree-ferns*, *Scitamineæ*, and *Palms*, stamp on them the real tropical character. The giants of the woods are the *Espavè* (*Anacardium Rhinocarpus*, DC.), *Corotù* (*Entherolobium Simbouva*, Mart.), and the *Cuipo* (a *Sterculiacea*), which attain a height averaging from 90 to 130 feet, and a circumference of from twenty-four to thirty-six feet. Occasionally a whole forest consists only of a single species of tree. Near Juan Lanas is a locality entirely clad with *Membrillos* (*Gustavia Membrillo*, Seem.): near the village of San Juan another with *Palos de velas* (*Parmentiera cereifera*, Seem.); while, in different parts of the country, may be seen vast groves solely composed of either *Maquenque*, *Palma real*, or *Palma de escoba*.

Mountains, exceeding 1,500 feet in elevation, situated principally in western Veraguas, possess a flora which resembles, in many respects, that of some parts of the Mexican highlands. *Alders* and *Blackberries* (*Rubus*, sp.) are common; evergreen oaks are intermingled with palms; and the genera *Styrax*, *Rondeletia*, *Salvia*, *Lopezia*, *Fuchsia*, *Centradenia*, *Ageratum*, *Conostegia*, *Lupinus*, *Hypericum*, *Freziera*, *Galium*, *Euphorbia*, *Rhopala*, *Equisetum*, *Tropæolum*, *Adiantum*, *Begonia*, *Clematis*, *Verbena*, *Inga*, *Solanum*, *Kellettia* (n. 1593), &c., are represented by one or more species; in fine, it is that vegetation in which the forms of the torrid are blended with those of the temperate region.

Such is a rough outline of the chief characters and general aspect of the Panamian flora. It is at present not my intention to go more into detail, but, in order to afford a still further insight into the vegetation of a country so little known, I will occupy the following paper with a short account of its principal productions, enumerating them according to their uses and properties.

(*To be continued.*)

Copy of a Letter addressed by Mr. Spruce *to* G. Bentham, Esq., *dated Santarem, Rio das Amazonas, Sept.* 10, 1850.

My dear Sir,—I have now another box of specimens to send you, and I hope it will be the last from Santarem; for, although I have no reason to complain of my collections during the last month, I am anxious to reach a new field, where I should undoubtedly find more novelty. The excursions are now long and painful, for I can hardly see a tree on the neighbouring campos, or in the forests, within a reasonable distance, whose flower and fruit I do not possess. I lose no opportunity of examining the coasts of the rivers and igarapés when I can get a boat and men. Boats I can at any time have at my disposal, but the "gente" to man them are difficult to catch. Should I need men but for a day, and ask them of the Capitao dos Trabalhadores, I must wait for a fortnight ere I can get them, for in all probability a detachment of soldiers has to be sent into the interior, to beat them up at their sitios. This delay is so annoying, that I prefer the uncertain chance of the loan of three or four men from the crew of any vessel which may be laid up in the port—a circumstance of rare occurrence.

I have quite had my edge taken off for ascending the Tapajoz; a merchant of Cuyabá, with whom I had some thoughts of going up in February, was wrecked by the unskilfulness of his pilot in ascending the first cataracts, and lost two of his men, besides all his cargo, valued at 8,000 milreis. Besides this, fever and ague have been fatal on this river, to an extent previously unknown; the number of deaths is variously stated at from 400 to 1,000, but the least of these numbers is a serious diminution of the scanty population. Even the coming of the dry season has not dispelled this formidable scourge. Two Frenchmen, who have just returned from a voyage of three weeks, found fever everywhere rife. They ventured to sleep on shore but once, in the house of the only English settler on the Tapajoz. He himself and all his household were ill, and he had already buried six of his people. We have accounts that the Madera and Rio Negro have had similar visitations. My countryman, Mr. Bradley, who is settled at the Barra, assures me that the true cause of the gradual depopulation of the Rio Negro is the now well-ascertained unhealthiness of its banks: he himself caught an ague up it, which held him six months. The cause assigned here is some supposed insalubrious property of the waters, but I cannot doubt that Humboldt's opinion as to the healthiness of the rivers of tropical America which run east and west, and the unhealthiness of those whose course lies north and south, is the true one. In the valley of the Amazon, for instance, we have the trade-wind blowing upwards nearly every day, while on the rivers above-mentioned the winds are variable and uncertain. Even here we have sometimes, at new and full moon, a day or two of what is called "vento da cima," or "wind from above;" and it is justly esteemed a "vento roim," for it brings with it colds in the head, toothache, and fever.

The specimens now sent are chiefly of plants of the "Gapó" (as it is called in *lingua géral*), or lands inundated by the rivers and lakes in winter, constituting a breadth of from twenty yards to several miles, according as the land is abruptly ascending or perfectly flat. I have got several more of the minute quasi-ephemeral plants, which spring up as the water recedes. The shores of the large rivers produce scarcely any of these; their waters beat on the sand with too much violence to allow of such frail things existing there; but by the small inland lakes connected with the Tapajoz, and near the creeks at the mouth of some of the igarapés, minute leafless *Utricularias, Eriocaulons, Alismas*, &c., cover the white sand in thousands. A *Utricularia*, which you will find under No. 1050 (*U. uniflora*, MS.), is surely the simplest in its structure of all its family. Stems of the size of an ordinary sewing-needle, fixed into the sand by a small cone of radicles, without leaves, but with a minute tubular 2-lipped bract a little below the flower, which is white and comparatively large, complete the description of its outward aspect. I have often been struck with the wonderful contrast in size which is presented here in both the animal and vegetable world. Under a gigantic *Castanheira*, or a *Caryocar*, may occasionally be seen an almost microscopic *Cyperacea*; and the same lake which produces this fairy *Utricularia*, bears on its bosom the queen of the waters, *Victoria regia*. Another *Utricularia* (Coll. no. 1053, *U. quinqueradiata*, MS.) has a peculiarity of structure to me quite novel, though you may have met with it before. It is a small species, with submerged stems and bladdery leaves, but the pedicels, which are about two inches long, have about midway a large horizontal involucre of five rays, resembling the spokes of a wheel. This floats on the water, and supports the upper part of the pedicel in an erect position; the whole recalling a sort of floating lamp I have seen, especially as the large yellow flower may be considered to represent the flame. The rays are half an inch long, clavate, not hollow, but composed of about six series of large diaphanous cellules. The cellules are convex on the surface, giving the rays a papillose appearance, hexagonal, pale green, with pink interstices. The rays are trifid at the extremity; segments short, twice dichotomous, the last divisions capillary, rarely sacciferous.

I said above that the Amazon and Tapajoz have scarcely any of these fugitive plants; but on the shore of the former, a little below Santarem, I have been delighted to meet with a couple of *Podostemaceæ*, growing on scattered stones. This is almost the only place where as yet I have seen stones by the Amazon; but as I ascend higher, I hope to meet more frequently with localities suitable for this interesting tribe of plants. The few igarapés which run into the rivers near Santarem have all either gravelly or muddy bottoms: it is true that decayed logs are frequent in them, but I have in vain searched these for *Podostemaceæ*.

The grasses which were so numerous in the months of February and March, are now withered up, and scarcely a grass is to be met with; but in their stead I have obtained the flowers of several fine trees, which will probably be more valuable in your eyes. I have now got most of the plants on the volcanic serras south of Santarem, but the scattered trees and shrubs on these hills comprise, when more closely examined, but very few species. The most abundant are a small Lythraceous tree (a *Lafoensia* ?) and a handsome *Vochysiacea*, probably a *Salvertia*, with white hexapetalous flowers. The last exhales a most delicious odour, and walking through a grove of these *Salvertias* reminds me of traversing large beds of Lily of the Valley, only the perfume is stronger: in drying they assume the still richer odour of the Violet.

Nearly south-west from Santarem, and communicating with the Tapajoz by a short channel, is a small lake called Maracauá-miré. In the height of the dry season it was a walk of an hour and a half to reach this lake, by the broad beach of the Tapajoz; but now, when the mouth of an intervening igarapé is swollen to half a mile in width, this is impracticable, and it is necessary to cross the igarapé about two miles up, and then penetrate the mato extending along its banks, to a campo, which stretches beyond to the shores of the lake. We first undertook to reach it by this route on the 15th of August, and we were not very certain of the way. We crossed the igarapé, and then attempted to pierce the mato; but the track by which we entered the latter ceased after we had followed it for a time, and we had then to cut our way through *Sipós* and *Pindoba Palms*, in the direction of the campo, steering by compass. While thus progressing slowly and with difficulty, I heard a distant roar, very much like that of an onça; but as I knew there had been several cattle on the opposite side of the igarapé, I was willing to suppose it might be one of them. Shortly afterwards the same sound was repeated, and a little nearer; and in a few minutes more it was repeated, so loud and so near, that it brought us both to a stand-still. Mr. King had heard the two former growls, but, like myself, he had not spoken. We were armed only with trésados, and had barely arranged our plan of defence, when we heard a tremendous crash among the underwood at thirty paces from us, which I confess made the blood run cold in my veins. After this, however, we heard no more. When we afterwards recounted the adventure to some Indians, they told us that the crash we had heard was undoubtedly the tiger, either springing on some deer, of which he had been in chase, or, arriving in sight of us, and doubting his capacity to overcome us, betaking himself to flight.

Rarely are tigers seen so near Santarem, yet a few years ago an engagement took place between three men and a tiger, in the very same valley. One of these men was armed with a musket, another with a trésado, and the third (a tall powerful man) was quite unarmed. It was upon the last that the tiger made his first attack, springing upon him out of a bush, and he had fortunately sufficient activity and presence of mind to seize the tiger by the fore-paws, one of which he secured by the wrist, and the other lower down, and consequently less firmly. They struggled until the tiger released this paw, and, making a claw with it at the man's crown, tore his scalp completely over his eyes. He is now living at Santarem, and constantly wears a black skull-cap, his head being still very tender. At the moment of the attack the man with the musket was some distance in the rear, but the one with the trésado flew to his friend's assistance, and the tiger, leaving the latter, turned on his new assailant, whom, also, he succeeded in wounding severely. He then sat down midway between them, eyeing first one and then the other, and looking, I dare say, as amiably as a cat might be supposed to do between two disabled mice, uncertain which to devour first. At this critical conjuncture the third man came up, and the contest was renewed, resulting finally in the death of the tiger, but not until he had wounded all his assailants.

In this strip of forest were numerous trees of the *Melastomacea* I have sent you under No. 819, its slender trunks rising to the height of fifty feet before branching. I might have passed it unnoticed, but that the ground beneath was strewed with its ripe fruits, which more resemble the Apple in flavour than any other fruit I have met with on the Amazon: they are, however, more insipid, and the flesh less firm. In shape and in the way they grow along the naked branches, they are not unlike the *Medinilla macrocarpa* figured in Lindley's 'Vegetable Kingdom,' but the 12-celled ovary and the 12-lobed stigma are characters I have not met with in any other of the Order.

We found the shores of the lake rich in the little plants above-mentioned, and I was glad to find a second species of *Mayaca*, with small white flowers, the petals barely equalling the sepals, and short pedicels reflexed in fruit; while in the one formerly sent (no. 375), the rose-coloured petals are twice the length of the sepals, and the long pedicels are erect in fruit: there are also slight differences in the leaves. I have since found them growing together abundantly near the Indian village of *Mayaca* (or Mahica). You will have noticed that the anthers of *Mayaca* terminate in a tube one-third the length and diameter of the anther itself. In these two species the filament is short and tolerably stout. It has not been noticed that the calyx has a very short tube adherent to the ovary; its segments are truly valvate. The parts of the flower have assuredly the normal position: viz. the stamens opposite the sepals, and the carpels opposite the petals.—The habit of *Mayaca* is certainly different from that of the *Commelynaceæ*; when out of flower its resemblance to *Polytrichum juniperinum* is most striking; yet its characters hardly suffice to sustain it in a separate Order. I have already sent you a true *Commelynacea*, with anthers opening at the apex.

What I am about to relate will illustrate the utility of the large marsh grass, called *Canna-rana* (wild cane), of which I have already sent you specimens. As beef is only half the price at Santarem that it is at Pará, a considerable traffic is kept up in cattle all the year round. A large portion come from the fazendas on the Furo de Sapucuá (near the mouth of the Trombétas), and the rest from the vicinity of Obidos, Almeirim, and Santarem. No supply of dry food is taken for their support during the voyage to Pará, but in leaving Santarem a quantity of *Canna-rana* is cut from the Ponta Negra, sufficient to last for two days, and all the way down a supply can be procured on the banks when required, until reaching the entrance to the channels below Gurupá, where there is no more *Canna-rana*, and the cattle subsist on the fronds of the Assai Palm until arriving at the city. The only objection to this latter food is that it causes the cattle to drink an inordinate quantity of water. So long as there is a sufficient supply of the *Canna-rana* in the marshes near Santarem, the cattle thrive well; but when they are reduced to eat other succulent grasses, they are constantly attacked with diarrhœa, and the cattle-keepers have not got into the way of giving them occasionally a little dry food.

As I mentioned lately to Sir W. Hooker, *Canna-rana* is a chief constituent of the floating islands on the Amazon. Those we have seen passing Santarem in such numbers come mostly from a great distance. A few masses have been detached from the Ponta Negra, but fortunately they floated by without coming in contact with any vessels in the port. I am told that in 1836, the year following the rebellion in Pará, five sloops of war were sent from Pará to receive the submission of the various towns on the river, and that whilst lying at anchor in the port of Santarem, an Ilha de Capim, of some acres in extent, was one night detached from the Ponta Negra, and coming full upon these vessels, tore them all from their anchorage, and carried them bodily down the river a distance of half a day's journey. A strong body of soldiers, blacks, and Indians, amounting to some hundreds, were despatched to liberate the ships, and it cost them several hours' labour with axes and trésados to effect it, for the mass was some yards in thickness. They found in it, and killed, several large snakes (boas and sucurijus), and even some peixe-boys (cow-fishes).

These formidable floating islands of grass seem to be liberated in three ways: first, when an island or a point of the mainland is detached by the force of the waters, as happened last year with a small island a little above Santarem, in the main channel of the Amazon. This island had on it trees of considerable size, a house, and a fazenda of cattle, yet it was carried away in a mass by the furious stream, in the height of the "eucheute." Secondly, the earth may be gradually washed away from the roots of a flat of Capim, until, having no longer anything to retain it in its place, the loosened mass is detached from the shore, and floats down the stream; or, thirdly, the lower part of the stem may actually be decayed, as is the case with most stems I have drawn up out of deep water, and thus have so slight a hold on the bottom, as to be readily dislodged by the swelling stream; and as the stems are much entangled in one another, it is only in masses they can be liberated. So far as my observations have extended, the Ilhas de Capim on the Amazon chiefly owe their origin to the concurrence of these two latter causes.

I must not close my letter without adverting to a melancholy circumstance that occurred here on the 7th of last month—no less than an attempt to take the life of our excellent friend, Captain Hislop. He had been robbed a short time before of 470 milreis, by a black girl, the slave of a woman of not very good character, and by the aid of the police recovered 280 milreis of the same. Probably in revenge for this, and in the expectation of acquiring more plunder, a man one night concealed himself in Mr. Hislop's sleeping-room (no very difficult task), and in the dead of night stabbed him in the breast, as he lay in the hammock, and then attempted to make off with a large trunk, but falling with it over a chair in the dark, Mr. Hislop's servants were aroused and ran to his assistance. The scoundrel had, however, taken the precaution of setting open the doors before commencing operations, and when the servants reached the spot he had effected his escape. The knife he had used was found on the floor; it seemed to have been made on purpose for the deed, of a piece of iron hoop, beaten into the shape of a poinard, and sharpened only at the point. Although in possession of this, all the efforts of the police to discover its owner have been unavailing. The wound had been intended "to serve;" but the knife had pierced the breast-bone without reaching any vital organ. It is now quite healed, to external appearance; but Mr. Hislop, who is some seventy years old, though up to this period hale and hearty, has received a shock, from the effects of which he will never recover.

October 7th.—Since last date we have been waiting in almost daily expectation of starting for the Barra, and I have consequently been able to leave home very little. On the shores of a small bay of the Tapajoz, I found, a few days ago, a diminutive *Isoetes*, much resembling the British *I. lacustris*. If I may trust to Endlicher, this is the first *Isoetes* that has been found in South America. Near it grew the smallest *Scrophulariacea* I ever met with; its stems creeping and buried in the sand, sending up short branches, with one or two pairs of minute fleshy leaves, and a comparatively large purple flower. In character it comes near *Herpestes*. I have added another *Podostemacea* to my collection, gathered on stones by the Amazon. This is a foliaceous species, and had I found it without flowers, I should probably have taken it for a *Jungermannia*. It is singular that the shores of the Tapajoz produce no *Podostemaceæ*, though more stony than the Amazon, and even in some parts rocky; but the rocks and stones, instead of bearing plants, are constantly encrusted by some *Polypidon*.

I have no description of the Gutta Percha tree, but from what I recollect of a figure I have seen, the *Sapotacea* now sent under No. 296 should come very near it. This tree is not unfrequent in sandy and gravelly situations near Santarem. Its milk is sweet and wholesome, and rapidly becomes hard; but it is not poured out in such quantities as in the *Masseranduba*, or Milk-tree. Several *Apocyneæ* here give out abundance of milk, which, from its tenacity and hardness when dry, I suppose may one day come to be used like Gutta Percha.

My specimens of the *Leguminosæ*, nos. 920 and 960, were gathered from young trees, and I now find that these two species are amongst the largest trees of the "Gapó." I lately found on the leaves of no. 960, what I at first took for a new *Rafflesiacea*, and the resemblance to some *Apodanthus* is indeed most striking: it was only by careful examination that I satisfied myself it was really produced by an insect. The perianth (for such it seems) is green in the earliest stage, changing to pink, and afterwards to dull purple, tubular from an oval base, from one to two lines long, and the tube a third of a line broad, hairy within and without with spreading white hairs (though the leaves are nearly smooth), the mouth expanded, 2-5-lobed, sometimes dimidiate. Ovary inferior, 1-celled, with one or two pendulous ovules. But these ovules are the true eggs of an insect, for, by examining individuals in progressive states of development, I have traced the formation from the egg of, first, a minute fusiform annulate body, and, ultimately, of a perfect insect with legs and wings. To make the resemblance to a flower more striking, there appears, beneath what I have called the perianth, what seems to be a calyx of four or five erect triangular brownish sepals; but these are really only the torn cuticle by the protrusion of the perianth.

To explain the form assumed by these excrescences, may we not suppose there has been an attempt to reproduce the tubular 5-lobed calyx of the species (which belongs either to *Inga*, or to some allied genus)? The juices of a plant, when diverted from their ordinary channels, must still go on forming tissue according to some law originally impressed on the species; and I have seen modes of development follow the puncture of an insect, such as in general only long cultivation calls forth. On the same leaves were the nidi of another insect. These were scarcely a line long, globoso-urceolate, regularly 20-striate, containing eggs in the concavity as in the other. They might easily be mistaken for some epiphyllous fungus. I enclose specimens of these productions, and I will afterwards send you a larger species.

The curious leaf-gall from the island of Cuba, figured on page 82 of Lindley's 'Vegetable Kingdom,' shows great resemblance to the flower of *Ochnaceæ*, to which Natural Order the plant bearing it belongs.

In its mode of growth it corresponds to the first of those above mentioned.

We are to embark to night (Monday), and if we have a continuation of the excellent winds which have blown for some time, we may hope for a speedy passage. RICHARD SPRUCE.

LICHENES ARCTICI; *collected in* 1848, *by* Mr. Seemann *of the Expedition of* Capt. Kellet, *in H.M.S. Herald, in search after Sir J. Franklin; by the* Rev. CHURCHILL BABINGTON, M.A., *St. John's College, Cambridge.*

From Avatscha Bay. August 1848.

Ramalina farinacea, *Ach.* A fragment on fir-bark.
Parmelia saxatilis, *Ach.* (P. sulcata, *Tayl.*!) On fir-bark.
—— olivacea, *Ach.* On fir-bark.
—— varia β symmicta, *Fries.* On fir-bark.
Cladonia digitata β viridis, *Schær.*! (n. 46). On rotten wood; in a bad state.
—— pyxidata? *Hoffm.* Barren; growing on the ground among *Polytricha.*

From Norton Sound. September 1848.

Evernia divergens, *Fries.* The specimens agree with examples from Lapland, collected by Wahlenberg.
Cetraria cucullata, *Ach.*
—— Islandica, *Ach.*
Cladonia rangiferina *v.* sylvatica, *Fries* (C. sylvatica, *Auct.*)

All the above species grow matted amongst each other, mixed with dead stems of herbaceous plants and with *Vaccinium Vitis-Idæa*, L.

From Kotzebue's Sound.

Evernia divergens, *Fries.*
—— ochroleuca, *Fries* (Cornic. ochr. *Auct.*). This species and *E. divergens* grow together among other lichens and mosses.
Cetraria cucullata, *Ach.* ⎫ Both fertile; mixed together and accompanied
—— Islandica, *Ach.* ⎭ by other lichens and by grasses.
—— glauca, *Ach.*; substraminea, *Bab.*; thallo substramineo. Barren;

growing among sticks, grasses, and *Polytrichu...* The yellowish
colour is probably accidental; the less exposed parts of the
thallus incline to glaucous.

Nephroma polaris, *Ach.* Barren, growing among *Hypna* and *Dicranum
scoparium*: the leaves of grasses and a *Betula* (*B. nana*...
intermixed. The thallus agrees with European and ...
American specimens. I suspect that the ...
(*Hook. fil.! Crypt. Ant.* p. 213) is not the same.

Peltidea aphthosa, *Ach.* Fertile, mixed with *Hypna Sphagna* ...

—— horizontalis, *Ach.* Mixed with leaves of *Betula nana* ... *S.
octopetala*, mosses, &c. The apothecia ...
but there is little doubt that the plant ...

—— canina, *Ach.* Fertile, among *Hypna*, *Vaccinium* ...

—— venosa, *Ach.* On earth, fertile.

Sticta pulmonacea, *Ach.* Barren; among ... *Hypna* ...
leaves of a tree (a *Salix* apparently) ...
neatly crisped, pale ferruginous below.

Parmelia saxatilis, *Ach., Tayl.!* On rocks ...

—— parietina, *Ach.* Particles of ...
cimens of the preceding.

—— pallescens *v.* Upsaliensis, *Fries !* ...
of some coarse grass.

—— tartarea β frigida, *Fries* ...
sum), which it has ...

—— oculata? *Fries, F. ...*
with the figure ...

Cladonia uncialis, *Fries* ...
five inches long ...
specimens ...
axils. The ...
(Enum. of ...
suppose ...
they are ...

—— rangiferina, *Hoff.*

—— digitata, *Hoff.* ...

—— deformis, *Hoff.* ...
thecia ...

—— coccocephala, *F.* ...

Cladonia gracilis β hybrida, *Schær.*! Intermixed with mosses, *Hepaticæ*, and lichens. Specimens of a very strong growth, much curved, fertile.

Stereocaulon tomentosum, *Fries*. Barren, but the specimens apparently belong to this species. Growing loosely among masses of *Jungermanniæ* and mosses, mixed with leaves of *Vaccinia*, also growing on sandy ground with *Hypna*.

Sphærophoron coralloides? *Pers.* A barren fragment attached to a specimen of *Parmelia saxatilis*. All the European species of this genus are Arctic (Tuckerm. Enum. p. 82): and the present fragment is too minute to determine satisfactorily.

BOTANICAL INFORMATION.

Death of M. REQUIEN.

We regret to find the death of M. Requien, of Avignon, announced in the French papers. He was much beloved, both in his native city, where he had always taken a very active part in the management of benevolent and charitable, as well as scientific, institutions, and by his numerous friends at Montpellier and other towns in the south of France. He had been on a visit to Montpellier early in April, from whence he went to Corsica, where he was carried off by apoplexy on the 30th of May, having lately completed his sixty-third year. His rich collections, chiefly of European plants, shells, and other departments of Natural History, were engaged during his life-time to be transferred on his death to a public institution, we believe in Avignon. He was in every respect a generous botanist; and his liberality in distributing from his vast stores of French plants is known to all botanists who have visited the south of France, and to all who were in any way in correspondence with him.

Death of J. E. BICHENO, Esq.

We have the painful duty to record the recent account of the death of a zealous Fellow of the Linnean Society while resident in his native country and long its Treasurer, J. E. Bicheno, Esq. This took place at Hobart Town, Van Diemen's Land, where he had been for some years Colonial Secretary, and where he was no less respected and beloved than he was by all who had the privilege of his acquaintance in England. The botany of Great Britain had chiefly engaged his attention. In the Linnean Society's Transactions, he published " On Systems and Methods in Natural History," and a "Monograph of the British species of *Juncus*."

Cereus triangularis.

Lieut. Agassiz has sent us a drawing of a plant of *Cereus triangularis* which has recently blossomed in a stove in this country, bearing flowers of a most extraordinary size. The length of the flower, accurately measured, is fourteen inches from the base of the ovary to the tip of the petals; the breadth, from tip to tip of the calyx, fourteen inches: the petals form a cup and are nearly erect, but they measure eight inches across the mouth of the cup, and the petals are seven inches long, including their curve: the calyx-tube eight inches long. The weight is nearly a pound. The stem was of the ordinary size. This species is well figured in the 'Botanical Magazine,' tab. 1884, in the year 1817, and in the 'Botanical Register,' tab. 1807, in the year 1837, although the author of the latter work says it had never previously been represented from an European specimen. But the fact is, it was first figured more than a century ago, in the Acta Naturæ Curiosorum, from a plant which flowered at Altorf.

NOTICES OF BOOKS.

De Vriese (Dr. and Professor); *Descriptions et Figures des* Plantes Nouvelles *et* Rares *du Jardin Botanique de l'Université de* Leide *et des principaux Jardins du Royaume des Pays Bas. Ouvrage dédié à sa Majesté la Reine. Livraison* II. Imp. folio. 1851.

We gave, as we had most just reason for doing, in the 7th volume of the 'London Journal of Botany,' a highly favourable notice of the first livraison of this splendid work, a publication as remarkable for the beauty of the Plates, which are coloured, as for the scientific value of the descriptions. The first figure in the present fasciculus represents a new species of *Hymenocallis, H. Borskiana*, De Vriese, from

Venezuela. The second and third are devoted to noble figures of *Cycas Rumphii*, Miq.; and the two last plates, one to the male, the other to the female, inflorescence of *Cycas circinalis*. The descriptive part includes some valuable remarks on the internal structure of the male cone of the latter species.

Dr. WIGHT: ORCHIDEÆ *of the Neilgherries*.

We have this moment the satisfaction of receiving Vol. V. of Dr. Wight's 'ICONES PLANTARUM INDIÆ ORIENTALIS,' containing Plates 1622 to 1762 (with specific characters), which are, all but one (the last) plate, devoted to admirably executed outline lithographic figures of the ORCHIDEÆ of the Neilgherries. We have been for some time anxious to record the invaluable and indefatigably continued labours of our excellent friend in the field of Indian Botany, but we have waited for one of the Parts of the fourth volume (Part II.) which, through some unaccountable accident, has never yet been received in England. The very mention above of the numbering of the plates will show the amount of species illustrated in this work alone. Equally meritorious is the coloured work, entitled 'Illustrations of Indian Botany, or Figures illustrative of each of the Orders of Indian Plants, described in the Author's Prodromus Floræ Peninsulæ Indiæ orientalis,' and extending to 182 plates, with copious and very instructive letter-press. We trust soon to have it in our power to notice both these works more fully.

JOHN SANDERS: *A practical Treatise on the Culture of the* VINE, *as well under glass as in the open air*. London: Reeve and Benham. 8vo. 1851.

This is a plain, intelligible, well-written pamphlet of between thirty and forty pages, devoted, 1st, To the culture of the Vine under glass; 2ndly, Culture of the Muscat Grape; 3rdly, Culture of the Vine in the open air; and 4thly, Culture of the Vine in pots. It is accompanied by nine plates : the first four illustrating " a double-fronted wall-house for grapes and other purposes." Plates V. and VI. represent a house for Muscat grapes. Plate VII. shows Vines as grown in the open air; and Plates VIII. and IX. exhibit plans of a house for growing Vines in pots and troughs. No vine-grower should be without this little book.

We have received the following PROSPECTUS *of a* FLORA GRÆCA EXSICCATA ; *ou Collection de plantes rares et intéressantes* DE LA GRÈCE *Printed at* ATHENS *in French, German, and Greek.*

La Flore grecque, qui renferme des espèces offrant un grand intérêt à la science, n'est pas encore assez connue. Si la Grèce a été tour à tour l'objet des recherches et des travaux de plusieurs botanistes distingués, tels que Tournefort, Sibthorp, Sieber, Friedrichsthal, Bory de St. Vincent, Boissier, Link, Grisebach, Fraas, Heldreich, Sprunner, Sartory et autres, il n'en est pas moins vrai que ce pays possède encore des richesses qui sont à explorer.

Au point de vue du règne végétal, la Grèce offre, par sa constitution physique, des variations qu'il est rare de retrouver ailleurs […] un point si circonscrit. Des groupes de montagnes plus ou moins élevées, entrecoupés par des vallées plus ou moins pro[fondes …] des différences considérables de température et d'h[…] qui se reflètent nécessairement dans les pr[oductions …] Ajoutez à ces inégalités du terrain, ce grand d[…] couvert souvent de plages et de marais, des […] les dattiers, les opuntia, les hesp[érides …] nombreuses, dans des latitudes plu[s …] idée approximative de la vaste […] richir la science.

Les savans qui ont recueilli […] talens étendus et leur zèle […] n'ont pu atteindre complè[tement …] ignorant la langue, étant […] leurs guides, et n'ayant […] localité, n'ont fait que tra[…] d'une saison quelc[onque …]

Entièrement vou[é …] été confiée, et habit[ant …] que les espèces de […] étudiées et connues […] tant d'autres, pu[…] avancement.

Pour arriver à […] critiques de […] communes […]

tions des savans et des amateurs de la botanique. Par ce moyen, les erreurs que je puis involontairement commettre, soumises au contrôle de juges compétens, n'iront pas augmenter la confusion qui malheureusement pénètre de jour en jour dans la science, et la détermination des plantes de la Grèce opérée ou confirmée par nos illustres maîtres, acquerra une valeur incontestable.

Résolu à étendre de plus en plus l'horison de mes excursions botaniques, j'entreprends la publication par centuries des plantes rares que j'ai déjà recueillies en Grèce, sans m'arrêter aux dépenses et aux fatigues que cette entreprise entraîne.

La première et seconde centurie comprennent les plantes brièvement décrites par Smith dans le *Prodromus Floræ Græcæ*, ainsi que plusieurs espèces découvertes récemment par MM. Boissier, Sprunner et Heldreich, et décrites par le premier dans son excellent ouvrage : *Diagnoses plantarum orientalium novarum*. Ces deux centuries seront publiées le printemps prochain, et les espèces qui en font partie, ont été cueillies au Lycabète, à l'Hymètte, au Pentélique, au Parnès, au Corydale (Attique), aux environs de Nauplie, au Parnon appelé par quelques voyageurs Cronion et vulgairement appelé Malevo (Péloponnèse) et à l'île de Syra.

La troisième centurie contiendra les plantes du Parnasse et du Cyllène ; la quatrième celles du Taygète, du Tymphreste et de l'Oeta ; la cinquième celles du Mont Athos, du Pélion, de l'Ossa et de l'Olympe ; la sixième et septième les plantes de l'Asie-mineure : et les centuries suivantes celles de l'île de Crète, des îles Ioniennes et d'autres points de l'Est de la Méditerranée. La succession de ces voyages pourra être intervertie par suite des circonstances.

Les personnes qui comptent faire l'acquisition de la *Flora Græca exsiccata* sont priées de faire parvenir leurs demandes d'une manière lisible à l'adresse du soussigné.

THEODORE G. ORPHANIDES,
Professeur de Botanique à l'Université-Othon, Rue d'Eole à Athènes.

Athènes, le 15 Septembre, 1850.

FLORULA HONGKONGENSIS : *an Enumeration of the Plants collected in the Island of Hong-Kong*, by Capt. J. G. Champion, 95*th Reg.*, the determinations revised and the new species described by GEORGE BENTHAM, ESQ.

Capt. Champion's labours in the investigation of the Flora of Hong-Kong, during his visit to that island with his regiment, have already been alluded to, and some of the new species he first sent over were described by the late Mr. Gardner, in the first volume of this Journal (pp. 240, 308, 321). He remained in the island three years, returning to Europe in 1850 with a collection of between five and six hundred phænogamous plants, comprising, with scarcely any exceptions, all those previously found in the island by Mr. Hinds, and enumerated in the 'London Journal of Botany' (vol. i. p. 482), and a considerable number either entirely new or not hitherto found on the Chinese coast. He has placed a set of specimens in the hands of Mr. Bentham, who has re-examined the whole, and in whose herbarium are deposited the original specimens here described.

RANUNCULACEÆ.

1. Clematis *uncinata*, Champ., sp. n.; scandens, glaberrima, glaucescens, foliis pinnatisectis, segmentis (quinis) petiolulatis ovato-lanceolatis uncinato-acuminatis, panicula laxa floribunda, carpellorum caudis plumosis.—*Folia* inflorescentiæ et calyces siccitate nigrescunt. *Petioli* et *petioluli* subcirrhosi. *Segmenta* 1½–3-pollicaria, basi trinervia, reticulato-venosa, subtus pallida, basi obtusa, apice in acumen recurvum interdum fere cirrhiforme desinentia. *Panicula* ultrapedalis, opposite ramosa, basi foliata; ramuli superiores et ultimi bracteis brevibus subulatis subtensi. *Sepala* 4, aperta fere 6 lin. longa, lanceolata, setaceo-acuminata, dorso glabra et nigricantia, anguste lanato-marginata (intus colorata?). *Stamina* numerosa, antheris linearibus filamento æquilongis longioribusve. *Ovaria* 8–10, glabra. *Styli* longe plumosi, summo apice glabri, recurvi, intus stigmatosi.

Gathered July 10, 1848, in a ravine behind Mount Parker, near Saywan, and not seen in other parts of the island. It is a very distinct species among the pinnately divided paniculate *Flammulæ*.

2. Clematis *parviloba*, Gardn. et Champ., Kew Journ. Bot. vol. i. p. 241.

Also a rare species in Hong-Kong. Gathered in the Happy Valley, near the Waterfall, and towards West Point under the Buddhist Cave, in April 1848 and March 1849.

3. Clematis *Meyeniana*, Walp., Pl. Meyen. p. 277.

Of common occurrence, and in great abundance in almost every ravine in the island. The flowers are nearly pure white, very pretty, and slightly sweet-scented. It flowers during May, and occasionally again in autumn, even so late as December.

DILLENIACEÆ.

1. Delima *sarmentosa*, Linn.

This is, at any rate, the Canton plant referred to this species in the 'Botanical Magazine,' t. 3058, and is the same as the one distributed in Fortune's Chinese collection, n. 127. It agrees also very well with Burmann's indifferent figure of the original Ceylonese plant, but requires further comparison with specimens from Ceylon absolutely to establish its identity. The Penang plant referred to the same species in the 'Botanical Magazine' is the *D. hebecarpa*, DC.

Common in ravines all over the island, flowering in summer.

MAGNOLIACEÆ.

1. Talauma *pumila*, Bl.—*Magnolia pumila*, Andr.—Bot. Mag. t. 977.

Shrubby and rather scarce on Victoria Peak, more common and subarboreous in the woods of the Happy Valley, flowering in summer. It is much cultivated in flower-pots by the Chinese on account of its oppressively fragrant flowers, which bloom towards evening, falling off the next morning.

ANONACEÆ.

1. Unona *discolor*, Vahl.—DC. Prod. vol. i. p. 91.

Of very common occurrence low down in ravines, where it is usually a humble shrub. The flowers have a strong perfume, and vary much in size according to the dryness or wetness of the weather about the flowering season. The same thing occurs in *Artabotrys* and others of this family.

2. Uvaria *microcarpa*, Champ., sp. n.; foliis breviter petiolatis obovatis oblongisve basi oblique cordatis supra sparse subtus densius stellato-pilosulis subtus ad venas ramulis pedunculisque rufo-tomentosis,

pedunculis terminalibus a...ctis, ...ails ...
tosi- carre ... ..., ha... globo-...
dispersa. ovule gemma... —Ramus ... ... primo intuitu ...
... ... ovaria ... ... et fructus ... ...
Folia ...-petiolata, ovato-...oblongo-acuminata, petiolis ... ...
rufo-tomentosa, ..., ramulo ... ...
... ... *Pedicelli* ...-pedicellat..., supra medium ... ...
articulatorum teretes, ... 3 lin long, ...
divisus ... membra ... ... reticulatas obtusos. *Petala* ... ...
circa 7 in diametro, inter se æqualia v inter ... ...
utrinque crassiuscula, pube minuta velutina ... ...
Stamina numerosa, filamentis linearibus ... ...
intror... ... ...; *antheræ* continuæ, locul... ... ...
... ... *Ovaria* numerosa, linearia, ferrugin... ...
... ... plurimis. *Stylus* brevi..., glaber ...
... ... *Carpella* globosa, 3 lin. diametro, ...
rubra, stipite carpello breviore, ...
transverso separata, in specimin... ... ...
Of very common occurrence, and ... ...

3. Uvaria *platypetala*, Champ ... ...
    basi cordatis supra præter ... ...
    tibus et ad venas ramul... ... ...
    subæqualibus ampl... or... ... ...
    membranaceas paull... ... ...
    grandifloræ, Roxb., ...
    submembranaceum, ...
    incrassatus. *Caly...* ... ...
    gulariter fiss... ... ...
    grandiflora d.ª ... ...
    I was unw... ... ...
dissecting the ... ...
belong to this ... ...
with a ta... ... ...
between the ... ...
    Found ... ...
    4. A... ... ...
    Mon... ... ...
woods ... ...

5. Artabotrys *odoratissima*, Bl. Fl. Jav.

This is without doubt the Chinese *Anona hexapetala*, Linn., usually referred to the *Artabotrys odoratissima*: it appears to have rather larger flowers than the common East Indian form, but that, as observed by Capt. Champion, is variable, and I can find no other difference.

In similar places to the last species, but less common.

SCHIZANDREÆ.

1. Kadsura *Japonica*, Dån.; apparently identical with the Japanese species.

Rare, in a ravine below Victoria Peak, flowering in autumn. The ripe carpels, found in January, are blackish-purple, juicy, sweet, and pleasant to the taste.

LARDIZABALEÆ.

1. Stauntonia *Chinensis*, DC. Syst. Veg. vol. i. p. 514.

Trailing on rocks in ravines in various parts of the island. The flowers, in January and February, have a nauseous smell.

MENISPERMACEÆ. (By J. Miers, Esq.)

1. Cyclea *deltoidea*, Miers, sp. n.; ramis gracilibus (demum glabris?), foliis deltoideo-rotundatis apice obtusis v. rarius acutiusculis mucronulatis utrinque glaberrimis infra sordide glaucis 5-nervibus margine subrevolutis, petiolo gracili limbo dimidio breviore, racemo fœmineo axillari subpaniculato glaberrimo folio longiore, baccis glabris.

A single female specimen, from a ravine on Mount Victoria.

This differs from other species in the smaller size of its leaves, which are only about $1\frac{3}{4}$ inch long, $1\frac{1}{2}$ inch broad, on a petiole $\frac{3}{4}$ inch, the peltate insertion of which is generally about 3 lines (rarely 1 line only) within the margin. The somewhat slender raceme is subflexuose, about 2 inches long; the pedicels, 3 or 4 in each alternate fascicle, are approximate and almost umbellate, and $\frac{3}{4}$ line in length. The nut, though small, corresponds in structure with the peculiar form common to the genus.

2. Hypserpa *nitida*, Miers, sp. n.; ramulis striatulis sparse ferrugineotomentosis, foliis ovatis subacutis v. obtusiusculis mucronatis supra nitidis utrinque glaberrimis 3-nervibus, nervis venisque reticulatis supra immersis et cum nervis 2 alteris marginalibus subtus promi-

nentibus, petiolo apice incrassato superne pubescente inferne glabro, racemo fœmineo petiolo breviore paucifloro.

In a ravine of Mount Victoria.

The leaves are 1½ inch long, 10–11 lines broad, on a petiole of 4 or 5 lines. The berry is fleshy, about ¼ an inch in diameter. The specimen is in fruit only, but the plant evidently belongs to the above genus, the type of which is the *Cocculus cuspidatus*, Wall. (See Ann. Nat. Hist. N. S. v. 7. p. 36.)

3. Nephroica *pubinervis*, Miers, sp. n.; ramulis striatulis retrorsum pubescentibus, foliis oblongis a basi sursum angustioribus apice obtusiusculis mucronatis reticulatis 3-nervibus cum nervis 2 alteris marginalibus adjectis supra nitidulis, utrinque in nervis venis prominulis petioloque molliter pilosis, racemulo axillari paniculato paucifloro petiolo sublongiore, pedunculo bracteisque pubescentibus, pedicellis floribusque glabris.

Found with the preceding, in a ravine of Mount Victoria.

Intermediate, as a species, between *N. ovalifolia* (*Cocculus*, DC.) and *N. cynanchoides* (*Cocculus*, Presl). Its leaves are 2–2¼ inches long and 1–1¼ inch broad, on a petiole 5–6 lines in length; the younger branches are floriferous, and from each axil there springs a single raceme out of a woolly tuft, the peduncles being 3 lines, the pedicels 1 or 2 lines in length.

PAPAVERACEÆ.

1. Argemone *Mexicana*, Linn.

On the sea-shore and waste places, as in other parts of the warmer regions of both the old and the new world.

CRUCIFERÆ.

1. Cardamine *hirsuta*, Linn. The only representative of this large family, and itself naturalized only as a garden-weed.

CAPPARIDEÆ.

1. Capparis *membranacea*, Gard. et Champ. Kew Journ. Bot. vol. i. p. 241.

Victoria Peak and Happy Valley woods. Flowers in April and May. One specimen has much narrower leaves, cuneate at the base, and long acuminated, but is apparently a mere variety produced by station.

2. **Capparis** (Eucapparis) *pumila*, Champ. sp. n.; inermis, foliis ovalibus v. ovato-oblongis acuminatis glabris, inflorescentia minute puberula, corymbis pedunculatis ad apices ramorum paniculatis folia subsuperantibus, floribus glabris, ovario ovoideo acuminato glabro, bacca globosa.—*Ramuli* tenues, novelli pulveraceo-puberuli, mox glabrati. *Stipulæ*, saltem in parte superiore ramorum, omnino inconspicuæ. *Folia* breviter petiolata, 2–2½-pollicaria, basi rotundata, consistentia fere laurina, costa media subtus valida prominente, cæterum tenuiter penninervia et tenuissime reticulato-venulosa, marginibus subrecurvis. *Paniculæ* terminales, parce ramosæ, pauciflora, ramis supra medium irregulariter corymbiferis. *Pedicelli* 2–3 lin. longi. *Flores* quam in *C. sepiaria* paullo majores. *Sepala* viridia, ovata, concava, 2 lin. longa. *Petala* alba, sepalis subæquilonga, oblonga, obtusa, interdum margine leviter ciliata. *Stamina* circa 20. *Bacca* subglobosa, 3 lin. diametro, abortu monosperma, stipite 4½ lin. longo.

Only once seen in a precipitous ravine in the Black Mountain, when in company with Mr. Fortune, in January 1850.

FLACOURTIACEÆ.

1. Phoberos *Chinensis*, Lour.—Wight et Arn. Prod. Fl. Penins. Ind. Or. vol. i. p. 30.

Very common in the Happy Valley woods, often arboreous, and varying in being sometimes thorny, sometimes unarmed.

VIOLACEÆ.

1. Viola *Patrinii*, β *Chinensis*, DC. Prod. vol. i. p. 293.
2. Viola *tenuis*, Benth. in Lond. Journ. Bot. vol. i. p. 482.

These two species are both common on the summits of the hills, flowering from January till March. The *V. tenuis* is a pretty species from its light-coloured flowers, but they have no scent. It has a great tendency to become stoloniferous.

3. Viola *confusa*, Champ., sp. n.; subacaulis, pilosula, stipulis vix dentatis, foliis imis orbiculato- cæteris ovato-cordatis obtusis crenatis, calycis laciniis lanceolatis acutiusculis, corollæ calcare obtuso calyci subæquilongo, antheris inferioribus appendiculatis, stylo recto hinc fisso, stigmate truncato-crostri leviter dilatato.—Habitus et folia speciminum minorum *V. hirtæ* nostræ. *Pedunculi* florum perfecto-

rum foliis longiores, tenues, infra medium bibracteolati. *Flores* iis *V. hirtæ* minores, violacei. *Calycis laciniæ* 2 lin. longæ, basi inæqualiter appendiculato-productæ, appendicibus angustis, majoribus sæpe lineam longis. *Petala* 5 lin. longa. *Calcar* inferiorum 2 lin. In speciminibus flores fertiles apetalos ferentibus folia majora, longius petiolata, pedunculi petiolis breviores. *Calyces* parvi, laciniis angustis basi longiuscule appendiculatis. *Ovarium* ovoideum, stylo uncinato acuminatum. *Capsula* elliptico-trigona.

Gathered probably on Mount Parker, but mixed up with specimens of *V. Parkeri*, which abounds there; also on the opposite hill of the China coast. The late Dr. Gardner considered this to be Blume's *V. inconspicua*, said to have been introduced into the Buitenzorg garden from China. It is evidently nearly allied, and Blume's character would agree well with the apetalous state of our plant, but that he says, "staminibus inappendiculatis," whereas I find them appendiculate in the petaliferous flowers, and deficient in the apetalous ones.

All the above species, like most others of the same groupe in Europe and North America, appear to bear, in summer, minute apetalous yet fertile flowers.

POLYGALACEÆ.

1. Salomonia *Cantonensis*, Lour.—DC. Prod. vol. i. p. 334.

In marshy fields, abounding near the Albany Barracks. Flowers in summer.

2. Polygala *arillata*, Hamilt.—Wight et Arn. Prod. vol. i. p. 39.

Very rare. A shrub on Mount Gough, and another on Mount Victoria (since destroyed by the Chinese), were all that had been seen.

3. Polygala *glomerata*, Lour., as described by De Candolle (Prod. vol. i. p. 326), who saw Loureiro's specimen, although it scarcely agrees with Loureiro's own character. It is the same as the one since described by Blume under the name of *P. densiflora*, and is perhaps a mere variety of the common East Indian *P. arvensis*.

Gathered on Mount Victoria.

4. Polygala *Loureiri*, Gardn. et Champ. Kew Journ. Bot. vol. i. p. 242.

On Mount Victoria.

DROSERACEÆ.

1. Drosera *Loureiri*, Hook. et Arn. Bot. Beech. p. 167. t. 31.

Common in a ＊Hong-Kong.

PITTOSPORACEÆ.

1. Pittosporum *glabratum*, Lindl. Journ. Hort. Soc. vol. i. p. 230.

Some specimens have the umbellate inflorescence, and agree in every respect with Fortune's plant described by Lindley; others have solitary flowers with rather longer calyces, coming nearer to *P. pauciflorum*, Hook. et Arn., but with perfectly glabrous fruits. Captain Champion is, however, convinced that the Hong-Kong specimens all belong to one species, common on hills and in the woods of the Happy Valley. Captain Champion adverts to an affinity, not (as far as I am aware) hitherto pointed out, between *Pittosporaceæ* and *Grossularieæ*, considering that they bear the same relation to each other which *Loganiaceæ* do to *Rubiaceæ*. There is, however, an additional difference of some importance in the hypogynal staminal insertion of *Pittosporaceæ*.

CARYOPHYLLEÆ.

1. Stellaria *uliginosa*, Linn.
In rice-fields.

GERANIACEÆ.

1. Oxalis *corniculata*, Linn.
On road-sides, &c.
2. Linum *usitatissimum*, Linn.
In a marsh near East Point, not seen elsewhere.

MALVACEÆ.

1. Urena *sinuata*, Linn.
Common in the island.
2. Paritium *tiliaceum*, A. de St. Hil.
On banks and enclosures.
3. Abelmoschus, apparently a new species allied to *A. rugosus*, Wall.; but unfortunately the single specimen, gathered on the summit of Mount Victoria, is insufficient for description.
4. Sida *rhombifolia*, Linn.
5. Sida *cordifolia*, Linn.

These two common tropical species are also frequent in Hong-Kong on road-sides, &c. A third species, *S. humilis*, Willd., also a common East Indian plant, was gathered in Hong-Kong by Mr. Hinds, but is not in Captain Champion's collection.

STERCULIACEÆ.

1. Helicteres *angustifolia*, Linn.
One of the commonest plants in the low grounds, flowering all summer.

2. Reevesia *thyrsoidea*, Lindl. Bot. Reg. t. 1236.
Common in the Happy Valley woods; rare on Mount Victoria. Its flowers are pretty, white, and very sweet-scented, blowing in May.

3. Firmiana *platanifolia*, Br. Pl. Jav. Rar. p. 235.
Not indigenous to Hong-Kong, but naturalized and planted all about Victoria, having been brought from the opposite coast.

4. Sterculia *lanceolata*, Cav.—Lindl. Bot. Reg. t. 1256.
Common in ravines and woods, especially about the Buddhist temple at East Point.

(*Waltheria Americana*, Linn., a common tropical weed, was gathered in Hong-Kong by Mr. Hinds, but is not in Captain Champion's collection.)

5. Buettneria *aspera*, Colebr. in Roxb. Fl. Ind. ed. Car. et Wall. vol. ii. p. 383.
An extensive creeper over rocks in ravines; rather local in Hong-Kong, but covering much ground in some parts of the China coast.

6. Pterospermum *acerifolium*, Willd. — Wight et Arn. Prod. vol. i. p. 69.
Woods near the Buddhist temple.

TILIACEÆ.

1. Corchorus *acutangulus*, Lam.
A weed in rice-grounds.

2. Corchorus *capsularis*, Linn.
Cultivated in fields.

3. Triumfetta *angulata*, Lam.
Road-sides near the town of Victoria.

4. Triumfetta *cana*, Bl.? Bijdr. ex Walp. Rep. vol. i. p. 355.
Road-sides near the town of Victoria.

5. Grewia *microcos*, Linn.—Hook. et Arn. Bot. Beech. p. 170.
Rather scarce in Hong-Kong. Found on the hill upon which stands the Fever Hospital.

6. Elæocarpus *serratus*, Linn.—Wight et Arn. Prod. vol. i. p. 82.
Woods in the Happy Valley. Flowers in May.

7. Friesia *Chinensis*, Gardn. et Champ. in Hook. Kew Journ. Bot. vol. i. p. 243.

Woods about the waterfall in the Happy Valley. To Dr. Gardner's description, quoted above, may be added:—Female plant, with the racemes fewer-flowered than in the male. Calyx and petals as in the male. Ovary pubescent, ovato-globose, on a crenated disc, and crowned by a long style. Stigmas short, acute. Ovary 2-celled, each cell with two pair of collateral ovules suspended by pairs from the central axis. Rudiments of stamens none. I find the stamens of the male flowers to be 8. I have not seen the fruit. This species grows to a much-branched under-tree; the male plants, loaded with clusters of pea-green flowers, are exceedingly attractive to bees: they blow in April. (J. G. Champ.)

8. Heptaca? *latifolia*, Gardn. et Champ. Kew Journ. Bot. vol. i. p. 243.

(*To be continued.*)

Sketch of the VEGETATION *of the Isthmus of* PANAMA; *by* M. BERTHOLD SEEMANN, *Naturalist of H. M. S. Herald.*

(*Continued from p.* 239.)

The Isthmus is rich in medicinal plants, many of which are only known to the natives, who have ably availed themselves of their properties. As febrifuges, they employ *Chicoria* (*Elephantopus spicatus*, Juss.), *Corpachi* (*Croton pseudo-China*, Cham. et Schlecht.), *Guavito amargo* (*Quassia amara*, Linn.), *Cedron* (*Simaba Cedron*, Planch.), and several *Gentianeæ*, also herbaceous plants, which are known to the inhabitants by the name of *Canchalaguas*, and which belong to the genera *Eustoma*, *Coutoubea*, and *Schultesia*. As purgatives are used, *Ninno muerto*, or *Malcasada* (*Asclepias Curassavica*, Linn.), *Frijolillo* (*Cassia occidentalis*, Linn.), *Cannafistola de purgar* (*Cassia Fistula*, Linn.), *Laurenno* (*Cassia alata*, Linn.), *Savilla* (*Hura crepitans*, Linn.), and *Coquillo* (*Jatropha Curcas*, Linn.). Emetics are obtained from *Garriba de penna* (*Begonia* sp.), and *Frailecillo* (*Jatropha gossypifolia*, Linn.). As vulneraries they use *Chiriqui* (*Trixis frutescens*, P. Br.), and *Guazimillo*, or *Palo del soldado* (*Waltheria glomerata*, Presl), and *Cope chico de suelo* (*Arrudea clusioides*, St. Hil.). Anti-syphilitics are, *Cardo santo* (*Argemone Mexicana*, Linn.), *Zarzaparilla* (*Smilax* sp. pl.), and *Cabeza del negro* (*Dioscorea* sp.). Cooling draughts are prepared from the Ferns, *Calahuala* (*Goniophlebium attenuatum*, Presl) and *Doradilla de palo* (*Goniophlebium incanum*, Swartz). Antidotes for the bites of snakes are found in the stem and leaves of the *Guaco* (*Mikania Guaco*, H.B.K.) and the seeds of the *Cedron* (*Simaba Cedron*, Planch.). Cutaneous diseases are cured by applying the bark of the *Palo de buba* (*Jacaranda Bahamensis*, Brown), and *Nanci* (*Byrsonema cotinifolia*, H.B.K.), and the leaves of the *Malva* (*Malachra capitata*, Linn.).

The most dreaded of the poisonous plants are the *Amancay* (*Thevetia neriifolia*, Juss.), *Cojon del gato* (*Thevetia nitida*, De Cand.), *Manzanillo de playa* (*Hippomane Mancinella*, Linn.), *Florispondio* (*Datura sanguinea*, Ruiz et Pav.), and *Bala* (*Gliricidia maculata*, Kunth). It is said of the *Manzanillo de playa* that persons have died from sleeping beneath its shade; and that its milky juice raises blisters on the skin, which are difficult to heal. The first of these statements must be regarded as fabulous, and the second be received with a degree of modification. Some people will bear the juice upon the surface of the body without being in the least affected by it; while others do experience the utmost pain: the difference seeming to depend entirely upon the state of a man's constitution. Great caution, however, is required in protecting the eyes, for if the least drop enters them, loss of sight and the most acute smarting for several days are the consequence. The smoke arising from the wood produces a similar effect; and I remember that, while surveying on the coast of Darien, a whole boat's crew of H.M.S. Herald was blinded from having kindled a fire with the branches of this tree. Whenever the natives are affected by the poison they at once wash the injured part in salt water. This remedy is most efficacious, and, as the *Manzanillo* is always confined to the edge of the ocean, of easy application. It has been stated that the Indians of the Isthmus dip their arrows in the juice of the *Manzanillo*. There are, however, various reasons for doubting this assertion; firstly, because the poison is, like that of all *Euphorbiaceæ*, extremely volatile, and, however virulent when first procured, soon loses its power; secondly, because its effect, even when fresh, is by no means so strong as to cause the death of human beings, it not even producing, as has already been stated, the slightest injury on some constitutions. We may, therefore, consider the statement as an inaccuracy, and rather suppose that the Indians, like those of Guayana, obtain their poison from the *Strychnos toxifera*, Bth., and *S. cogens*, Bth., two plants very common throughout Panamà and Darien. The fruit of the *Amancay* (*Thevetia neriifolia*, Juss.) is also considered very poisonous, but its dangerous qualities have probably been overrated. I knew a gentleman in Panamà who, when a boy, ate four of these fruits, without experiencing any other effect than that of mere griping. The leaves of the *Bala*, or, as it is also called, *Madera negra* (*Gliricidia maculata*, Kth.), are used to poison rats. The *Florispondio* (*Datura sanguinea*, Ruiz et Pav.) appears to have always played, and still continues to play, a prominent part in the superstition of tropical America. The Indians of Darien, as well as those of Chocò, prepare from its seeds a decoction, which is given to their children to produce a state of excitement in which they are supposed to possess the power of discovering gold. In any place where the unhappy patients happen to fall down, digging is commenced; and, as the soil nearly everywhere abounds with gold-dust, an amount of more or less value is obtained. In order to counteract the bad effect of the poison, some sour *Chicha de Maiz*, a beer made of Indian corn, is administered.

Many indigenous plants bear eatable fruits, some of most delicious flavour. The principal are: *Algarrobo* (*Hymenæa splendida*, Vog.), *Boca vieja* (*Posoqueria longiflora*, Aubl.), *Cannafistola* (*Cassia Brasiliana*, Lam.), *Cerezo* (*Bunchosia glauca*, H.B.K.), *Coco* (*Cocos nucifera*, Linn.), *Coronillo* (*Scheeria Coronillo*, Seem.), *Espavè* (*Anacardium Rhinocarpus*, De Cand.), *Fruta de pava* (*Ardisia coriacea*, Swartz), *Granadilla* (*Passiflora quadrangularis*, Linn.), *Guayavo di savana* (*Psidium polycarpon*, Lamb.), *Guayavo* (*Psidium pyriferum*, Linn.), *Guavo* (*Inga spectabilis*, Willd.), *Icaco* (*Chrysobalanus Icaco*, Linn.), *Jagua* (*Genipa Caruto*, H. B. et Kth.), *Juan Bernardo* (*Conostegia lasiopoda*, Bth.), *Jobito de puerco* (*Spondias spinosa*, Seem.), *Marannon* (*Anacardium occidentale*, Linn.), *Madronno de comer* (*Sabicea edulis*, Seem.), *Membrillo* (*Gustavia Membrillo*, Seem.), *Nance* (*Byrsonema cotinifolia*, H. B. K.), *Nispero* (*Sapota Achras*, Mill.), *Panamà* (*Sterculia Carthagenensis*, Cav.), *Papayo Cimarron* (*Carica* sp)., *Pinna* (*Ananassa sativa*, Lindl.), *Pita di zapateros* (*Bromelia* sp.), *Sastra* (a *Guttifera*), *Pinajita* (*Pentagonia Pinajita*, Seem.), and *Zarzamora* (*Rubus trichomellus*, Schlecht.).*

* With a few exceptions, the Spanish names, when ending in *o*, denote the tree; in *a*, its fruit. For instance, *Naranjo*, an orange-tree, *Naranja*, an orange; *Manzano*, an apple-tree, *Manzana*, an apple.—B. S.

(a tree botanically unknown), *Cedro espinoso* (a *Sterculiacea*), *Caoba* (*Swietenia Mahagoni*, Linn.), *Espavè* (*Anacardium Rhinocarpus*, DC.), *Guachapali* (*Schizolobium excelsum*, Vogel), *Guavito cansaboca* (*Pithecolobium ligustrinum*, Bth.), *Guayacan* (*Tecoma flavescens*, Mart.), *Guazimo colorado* (*Lühea rufescens*, St. Hil.), *Laurel* (*Cordia Gerascanthus*, Jacq.), *Macano* (*Diphysa Carthagenensis*, Jacq.), *Maria* (a *Guttifera*), *Nance* (*Brysonema cotinifolia*, H. B. et Kth.), *Naranjo de monte* (*Swartzia triphylla*, Willd.), *Nispero* (*Sapota Achras*, Mill.), *Peronil* (a Papilionacea, n. 1673), *Quira* (*Machærium Schomburgkii*, Benth.), *Roble* (*Tecoma pentaphylla*, Jacq.), *Terciopelo* (*Bixacearum* gen. nov.), and *Corotù* (*Enterolobium Timboüva*, Mart.). From the *Roble* and *Guayacan* the most durable wood is obtained: some beams of the latter are still to be seen among the ruins of the cathedral of Old Panamà, where they have been exposed to the influence of the climate ever since the destruction of that city, a period of nearly 200 years. The *Nazareno*, a beautiful bluish fancy-wood, the produce of a scientifically unknown tree, would fetch a high price in Europe. The *Quira*, remarkable for its black and brown streaks, is the *Itaka* of commerce. The *Corotù* and *Espavè* supply the natives with materials for constructing their canoes, and no better estimate can be formed of the magnitude of these giants of the forest, than by an inspection of the Port of Panamà, where vessels of twelve tons burden, made of a single tree, are seen riding at anchor.

Dyes are numerous: a yellow one is obtained from the wood of the *Macano* (*Diphysa Carthagenensis*, Jacq.), a scarlet from the leaves of the *Hojita de tennir* (*Bignonia? Chicha*, H. et B.), a blue from the foliage of the *Annil silvestre* (*Indigofera Anil*, Linn.), a violet from the fruit of the *Jagua* (*Genipa Caruto*, H. B. K.), a red from the pulp of the *Bija* or *Achotte* (*Bixa Orellana*, Linn.), and a black from the seeds of the *Ojo de venado* (*Mucuna urens*, DC., and *Mucuna altissima*, DC.). A brown colour might be extracted from the *Clava*, a *Cyperacea*, (n. 147), which abounds in the savanas, and produces on cotton and linen a stain very much like that caused by the rusting of an iron nail, hence the vernacular name, *Clava*, a *nail*. The Indians of Southern Darien paint their faces with the colour obtained from the *Bixa Orellana*, Linn., or, as they themselves term it, *Bija*. The scarlet dye, observed in the hammocks of Veraguas, is not given with the purple shell (*Purpura patula*, Lam.), as the people of Panamà assert, but with the leaves of the *Bignonia Chicha*.

The cordage, which the Isthmians use, is solely procured from indigenous plants. The best and whitest rope is made from the fibre of the *Corteza* (*Apeiba Petaumo*, Aubl.). A brownish-looking rope, easily affected by dampness, probably because the tree from which it is taken contains saline properties, is manufactured from the *Majagua de playa* (*Hibiscus arboreus*, Desv.), and a third is obtained from the *Barrigon* (*Bombax Barrigon*, Seem.). The *Xylopia sericea*, St. Hil., also yields a fibre fit for ropes. It is on that account named *Malagueto hembra*, or *Female Malagueto*, in order to distinguish it from the *Malagueto macho*, or *Male Malagueto* (*Xylopia grandiflora*, St. Hil.), which is destitute of such a quality. The far-famed hammocks of Veraguas consist of the fibres of the *Cabuya* (*Agave* sp.), and those of a Palm called *Chonta*. A strong fibre is contained in the leaves of the *Pita de zapateros* (*Bromelia* sp.), which is prepared like flax, woven into bags, or Chacaras, by different Indian tribes, and extensively used by shoemakers for sewing. The fibre surrounding the wood of the *Cucua* or *Namagua* (*Brosimum Namagua*, Seem.) forms a close texture of regular natural matting, which the natives soak in water, beat and make into garments, beds, and ropes, or use as sails for their canoes. The mats which the poorer classes use to sleep upon, are manufactured from the fibre of *Plantain* leaves (*Musa paradisiaca*, Linn.).

Numerous vegetable substances are applied to miscellaneous purposes. An infusion of the leaves of the *Tè* (*Corchorus Mompoxensis*, H. B. K.) is drunk instead of tea, and a similar preparation is now made from those of the *Freziera theoides*, Swartz, a shrub common on the volcano of Chiriqui. The aerial roots of the *Zanora* (*Iriartea exorrhiza*, Mart.), being clad with numerous spines, are used as graters, and although they are not so fine as those supplied by art, yet in a country where, from the humidity of the climate, tin ones soon get rusty, they are almost preferable. The natives chiefly employ them when grating Cocoa-nuts, which, boiled with rice, compose one of their favourite dishes. The leaves of the *Papayo* (*Carica Papaya*, Linn.) are a substitute for soap. The wood of the *Balsa* (*Ochroma Lagopus*, Swartz), being soft and light, like cork, is used for stopping bottles: the never-sinking rafts, which, at the discovery of South America, caused such surprise among the early adventurers, were then constructed of it and are so still. The prevalence of the *Balsa* along the coast of Western America has hitherto, it seems, not been sufficiently appreciated by historians; but the nature of such a tree might indeed account for much intercourse, and many an early migration, which, under other circumstances, may almost appear inexplicable. The wool of various *Sterculiaceæ*, the *Balsa* (*Ochroma Lagopus*, Swartz), *Ceiba* (*Eriodendron Samauma*, Mart.), and *Barrigon* (*Bombax Barrigon*, Seem.), is employed for stuffing pillows, cushions, &c. Hedges are made of the *Ortiga* (*Urtica baccifera*, Linn.), *Poroporo* (*Cochlospermum hibiscoides*, H. B. et Kth.), *Pitajaya* (*Cereus Pitajaya*, De Cand.), and *Pinnuela* (*Bromelia* sp.). The hard shells of the *Crescentia cuneifolia*, Gardn., *C. Cujuta*, Linn., and *C. cucurbitacea*, Linn., are turned into bottles, sieves, pails, spoons, and various other household articles. In catching fish by stupefaction, the natives avail themselves of the juice of the *Manzanillo de playa* (*Hippomane Mancinella*, Linn.), the bark of the *Espavè* (*Anacardium Rhinocarpus*, De Cand.), and the leaves of the *Barbasco* (*Piper* sp.). Oil is obtained from the fruit of the *Corozo colorado* (*Elaïs melanococca*, Gærtn.), and wine, vinegar, food, habitations, clothing, and numerous other necessaries of life, from the different palms which inhabit the country. The leaves of the *Chumico* (*Curatella Americana*, Linn.), and *Chumico bijuco* (*Tetracera volubilis*, Linn.), are used for cleaning iron, polishing and scouring wood; indeed, they serve all the purposes of sand-paper.

(*To be continued.*)

Journal of a Voyage from SANTAREM *to the* BARRA DO RIO NEGRO; *by* RICHARD SPRUCE, ESQ.

We left Santarem for the Barra do Rio Negro, on Tuesday the 8th of October (1850), in an igarité belonging to Mons. Gouzennes, a French gentleman, who has been many years established at Santarem, and is in the habit of sending several small vessels up the Amazon every year, to procure pirarucú, turtle-oil, castanhas, &c. Our vessel was a very small one, of little more than 300 arrobas burden, and my baggage half-filled it. For want of room we were put to much inconvenience in drying our plants; and what was still worse, the palm-leaf tolda was so ill-constructed that every heavy rain penetrated it, and gave us afterwards much trouble in drying our soaked clothes and paper. However, there was no remedy, and for this conveyance, wretched as it was, I had waited nearly three months.

We had light winds to Obidos, and did not reach it until towards evening of the 14th. By night there was seldom any wind sufficient to enable us to stem the current, and our usual practice was to anchor near shore or to tie to the projecting branch of some submersed trunk. In such situations I had sufficient opportunity (sorely against my will) of further cultivating my acquaintance with the carapanás, and I received ample proof of a fact which I had previously doubted, namely, that with the wind *off shore*, vessels are not persecuted at all by carapanás, but the contrary when the wind blows *on shore*; and the reasons assigned for this by the Indians seems to be the correct one, namely, that in the former case these blood-thirsty creatures scent their prey in the wind, which they cannot do in the latter. It is rarely that there is perfect silence on the shores of the Amazon;—even in the heat of the day, from 12 to 3 o'clock, when birds and beasts hide themselves in the recesses of the forests, there is still the hum of busy bees and gaily-coloured flies, culling sweets from the flowering-trees that line the shore, especially from certain *Ingas* and allied trees; and with commencing twilight innumerable frogs in the shallows and among the tall grasses chant forth their "Ave Marias," sometimes simulating the chirping of birds, at others the hallooing of crowds of people in a distant wood. About the same hour, the carapaná begins its night-enduring song, more annoying to the wearied voyager than even the wound it inflicts. There are besides various birds which sing, at intervals, the night through, and whose native names are uniformly framed in imitation of their note; such are the Acuráu, the Murucututú, a bird of the owl tribe, and the Jacurutú, whose song is peculiarly lugubrious. A sort of pigeon, which is heard at 5 o'clock in the morning, is called, and is supposed to say, "Maria, já he dia!" ("Mary, it is already day!")—a name which reminds me of "Milk the cow clean, Katty," a common Yorkshire appellation of the stockdove. Amongst the birds which most amused us by day, may be mentioned the "Bem, te vi!" ("Well, I saw thee!") and "Joaô, corta páo!" (John, cut the stick!")

I one night much amused the sailors by inquiring what bird it was that was making a croaking noise in an opposite cacoal. I was informed it was a small quadruped the size of a rat, which had its residence in the cacoals and lived on the fruit. It is one of the animals specially resorted to by the Indian "feiticeiros," and great importance is attached to its replies, which are merely a repetition of its note, "Torô, Torô" (sounding almost exactly like the French word "*trou*") for an affirmative, and perfect silence for a negative. I was in my turn diverted by one of the crew commencing a conversation with the Torô, of which what follows is a nearly literal translation.—

"Your worship sings very sweetly all alone by night in the cacao-tree."—"*Toró! Toró!*"

"Your worship seems to be enjoying your supper on the delicious cacao."—"*Toró! Toró!*"

"Will your worship tell me if we are to have a favourable wind in the morning?"—Torô respondeth not.

"Your worship, do me the favour to say if we shall arrive at Obidos to-morrow?"—Again no reply.

"Your worship may go to the d—l!"—An insult of which Torô taketh not the least notice; and so ends the dialogue, the Indian being too angry to interrogate further.

When lying-to for a wind by day I gathered a few plants unobserved the preceding year; and when slowly beating against the furious current below Obidos I twice swam on shore to gather a stout Mimoseous twiner that adorned the shore for miles with its thick secund panicles, sometimes a foot long, of minute pale yellow flowers.

Obidos seems unlucky for travellers. Here Spix and Martius had to repair their helm, and we ourselves had scarcely embarked, early on the morning of the 15th, when our canoe took the ground in a stony place, and part of the ironwork of the helm was broken by the shock. It took a blacksmith the whole of the day to make a new *femea*, and it was not until 10 of the following morning that we got it fastened on and again set forth on our voyage. Below Obidos jacarés are only occasionally seen, but we now began to meet with them in numbers. When anchored on the night of the 16th in the still bay at the mouth of the Trombétas we were surrounded by them, mostly floating nearly motionless on the water and only distinguished from logs by the elevations and depressions of the back. Their grunt is something like what a pig would make with his mouth shut: our people imitated it and thus drew several of them near us, but I did not consider it worth while wasting powder and shot on them.

The coast from Obidos to Villa Nova is flat and uninteresting, until a little below the latter town its tameness is somewhat relieved by a lowish wooded ridge, called Os Parentins, running close by the right bank. We had still only light winds, and we did not reach Villa Nova before late on the 24th. It is a miserable-looking town, the houses going sadly to ruin, and we found but a single small vessel in the port. It is seated on a small bay, skirted by a lowish cliff, up which are piled blocks of apparently volcanic rock, glazed on the surface and quite honeycombed. A little before reaching the town we had rounded a rocky point, of nearly horizontally stratified slaty sandstone. We went on shore and visited the Vigario—the "Padre Torquato" celebrated in Prince Adalbert's voyage up the Xingú—a young man, apparently under forty, good-looking and rosy. We found him exceedingly courteous in his manners, but delighting wonderfully to hear himself talk, and therefore not unlikely to be led into the relation of marvellous tales, *as true*, though himself sceptical respecting them. He seemed highly flattered to hear that the Prince had made mention of him in his travels.

We had now to leave the Amazon and enter the Paraná-mirí * dos Ramos, a furo which, commencing a little above the mouth of the Lago de Saracá, but on the opposite side of the river, joins the Amazon again near Villa Nova, the principal mouth being a little below the town. We, however, entered it by a narrow channel called the Paraná-mirí dos Limoês, the mouth of which we reached in an hour and a half's rowing above Villa Nova. Our captain's object in following this route was to collect some debts left on by Mr. Gouzennes the preceding year, and we expected to be detained only a few days; whereas, from entering the Ramos to again quitting it, we spent an entire month. In this time I might have made many interesting observations respecting the great country of guaraná and pirarucú, but I was unfortunately taken ill almost immediately after entering it. At the first sitio where we made any stay, I was wishful to obtain an astronomical observation, and for this purpose lay all night on deck, a thing which I had often done on the Amazon without feeling any inconvenience; but the night was very cloudy and, what was worse, so strong a dew fell

---

* *Paraná-mirí* (that is, "small river") is the general term for any narrow channel through which the waters of the Amazon or the Madeira pass, either in the rainy season only or throughout the year. Above Villa Nova it is constantly heard instead of *furo*—a term most in use in the lower part of the river. The Indian name for the Amazon is *Paraná-pitinga*, or "the white river:" more rarely *Paraná-açu*, "the great river."

that in the morning my blanket was soaked, and one of my arms quite stiff and benumbed. A feeling of general *malaise* throughout the day was followed by an attack of fever, from which I did not fairly recover until we entered once more the broad Amazon. During the whole of our stay in the Paraná-mirí we had constantly heavy dews, while on the Amazon the dews are none or scarcely perceptible. This difference is doubtless owing to the strong winds which almost daily sweep up the Amazon, the winds in these narrow channels being either light or in no certain direction; to which may be added, that in the dry season the channels are hemmed in by steep banks of twenty or thirty feet high, surmounted by dense and lofty forest.

The region included by the Paraná-mirí dos Ramos and the Amazon is literally sown with lakes, the outlets of which are narrow brooks, communicating, some with the former and some with the latter, but nearly all dried up in summer. On the south side of the Ramos there are also several lakes. All these are richly stored with pirarucú, and usually take their names from some animal or plant abundant in their waters or on their shores; as the Lago das Garças, Lago dos Jacarés, Lago do Arrozal (the Heron's lake, the Alligator's lake, the Rice lake), &c. In the height of the dry season, when the water of the lakes is low, numbers of fishermen resort to them for the purpose of taking pirarucú; including not only all the available population of the Ramos, but also fishing-parties from places as far distant as Pará and Macapú. When I had somewhat recovered from my sickness I managed, but with difficulty, to reach one of the lakes; to do which I had to thread an Indian track of an hour's length, through a dense forest, consisting chiefly of wild cacao-trees, castanheiras, and Urucurí palms. I found the lake nearly circular, of about a mile in diameter, and several fishing parties were at work on it. The general sleeping-apartment was a large palm-leaf house erected on poles in the lake, at a sufficient distance from shore to secure it from the visits of carapanás. This contrivance is resorted to on all the lakes, which are abominable places for *praga* of every description.

I was disappointed not to observe a single plant, save the rank grasses round the margin; but jacarés were laid in the water in almost countless numbers, resembling so many huge black stones or logs. What we had seen in the Amazon of these reptiles was nothing compared to their abundance in the Ramos and its adjacent lakes. I can safely say that at no one instant during the whole thirty days when there was light enough to distinguish them, were we without one or more jacarés in sight. Jacarés sometimes take the bait intended for pirarucú, and the line is strong enough to hold them.

The following is the mode in which pirarucú is prepared on the Ramos. When the fish is brought to shore, its head is first cut off; it is then skinned and the backbone taken out. The two halves of the back and belly are next each cut into two slices, so as to make in all four *postas* (as they are called), which are salted and laid on the skin spread out on the ground. Very little salt is used, and after remaining a few hours in pickle, the postas are suspended across poles, supported horizontally on a couple of forked sticks stuck erect in the ground, and here they remain, being occasionally turned, until quite dried in the sun. No part of the head is preserved, except the tongue-bone, which makes an excellent grater. A full-grown pirarucú weighs, when fresh, from two to three arrobas, and affords from half an arroba to an arroba of dried fish.

A little beyond the middle of the Ramos, we passed the mouth of the river Maué, upon which, at a distance of thirty hours in montaria, stands the town of Luzéa, anciently " Aldéa dos Maués." The river Maué enters from W.S.W., but a little way it turns to S., and its general course is said to be from S. to N. The direction of the Ramos at the junction is about N.N.E. Although Luzéa is not found on any published map, it is now a place of growing importance, and boasts of a church and chapel, with several white residents and a few shops. It was founded by the Portuguese in 1800, with 248 families of Maué and Mundrucú Indians, the government furnishing them with iron tools and building them a church. In 1808 the population amounted to 1627 souls, of whom 115 were whites.

The progress of Luzéa is entirely owing to its being the great depôt for guaraná, there being now large guaraná plantations (or *guaranals*) near the town, and the plant also being found wild on the Maué at some distance higher up. Formerly all the guaraná intended for the miners of Matto Grosso passed through Santarem, but it is now conveyed direct from Luzéa to Cuyabá, first in shallow canoes up the Maué-mirí nearly to its source, and then by a short portage to the river Tapajoz, above the first cataracts, where a depôt is established, and agents from Cuyabá are stationed to receive the guaraná from the Maué Indians, and forward it to its destination.

I was too ill to go to Luzêa in an open montaría, exposed to the burning sun and the dews of night, but I induced a woman to go and procure for me the fruit and the plant. Unfortunately, she had not understood the extent of my requirements, and brought me only a single bunch of fruit and a couple of leaves. These I hope to send you shortly. The fruit is about the size of the wild grape, and hangs in similar but smaller bunches. In form it is obovato-pyriform, apiculate, 3-lobed, the surface of each lobe being furnished with a midrib and reticulations like those of a leaf. The kernels are coated by a thin flesh of a yellow or slightly vermilion colour, which is picked off, and stains the hands of those who perform the operation. They are afterwards roasted, pounded, and made up into sticks, in the same way as chocolate.

The country of the Maués has never been accurately surveyed, and all the maps I have seen are much in error respecting it. A manuscript map in the possession of Dr. Campos of Santarem is the most correct, as regards the names and the relative positions of the rivers, &c.; but even this is very imperfect. Perhaps the Useful Knowledge Society's map is nearer the truth than that of Martius, for this particular part of Brazil, though the latter is correct in making the Maué-açú and the Maué-mirí branches of one and the same river. It is difficult to explain matters of this sort without a diagram; but I will suppose that it is required to reach the Madeira from Villa Nova by what is called the inside passage. The course will be to enter the Ramos by the Paraná-mirí dos Limoês, to sail up it as far as the mouth of the Maué, up which we must proceed to Luzêa, where we shall find a channel called the Paraná-mirí de Canomá (or sometimes the Furo de Urariá), taking us across into the Madeira a little below Borba. In winter the water of the Madeira enters the Rio Canomá, and it is then alone that this passage is practicable. In the same season the Amazon enters the Ramos by the upper mouth; hence from the latter to the mouth of the Maué we have Amazon water, and from the Maué to the lower mouth of the Ramos, Amazon, and Madeira water intermixed. The Madeira, like the Amazon, is a river of white water, which explains why in summer the waters of the Canomá and the Ramos are green, and in winter white.

On the two maps above referred to, the Ramos and the Canomá are erroneously represented as continuous, in the U. K. Society's map under the name of the Furo de Canomá, and in Martius's map under that of the Furo de Albacaxis. What is called the R. Mauhé on the latter is the Paraná-mirí dos Limoēs. I could hear of no "Ilha de Tupinambarana," though a river of this name is the most easterly tributary of the Ramos; nor is there any "Ilha dos Ramos," as indicated on Martius's map, on the north shore of the Amazon.

The Seringue-tree has long been known to exist abundantly on the Rio Madeira, but it is only during the present year that it has been found to grow on the Ramos in considerable quantity. About two months before our visit three small Seringals had been opened a little higher up than the mouth of the Mauć, and late on the evening of the 17th of November we reached one of these, belonging to Capitaô Pedro de Macedo of Saracá (or Silves, as it is called on the maps). A considerable opening had been made in the forest to erect the necessary huts, and to plant a few cabbages and water-melons. Amongst the trees was an enormous *Samaüma* (*Eriodendron Samaüma*, Mart.), divided from near the base into two trunks, of which the stoutest had been cut off at a height of about fifteen feet. In the morning I took a sketch of it, and measured its circumference, which was eighty-five feet at the lowest part, where the tape would ply of itself, that is, from one to three feet from the ground; but had the tape been applied to the recesses of the sapopemas (as the buttresses are called) the circumference would have been much increased.

We found the Capitaô a very hospitable and intelligent man, and were glad to accept his invitation to join him at supper and breakfast on game caught near his seringal, including Porco do mato, Macaco barrigudo, and Mutún—the last a bird much resembling a turkey, good eating, but rather dry; the monkey is rather insipid, and the pig very savoury, though with a thick tough skin. After breakfast he accompanied us into the forest, and showed us the Seringue-trees, and the mode of collecting the milk. A track had been cut to each tree, as also to adjacent flats of Urucurí palm (*Cocos coronata*, Mart.), which, curiously enough, is almost invariably found along with the Seringue, and whose fruit is considered essential to the proper preparation of India-rubber. A stout sipó is wound round the trunk of the Seringue, beginning at the base and extending upwards about as high as a man can reach, and making in this space two or three turns. This sipó supports a narrow channel made of clay, down which the milk flows as it issues from the wounded trunk, and is received into a small cuya deposited at the base. Early in the morning a man goes into the forest and visits in succession every tree, taking with him a terçado and a large cuya (called *cuyamboca*) suspended by a handle so as to form a sort of pail. With his terçado he makes sundry slight gashes in the bark of each tree, and returning to the same in about the space of an hour he finds a quantity of milk in the cuya at the base, which he transfers to his cuyamboca. The milk being collected and placed in a large shallow earthenware-pan, several large caraipé-pots with narrow mouths are nearly filled with the fruit of the Urucurí and placed on brisk fires. The smoke arising from the heated Urucurí is very dense, and as each successive coat is applied to the mould (which is done by pouring the milk over it, and not by dipping it into the milk), the operator holds it in the smoke, which hardens the milk in a few moments. The moulds now used are all of wood, and not of clay as formerly, and the one generally preferred is in the form of the battledores which English housewives use for folding linen, only thinner and flat on both sides, and the milk is applied only as far as to the insertion of the handle, the latter being held by the operator. When the requisite number of coatings has been applied and time has been allowed for the whole to stiffen, the seringue is withdrawn from the mould by slitting it along one side and end. In this state it is known in the Pará market as "Seringue em couro," or hides of India-rubber, and it is preferred to the bottle-rubber by purchasers. I send you one such "hide," from which you will see that Capitaô Pedro's manufacture is not despicable. If the bottle-moulds are used, or if a shoe is to be moulded on a last, a stick of two feet long is always inserted into the mould to guarantee the operator's hand from the milk and smoke. Some shoes we saw here had thirty coatings apiece of seringue. The Capitaô was getting about six milreis an arroba (32 lbs.) for his seringue, but in Pará it sells for as much as ten milreis. November is the season of ripe fruit of the Seringue, but the trees on the Ramos had been completely stripped by the Aráras, a sort of long-tailed parrot.

(*To be continued.*)

Contributions to the Botany of WESTERN INDIA;
by N. A. DALZELL, Esq., M.A.

(*Continued from p.* 233.)

Nat. Ord. LENTIBULARIEÆ.

UTRICULARIA. A. *Utriculiferæ.* § 5. Integræ.

1. U. *decipiens*; squamis raris acutis basifixis, floribus 6–7, bracteis ternatis, calycis lobis late ovatis superiore cuspidato-acuminato inferiore emarginato-bidenticulato paulo minore, corollæ violaceæ labio superiore *obovato-cuneato* emarginato erecto plano, inferiore *multoties ampliore* galeiformi orbiculari integro vel emarginato calcar dependens acutum æquante vel paulo excedente.

Radix circ. ½-poll., utriculifera. *Folia* decidua, obovato-spathulata, obscure 3-nervia, 2–4 lin. longa. *Scapus* teres, 3–9-poll., strictus vel volubilis. *Pedicelli* alato-marginati, 1½ lin., *fructiferi erecti*, bractea acuta basifixa 2-plo longiores. *Calycis* lobi 1¼ lin., post anthesin crescentes. *Corolla* a summo labii superioris ad apicem calcaris 3–3¼ lin. longa, labio superiore calycis lobo breviore v. longiore. *Capsula* intra calycem sacciformis. *Semina oblonga*, utrinque truncata, scrobiculato-reticulata.—*Herba* gracilis, erecta vel volubilis, primum 1–2-flora, postea elongata, 6–7-flora; floribus distantibus, labii inferioris marginibus horizontalibus.—*U. acutæ* (Benjamin in Linnæa, vol. xx. p. 314) proxima; differt labio superiore obovato-cuneato, inferiore multo longiore pedicellisque duplo brevioribus.—Crescit in locis humidis prope Vingorla, cum *U. nivea* associata; fl. temp. pluviali.

2. U. *albo-cærulea*; scapo tereti, squamis raris acutis basifixis, floribus paucis, bracteis ternatis, calycis lobis ovatis superiore acuminato inferiore acuto bidenticulato, corollæ læte albo-cæruleæ labio superiore orbiculari integro v. emarginato *marginibus reflexis* inferiore multoties ampliore galeiformi quadrato-orbiculari emarginato calcar descendens acutum duplo excedente.

Radix pollicaris, utriculifera. *Folia* decidua, spathulata, obscure 3-nervia, *utriculifera*, 2–3 lin. longa. *Pedicelli* alato-marginati, floriferi 3 lin., fructiferi reflexi, 5 lin. *Calycis* lobi 2 lin. *Corolla* a summo labii superioris ad apicem calcaris 6 lin. *Capsula* sacciformis, calyce aucto inclusa. *Semina* sphærica, scrobato-reticulata.—*Herba stricta, fragrans*, 4-6-pollicaris, *corollæ* palato macula pallida venosa antice 3-lobata notato; labio superiore albo.˙ *Flores U. reticulatæ* magnitudine, pulchriores et labio superiore 4-plo minore.— A *U. Smithiana*, R. Wight, cui proxima est, dignoscitur labio superiore et calcare multo brevioribus, calycisque lobo superiore *acuminato*.—Crescit in rupibus prope Vingorla; fl. temp. pluviali.

The three nerves, visible in the leaf of these two species, are formed by a spiral thread which traverses the fibres of the root, runs up the petiole, and divides into three branches in the lamina. It is from the lateral branches of this spiral thread that the utricles, when present on the leaf, take their origin, and the cellular tissue at the origin of the utricle assumes the form of a beautiful disc, formed by long cylindric cells radiating round a central spot.

Nat. Ord. ERIOCAULEÆ.

ERIOCAULON.

1. E. *rivulare*; caule simplici elongato submerso dense folioso, foliis linearibus planis apice in acumen setaceum attenuatis 7-nerviis vagina duplo longioribus, vaginis multistriatis glaberrimis apice laceratis vix fissis arctis, pedunculis umbellato-congestis subteretibus 10-sulcatis glabris 7-18-pollicaribus folio duplo longioribus, capitulis niveo-villosis, bracteis involucrantibus scariosis capitulo multo brevioribus glabris obovatis concavis apice laceratis; floribus masc. 6-andris, fœm. trigynis, bracteis flores stipantibus spathulato-acuminatis incurvis apicem versus pilis albis detergibilibus barbatis; sepalis masc. exterioribus in tubum fissum apice trifidum barbatum connatis, interioribus ovatis inæqualibus, filamentis brevioribus superne barbatis ibique glandula virescente notatis; filamentis interioribus longioribus, antheris *albis*; sepalis fœm. exterioribus linearibus barbatis, interioribus basi filiformibus apice spathulatis barbatis.

Receptaculum columnare. *Folia* 4-9 poll. longa, lineam lata.—Crescit in saxis demersis rivulorum provinciæ Malwan; in aquis stagnantibus nunquam obvenit.—Species unica a me visa *antheris albis*. Fl. temp. pluviali.

2. E. *odoratum*; acaule, foliis subulatis recurvis pollicaribus 7-nerviis, vaginis arctis apice fissis foliorum longitudine, pedunculis pluribus 5-angularibus 6-pollicaribus *tortis* filiformibus glabris, capitulis niveo-villosis 3 lin. diametro, bracteis involucrantibus brevissimis obo-

vato-cuneatis scariosis, bracteis floralibus rhombeo-cuneatis apice concavis pilis albis opacis comatis; floribus masc. 6-andris, fœm. trigynis; sepalis masc. exterioribus in tubum fissum apice truncatum connatis, interioribus lineari-lanceolatis glandula fusca notatis inæqualibus, omnibus apice pilis albis opacis comatis; filamentis æquilongis, antheris didymis olivaceis; sepalis fœm. exterioribus 2 carinato-navicularibus inæqualibus, interioribus angustioribus spathulatis, omnibus pilis albis opacis comatis.

Species insignis, odore *Anthemidis nobilis*, pedunculisque præter omnes gracillimis, vix ¼ lin. diametro.—Crescit in aqua stagnante prov. Malwan.

3. E. *cuspidatum*; acaule, foliis lineari-ensiformibus *obtusissimis* cuspidatis 7–9-nerviis subpellucidis glabris sesquipollicaribus 3 lin. latis vagina duplo triplo brevioribus, pedunculis umbellato-congestis 9–10-pollicaribus 7-angularibus glabris, capitulis hemisphæricis albido-villosulis, bracteis involucrantibus ovato-rotundatis glaberrimis capitulo brevioribus, bracteis flores stipantibus obovato-cuneatis apice incurvis ibique dorso pilis albis opacis confertis minutissimis obsitis marginibus glabris; floribus masc. 6-andris, fœm. 3-gynis; calycis masculi laciniis exterioribus 2 carinato-navicularibus dorso pilis albis opacis comatis marginibus glabris, interioribus 3-linearibus, 2 minutis, tertia multo majore, apice pilis *diaphanis* barbatis; antheris olivaceis, filamentis alternatim brevioribus; calycis fœminei laciniis exterioribus 2 carinato-navicularibus cucullatis dorso pilis albis opacis puberulis, interioribus 3 lineari-lanceolatis acutis inæqualibus apice glandula fusca notatis; ovario profunde 3-lobo albo. —Crescit in locis aquosis inter Vingorla et Malwan.

4. E. *pygmæum*; acaule, foliis planis linearibus acuminatis obscure 7-nerviis vagina duplo longioribus *pedunculos subæquantibus*, vaginis striatis glabris apice acuminatis fissis, pedunculis umbellato-congestis 3–4-angularibus glabris pollicaribus, capitulis hemisphæricis albidis, bracteis involucrantibus argenteis diaphanis lanceolatis acuminatis minutissime striatis *capitulo 3–4-plo longioribus* sub anthesi patentibus, bracteis flores stipantibus lineari-cuneatis truncatis pilis albis opacis comatis; floribus masc. 6-andris, fœm. 3-gynis; sepalis masc. exterioribus ima basi connatis superne liberis lineari-cuneatis truncatis pilis albis opacis comatis, interioribus ovatis glaberrimis inæqualibus; filamentis brevioribus apice glandula olivacea instruc-

tis, antheris fuscis; sepalis fœm. exterioribus filiformibus apice parce barbatis, interioribus spathulatis multo latioribus apice ciliatis ibique glandula fusca notatis.

Folia 8–12-lin. *Semina* lutea.—Crescit in locis uliginosis prov. Malwan; fl. Aug. et Sept.

Nat. Ord. ORCHIDEÆ-VANDEÆ.

MICROPERA.

1. M. *maculata*; subacaulis, foliis planis lineari-oblongis basin versus angustatis apice oblique emarginatis, cum mucronulo racemis basilaribus vel axillaribus simplicibus solitariis elongatis erectis multifloris a basi floriferis folio duplo longioribus, sepalis petalisque subæqualibus obovatis basi liberis, labelli albo et roseo picti utroque margine cornuto, cornubus reclinatis columnam æquantibus, calcare perianthio breviore saccato obtuso porrecto intus piloso laminæ 3-lobatæ brevicucullatæ supposito.

Capsula oblonga, sesquipollicaris, diametro 4–5 lin., cernua, 6-sulcata, angulis obtusis. *Flores* 4–5 lin. diametro. *Sepala* et *petala* lutea, medio macula purpurea notata; sepala lateralia patentia; petala cum sepalo supremo erecta, conniventia. *Pollinia* inæqualiter bipartibilia. *Folia* 4–5 poll. longa, 1 poll. lata.—Crescit in arboribus prope Tulkut, in montibus Syhadrensibus, lat. 16°; fl. Maio. Hujus generis, meo judicio, forma aberrans, dubia.

2. M. *viridiflora*; 4-pollicaris acaulis, foliis planis lineari-oblongis apice bilobis, racemis basilaribus brevissimis plurifloris folio brevioribus, sepalis petalisque conformibus æquilongis obovato-spathulatis pallide viridibus, petalis erectis conniventibus, sepalis lateralibus patentibus, labelli calcare perianthium æquante porrecto calceiformi laminæ cucullatæ submembranaceæ 3-lobatæ albæ supposito, laminæ lobis lateralibus rotundatis intermedio triangulari acuto, cornubus in labelli margine utroque parvis conicis erectis labelloque basi albo et roseo pictis.

Flores 5–6 lin. diametro, racemis 1–2-poll. *Folia* 3 poll. longa, 6–8 lin. lata.—Crescit cum præcedente, floretque eodem tempore. Differt a *M. pallida* caule nullo.

(*To be continued.*)

BOTANICAL INFORMATION.

Mr. Plant's Botanical Journey to South America, &c.

Mr. Plant, an active and intelligent naturalist, is about to travel for some years and employ himself as a collector in Botany and other departments of Natural History. He proceeds first to Rio Grande in South Brazil, and proposes to spend some time in Paraguay and other countries in the interior, then to cross the Pampas from Buenos Ayres to Mendoza, explore the eastern side of the Andes, and proceed thence into Chili, eventually probably to the Sandwich or other islands of the Pacific.

Mr. Samuel Stevens, 24, Bloomsbury-street, London, is prepared to give further information respecting this interesting journey, and to receive the names of any gentlemen who desire to subscribe to collections.

Death of Professor Ledebour.

Again we have to announce the death of a distinguished and energetic botanist, Dr. Ledebour, formerly Professor of Botany in the University of Dorpat, eminent as a traveller, and the author of Travels in the Altai, 'Flora Altaica,' and the 'Icones Plantarum novarum vel imperfecte cognitarum Floram Rossicam, imprimis Altaicam, illustrantes,' in 5 vols. folio, with 500 exquisitely executed and coloured plates. His most recent publication is the 'Flora Rossica,' still incomplete. This author died at Munich on the 4th of July. Our last fasciculus is X. of the 'Flora Rossica,' extending to the end of *Coniferæ* (following De Candolle's arrangement). The manuscript, however, is known to be very nearly complete, and it is said (Regensburg 'Flora') the " Russian subvention will be continued to secure the completion of the work."

NOTICES OF BOOKS.

Bulletin physico-mathématique de l'Académie Impériale des Sciences de Saint Pétersbourg.

The nos. 10, 11, and 12 of vol. ix. of this work, for May 1851, give the following information under the head of "Botanique :"—

La chaîne de l'Oural, à compter du 60me degré de latitude, au nord, jusqu'à la Mer Glaciale, n'avait jamais été foulée par le pied d'un naturaliste ; et cependant sa position, comme ligne de démarcation de deux parties de l'ancien continent, lui donnait, en tous temps, le caractère d'une haute importance géographique. Aussi notre Société de géographie débuta-t-elle, comme on sait, par une expédition qui, placée sous la direction du colonel Hofmann, a fourni, dans trois rudes campagnes, les notions les plus détaillées et les plus exactes de ce pays lointain et inhospitalier. La récolte botanique, rapportée par cette expédition, fut confiée à notre savant collègue, M. Ruprecht, qui en a fait l'objet d'un mémoire étendu, dont voici les résultats généraux :—L'Oural septentrional offre deux régions naturelles de plantes, très distinctes ; la région des forêts, au pied des montagnes, et la région alpine, sur les hauteurs. Dans les latitudes élevées, la région alpine descend jusqu'à la plaine ; plus au sud, elle monte successivement et finit par se circonscrire aux sommets. La région des forêts, sur les deux versants de l'Oural, est identique, d'un côté, avec celle du pays des Samoïèdes et des contrées adjacentes méridionales, jusqu'à la Mer Blanche ; et de l'autre, avec la région des forêts de la Sibérie occidentale, entre les marais du nord et les steppes du sud. Ce domaine vaste et continu se trouve donc coupé, dans l'Oural, par une bande de la flore alpine qui, s'avançant vers le sud, se retrécit de plus en plus, et finit par se dissoudre en points ou cimes isolées. La flore alpine se compose, en grande partie, d'espèces propres aux marais des Samoïèdes et d'un petit nombre de représentants du pays de Taïmyr, de l'Altaï et du Baïkal. Des espèces nouvelles, particulièrement propres à l'Oural, manquent entièrement, et l'on a tort d'admettre soit une flore particulière de l'Oural, soit, en général, une différence quelconque entre la végétation du N.E. de l'Europe et celle du N.O. de la Sibérie. Aussi, une flore de l'Europe, tant désirée par certains botanistes, ne saurait être écrite, qu'en établissant des limites artificielles, les limites naturelles n'existant pas.—Un second travail de notre botaniste, dont nous avons fait mention déjà dans notre dernier compte rendu, sa description des algues de la mer d'Okhotsk, a reçu, depuis, de nouveaux développements, relatifs surtout aux organes de fructification de ces cryptogames et à un nouveau système dont ces observations ont fourni la base. Ce travail, qui vient de quitter la presse, outre la partie descriptive proprement dite, a pour but de faire voir, d'une manière conforme à l'état actuel de la science, que cette végétation soi-disant forme réellement un règne à part, analogue à celui des plantes terrestres : bien que cette analogie, justement reconnue de tous temps, depuis les plus anciennes traditions des Grecs, par le simple instinct de la nature, ait dû nécessairement être rejetée par l'esprit de généralisation qui domine la science. M. Meyer a lu un mémoire sur *Astragalus asiaticus* de Pallas et ses rapports d'affinité avec d'autres espèces, et a soumis, dans un second mémoire, à une nouvelle révision les genres *Trinia*, *Rumia* et *Stenocoelium* de la famille naturelle des Ombellifères. M. Bunge de Dorpat a consacré un travail étendu à la description de la récolte botanique du voyage d'Alexandre Lehmann dans les steppes de l'Asie centrale, mémoire que l'Académie se fera un devoir de publier avec les autres résultats scientifiques du même voyage. M. Trautvetter de Kiev a déposé dans notre Bulletin une esquisse des classes et ordres du système naturel des plantes, et M. Bode, professeur à l'Institut forestier, a livré au recueil de MM. Baer et Helmersen une carte de la Russie européenne, accompagnée d'un texte explicatif et représentant les limites de la naissance des diverses espèces d'arbres, selon les données attentives et authentiques.

And under the head of "Rapports" we find as follows:

MM. Meyer et Ruprecht, rapporteurs, font un rapport très favorable sur le mémoire de M. le professeur Lehmann intitulé : *Alexandri Lehmanni Reliquiæ botanicæ*. Après avoir caractérisé le but et le contenu de cet ouvrage et exposé sommairement les résultats qu'il rapporte à la science, les commissaires émettent le vœu de le voir suivi bientôt de la seconde partie relative à la géographie des plantes. En attendant, ils pensent qu'on peut toujours procéder à la publication de cette première partie, soit dans le Recueil des savants étrangers, soit dans le recueil botanique (Beiträge, etc.) de M. Meyer. La classe y adhère, et, pour introduire plus de régularité dans la publication des posthumes, dont une partie déjà est imprimée dans les divers recueils, et dont on avait le projet de faire une édition à part, la classe charge MM. Baer, Meyer et Helmersen de se réunir en commission, afin de s'enquérir de l'état des éditions publiées déjà, et d'accorder avec celles-ci ainsi que de surveiller celles qu'il reste encore à publier.

Comptes rendus Hebdomadaires des Séances de l'Académie des Sciences.
Paris, Août 1851.

From the above-mentioned number of the 'Comptes rendus,' under the head of "Voyages Scientifiques.—Exposé des observations faites dans la Nouvelle-Grenade; par M. B. Lewy," we learn that during his residence in that country since 1847, he has been able to devote attention to botany. "Dans le règne végétal," he observes, "les collections que j'ai rapporteés comprennent plusieurs espèces de graines; un grand nombre d'imitations de fruits exécutées avec une rare perfection; plusieurs espèces de Palmiers, dont quelques-unes sont à peine connues et d'autres entièrement nouvelles; enfin une série de plantes utiles, en pleine végétation. J'ai déjà cité * l'arbre qui porte le *Cédron*; je puis y ajouter l'*Arracacha* qui, depuis quelques années en présence de la maladie des pommes de terre, excite à un haut degré la sollicitude des cultivateurs européens. Un botaniste français, Augustin Goudot, a trouvé la mort dans la tentative qu'il fit en 1847 pour enrichir son pays de cette précieuse racine alimentaire : plus favorisé que lui, il m'a été donné d'accomplir avec plein succès la mission à laquelle il s'était dévoué."—We may here observe that of the only two plants here named, the *Cedron* (*Simaba Cedron*, Planch.), is figured and fully described by us in the second volume of this work (Kew Garden Miscellany), p. 377, tab. XI.: and the *Arracacha* has been repeatedly imported into the English gardens, and was figured and described from specimens that flowered in this country, in Hook. Exotic Flora, tab. 152, under the name of *Conium Arracacha* nearly thirty years ago, and again in Botanical Magazine for 1831, tab. 3092, under the name of *Arracacia esculenta*, De Cand., where the properties of the root are fully detailed. Recent experiments have shown that it cannot be cultivated advantageously in Europe; and if it could, its value as an esculent to the European palate may well be called in question.

* P. 140. "J'espère être bientôt à même d'entretenir l'Académie de mes recherches sur les fruits fébrifuges du *Cédron* : j'ai été assez heureux pour rapporter en pleine végétation l'arbre qui produit ces fruits. Le *Cédron*, j'espère, sera une acquisition importante pour la thérapeutique."

Catalogue of Mr. Geyer's *Collection of Plants gathered in the* Upper Missouri, *the* Oregon Territory, *and the intervening portion of the Rocky Mountains; by* Sir W. J. Hooker, D.C.L., F.R.A., *and* L.S.

(*Continued from London Journal of Botany,* vol. vi. p. 256.)

LOBELIACEÆ.

1. Clintonia *elegans.* Dougl. in Bot. Reg. t. 1241. Hook. Fl. Bor. Am. vol. ii. p. 31.

Hab. Most abundant along the borders of Cœur d'Aleine Lake. The white and yellow variety rare. *n.* 665.

VACCINIEÆ, *De Cand.*

1. Vaccinium *cæspitosum,* Mich.—Hook. Fl. Bor. Am. vol. ii. p. 33. t. 126.

Hab. Grassy elevated Pine-woods of Upper Oregon; abundant in the highlands of the Nez Percez and Cœur d'Aleine, forming dense thickets about six inches high from the ground. Berries bluish-black. May. *n.* 285.

2. Vaccinium *ovalifolium,* Sm.—Hook. Fl. Bor. Am. vol. ii. p. 33. t.127.

Hab. Shady alpine woods, Cœur d'Aleine mountains, three feet high, forming thickets. April. *n.* 456.

MONOTROPEÆ, *Nutt.*

1. Pyrola *rotundifolia,* L.—Hook. Fl. Bor. Am. vol. ii. p. 46.

Hab. Moist rocks, shady alpine woods, defiles of the mountains in the Upper Columbia, with *Linnæa borealis* and *Pyrola secunda.* *n.* 427.

2. Pyrola *secunda,* L.—Hook. Fl. Bor. Am. vol. ii. p. 45.

Hab. Moist rocks, shady alpine woods in the mountains of Upper Columbia River, growing with *Linnæa borealis.* July. *n.* 428.

These specimens are larger and have broader leaves than any I have ever seen, some of the latter approaching to orbicular.

1. Pterospora *andromedea,* Nutt.—Hook. Fl. Bor. Am. vol. ii. p. 48.

Hab. On the roots of decaying Pine-trees, rare, Cœur d'Aleine mountains, often 3 feet high. May, July. *n.* 457.

APOCYNEÆ, *Br.*

1. Apocynum *androsæmifolium,* L.—Hook. Fl. Bor. Am. vol. ii. p. 51.

Hab. Volcanic plains and sandy Pine-woods, Spokan and Nez Percez country. July, August, very abundant. *n.* 449. This species had not been detected before to the west of the Rocky Mountains.

Asclepiadeæ, *Br.*

1. Asclepias *Douglasii*, Hook. Fl. Bor. Am. vol. ii. p. 52. t. 142.
Hab. Moist sunny valleys of Upper Missouri and Oregon territories, 2 feet high. Corolla pale rose-colour. Flowers very fragrant. July. *n.* 235.

Gentianeæ, *Juss.*

1. Gentiana *affinis*, Griseb. in Hook. Fl. Bor. Am. vol. ii. p. 56.
Hab. Fertile moist grassy meadows of Upper Oregon, with "*Castilleja miniata.*" August. Also in the Missouri territory. *n.* 84.
1. Frasera *thyrsiflora*, n. sp.; pentamera, caule elato crasso, foliis oppositis ternisve obovato-oblongis acutis radicalibus longe petiolatis supremis basi latioribus, cyma multiflora densissima racemiformi interrupta, calycis segmentis lineari-subulatis corollam cæruleam ¾ æquantibus, fovea solitaria elliptica.—F. *Caroliniana*, Hook. Fl. Bor. Am. vol. ii. p. 66, non alior.
Hab. Mountain valleys, Spokan and Kettle Falls, valley of the Columbia, *D. Douglas*. Mountain woods of the Nez Percez and Cœur d'Aleine, in moist open mossy places. Corolla litmus-blue. 3 feet high when in fruit. Succulent when young. Very rare. *Geyer*. *n.* 335.

A large species, with flowers equal to those of *F. Caroliniana*, and like them having but a solitary foveola at the base of each segment. Mr. Douglas's specimen is in fruit, and in other respects imperfect, and I was hence led erroneously formerly to refer it to the *Caroliniana*, which is peculiar to the east side of the Rocky Mountains.

2. Frasera *speciosa*, Dougl. in Hook. Fl. Bor. Am. vol. ii. p. 67. t. 153.
Hab. Ravines of Upper Platte and Colorado Rivers. From 4 to 8 feet high. June. *n.* 266.
3. Frasera *albicaulis*, Dougl. in Hook. Fl. Bor. Am. vol. ii. p. 67. t. 154.
Hab. Rare on the sandy denuded slopes of hills in the high plains of the Nez Percez Indians, along with *Clematis Douglasii*; also in the Kooskooskie valley. A foot high when in ripe fruit. Flowers lavander-blue. Leaves glaucous. May. *n.* 352.

These fine specimens show in its root-leaves to be numerous, spathulate, half as long as the slender stem. The flowers of this and *F. thyrsiflora* ex Pennington, and I fear the have as much claim to be ranked with *Swertia* as with *Frasera*.

1. Villarsia *cordata* Dougl. in Hook. Fl. Bor. Am. vo. 1. p. 76. t. 157 B.

Hab. Rocky loamy watercourses on the Coeur d'Alene mountains with *Platyspermum scopigerum*. April. n. 319.

POLEMONIACEAE Juss.

1. Polemonium *cæruleum*. L.—1. cæruleum. c. vulgare. Hook Fl. Bor. Am. t. 5. p. 71.

Hab. Swampy meadows, borders of rivulets. Spokan plains, at Tshimakeine, with "*Aster paniceus*." September 1; has the odour of "*Polanisia graveolens*." n. 536.

2. Polemonium *capitatum*. Eschsch.—Benth. in De Cand. Prodr. vol. ix. p. 317. P. Richardsoni. Hook. Bot. Mag. t. 2800. P. cæruleum, var., Lindl. Bot. Reg. t. 1303. P. cær. β humile. Hook. Fl. Bor. Am. t. 2. p. 71.

Hab. Gravelly table-lands, cataracts of Upper Lewis River, and abundant in the adjacent Pine-forests. September. n. 529.

3. Polemonium *micranthum*, Benth. in De Cand. Prodr. vol. ix. p. 318

Hab. Wet sunny rocks, very abundant at the Kettle Falls, Upper Columbia, with "*Collinsia minima*" and *Platycaryum scopigerum*. March, April. n. 463.

1. Phlox *speciosa*, Ph.—Lindl. Bot. Reg. t. 1351.—1. var. *latifolia*. 2. var. *linearifolia*; calycis tubo latiore, laciniis brevioribus.

Hab. Var. *latifolia*. Mountain meadows, plateaux of the Cœur d'Aleine and Nez Percez. Stem woody, with some evergreen leaves at the lower branches. A small globose shrub, covered with flowers. May. Rare. n. 375, "P. speciosa," Gey.—Var. *linearifolia*. Only in the valley of the Kooskooskie River and the adjoining plains. Prostrate or decumbent. Corolla clear, bright rose-colour, margin of the limb obsoletely erose: panicles mostly cymose. June. n. 340.

The narrow-leaved plant, here mentioned, quite agrees with several of the specimens of our *P. speciosa*, described in the 'Flora Bor. Am.' The broad-leaved plant has a different aspect, larger and more oval calyces, with shorter segments; but so variable is this species and its allies, that I dare not venture to consider the two distinct.

2. Phlox *Sibirica*, Gmel.—Hook. Fl. Bor. Am. vol. ii. p. 73.

HAB. Drift sand-hills of Lower Platte, with "*Psoralea tenuifolia*" and "*Rumex venosus.*" Flowers always white. June. *n.* 88. This quite agrees with my specimens from Nuttall, collected in the Rocky Mountains of the Platte.

1. Gilia (Leptodactylon, *Hook.*) *Hookeri*, Benth. in De Cand. Prodr. vol. ix. p. 316.—*Phlox Hookeri*, Dougl. in Hook. Fl. Bor. Am. vol. ii. p. 73. t. 159.

HAB. Elevated stony plains, Missouri and Oregon territories. A small depressed evergreen shrub, 1 foot high. Corolla white. July. *n.* 480.

2. Gilia (Eugilia, *B.*) *inconspicua*, Dougl. in Hook. Bot. Mag. t. 2833. Hook. Fl. Bor. Am. vol. ii. p. 74. Ipomopsis, *Sm.*

HAB. Gravelly banks of Upper Platte River. Corolla varying from reddish-blue to white. July. *n.* 42 and 25.

3. Gilia (Eugilia) *spicata*, Nutt. MS.; caule erecto simplici basi pilosulo superne viscido-pubescente, foliis caulinis anguste linearibus simplicibus trifidisque glabriusculis, floribus dense capitatis, capitulis breviter pedunculatis axillaribus longe virgato-thyrsoideis, bracteis linearibus, corollæ tubo calycem æquante. *Benth.*

HAB. Scott's Bluffs, Platte, *Nuttall,* in Herb. Hook. Rocky Mountains, *Fremont,* in Herb. Benth. Sunny but moist gravelly banks of Horse River of the Upper Platte, within the Black Hills, *Geyer, n.* 51.

Planta biennis videtur. *Caulis* crassiusculis, pedalis, a medio ad apicem florifer. *Folia* radicalia desunt; caulina $1-1\frac{1}{2}$-pollicaria, pilis paucis crispulis conspersa, utrinque lacinia brevi lineari aucta v. rarius integra, florialia breviora, pleraque simplicia. *Pedunculi* inferiores 3–6 lin. longi, viscoso-pubescentes, superiores breviores v. subnulli. *Capitula* densa, 6–8 lin. diametro. *Calyces* viscosi, $2\frac{1}{4}$ lin. longi, bractea eos æquante suffulti, laciniis acutis sed non mucronatis. *Corollæ* laciniæ $1\frac{1}{4}$ lin. longæ. *Antheræ* ad faucem filamento brevissimo insertæ. *Ovula* in loculis ovarii 6. *Benth.*

4. Gilia (Eugilia?) *iberidifolia*, sp. n.; humilis erecta ramosa, caule canescenti-pilosula, foliis profunde pinnatifidis pinnatisectisve segmentis anguste linearibus integris glabratis, cymis multifloris dense

capitatis terminalibus bracteatis, calycis laciniis mucronulatis corollæ tubum æquantibus, ovarii loculis uniovulatis.

HAB. The specimens mixed with *n*. 46 (*Gilia trifida*).

Herba annua? semipedalis. *Pubes* caulis et ramorum crispa, sub capitulis sublanata. *Folia* majora pollice paulo longiora; laciniæ utrinque 2-3 lacinia terminali breviores, omnes fere glabræ virides mucronula alba terminatæ. *Capitula* densa, fere sect. *Hugeliæ*, sed bracteæ et calycis laciniæ, etsi mucronatæ, non spinescunt. *Bracteæ* palmatim 3-4-fidæ, calycem vix superantes. *Calyx* 1½ lin. longus, pilis crispulis basi sublanatus. *Corollæ* laciniæ ovales, obtusiusculæ, lineam longæ. *Stamina* ad faucem inserta, breviter exserta. *Benth.*

This species connects the sections *Hugelia* and *Pseudocollomia* with *Eugilia*, and affords additional proof of the necessity of retaining the genus *Gilia* entire, as proposed in the 'Prodromus.' The ovules in the flower I examined were certainly solitary in each cell; but my specimen is very young, and I have not verified the structure in other flowers. *Benth.*

5. Gilia (Dactylophyllum, *B*.) *pharnaceoides*, Benth. in Hook. Fl. Bor. Am. vol. ii. p. 74. t. 161.

HAB. Grassy gravelly borders of Pine-woods, valley of Tshimakeinc. Corolla white, somewhat inclining to flesh-colour, yellow within. July. *n*. 535.

6. Gilia (Ipomopsis, *B*.) *pulchella*, Dougl. in Hook. Fl. Bor. Am. vol. ii. p. 174. G. aggregata, *Linn*.—*Br. Fl. Gard. n. ser. v.* 3. *t.* 218. *Ipomopsis elegans*, Lindl. Bot. Reg. t. 1281 (non Mich.).

HAB. Borders of dry Pine-woods, and on the loamy ferruginous slopes of the high plains of Upper Oregon. July, August. *n*. 435.

7. Gilia (Ipomopsis) *trifida*, sp. n.; humilis erecta laxe pilosula, foliis radicalibus integris trifidisve lobis brevibus obtusis, caulinis trifidis anguste linearibus, floribus subsessilibus demum dissitis, corollæ tenuis tubo exserto.

HAB. Sandy moist sunny spots encircled by the high precipitous rocks of Scott's Bluffs, Upper Platte River. *n*. 46.

Herba annua, semipedalis, ramosa, *G. inconspicuæ, arenariæ*, et *sinuatæ* affinis. *Pili* crispi albidi longiusculi in caule ramulisque. *Folia* fere glabra, radicalia crassiuscula 6-8 lin. longa, caulina pleraque longiora lobis angustis acutis divaricatis. *Flores* parvi, secus ramos solitarii v. ad apices ramulorum subglomerati, pedicello brevissimo

v. fere nullo. *Calyx* 2 lin. longus, laciniis setaceo-mucronatis. *Corollæ* tubus fere 3 lin. longus, limbi laciniæ oblongæ lineam longæ. *Stamina* ad faucem inserta, laciniis corollinis breviora. *Capsula* trilocularis, loculis 2–3-spermis (ovulis in quoque loculo 4 ?). *Benth.*

1. Collomia *grandiflora*, Dougl. in Bot. Reg. t. 1174. Hook. Bot. Mag. t. 2894. Hook. Fl. Bor. Am. vol. ii. p. 76.

HAB. Sandy Pine-woods of Spokan River, and sterile banks of streams, Upper Columbia region, growing with *Clarkia pulchella.* June, August. *n.* 441.

2. Collomia *linearis*, Nutt.—Hook. Bot. Mag. t. 2893. Fl. Bor Am. vol. ii. p. 76.

HAB. Open Pine-woods, banks of streams and waste places, common. June, July. *n.* 660 and 193.

3. Collomia *gracilis*, Dougl.—Gilia gracilis, *Hook. Fl. Bor. Am. v.* 2. *p.* 76.

HAB. Banks and denuded dry places, Upper Oregon. *n.* 871.

1. Navaretia *intertexta*, Hook. Fl. Bor. Am. vol. ii. p. 75.

HAB. Stony, clayey, and calcareous places, abundant in dry water-courses on the Spokan Plains and in the Missouri territory. August. *n.* 544.

CONVOLVULACEÆ, *Juss.*

1. Evolvulus *argenteus*, Ph. Fl. Bor. Am. vol. i. p. 187 (not Brown). Chois. in De Cand. Prodr. vol. ix. p. 443.

HAB. On the gravelly hills of the middle part of Platte River, growing with "*Astragalus hypoglottis*," "*Polygala alba*," and "*Mammillaria simplex.*" Corolla varying from pure white to bluish-lilac. June, July. *n.* 236.

1. Cuscuta *Americana?* L.? Hook. Fl. Bor. Am. vol. ii. p. 77?

HAB. Mouth of the Walla-Walla River; on the muddy borders; infesting the stems of "*Xanthium microcarpon.*" Sept. *n.* 674.

I am unable to determine this species at the present time, my entire collection of *Cuscuteæ* being in the hands of Dr. Engelmann, who is preparing a monograph of that family.

HYDROPHYLLEÆ, *Br.*

1. Hydrophyllum *capitatum*, Dougl. in Herb. Hook.—Fl. Br. Am. v. ii. p. 78.—H. densiflorum, *Nutt. MS. in Herb. Hook.*—β. *pumilum*, H. pumilum, *Gey. MS., minus, capitulis sessilibus.*

HAB. Shady swamps and rocky groves, springs and rivulets in the

Kooskooskie valley. Two feet high; very bushy and very succulent. Corolla pellucid, white. May. *n*. 401.—β. Pine-woods of Upper Oregon. Flowers pale lilac. March, April. *n*. 326.

This is exactly *H. capitatum*, Douglas, from the Columbia, and *H. densiflorum*, Nutt., from the same country; but it is, I think, very different from the Californian plant of Mr. Bentham, included under that species, and which is the *speciosum* of Nutt. in Herb. Hook.—β. is smaller, segments of the leaves less serrated or laciniated, and capitula sessile; but I dare not venture to consider it as more than a variety.

1. Nemophila *parviflora*, Dougl. in Hook. Fl. Bor. Am. vol. ii. p. 79. N. diffusa, *Nutt. MS.*

HAB. Alpine ravines of the Kooskooskie River. June. *n*. 646.

1. Ellisia *Nyctalea*, L.—Pursh, Fl. N. Am. vol. i. p. 141. A. De Cand. Prodr. vol. ix. p. 291. Br. App. to Frankl. Journ. p. 764. t. 27.

HAB. Marmot-burrows of the Prairie. June. *n*. 53.

1. Eutoca *Menziesii*, Benth.—Hook. Fl. Bor. Am. vol. ii. p. 79. E. multiflora, *Lindl. Bot. Reg. t.* 1180.

HAB. On the first range of trap mountains at the Upper Platte and Sweet-water Rivers, opposite to "Red Butter." July. *n*. 248; and sunny rocks of Kooskooskie and Cœur d'Alène Rivers. Biennial. Corolla diluted blue. *n*. 613. Young.

2. Eutoca *glandulosa*, Nutt. MS. in Herb. Hook; annua robusta, tota pubescenti-tomentosa viscosa, caulibus stellatim prostratis copiose foliosis, foliis lineari-oblongis petiolatis pinnatifidis lobis ovatis rigidis inferioribus remotiusculis, racemis copiosis digitato-corymbosis, floribus disticho-secundis, calycis parvi lobis linearibus corolla rotata subdimidio brevioribus, staminibus longe exsertis, capsula elliptica calycem paulo superante.

HAB. On decomposed bituminous slate rocks, hills of Upper Colorado. A very beautiful plant. Corolla deep azure or rather indigo blue. July. *n*. 93.

A most distinct and well-marked species, with copious, singularly stout, spreading stems and branches, and firm downy leaves. No one else appears to have met with it, except Mr. Nuttall, whose specimens, now before me, were also gathered in "Colorado of the Rocky Mountains." It is probably a very local plant.

1. Phacelia *circinata*, Jacq.—A. DC. Prodr. vol. ix. p. 298. β. hastata, *Dougl. in Hook. Fl. Bor. Am. v.* 2. *p.* 81.

HAB. Two states of this plant are found. 1. Stem-leaves entire, radical ones hastate or lyrato-pinnate. Stony, arid places, Kooskooskie valley. June. 1–4 feet high, erect. Almost burning like a nettle. Corolla lurid. Radical leaves more or less pinnated. *n*. 413.—2. All the leaves entire. Clayey saline water-courses, hills of the Upper Platte, at Scott's Bluffs. Corolla bluish. June. *n*. 164; and stony places along the banks of rivulets, Spokan Plains. *n*. 538.

BORAGINEÆ, *Juss.*

1. Myosotis *flaccida*, Dougl. in Hook. Fl. Bor. Am. vol. ii. p. 82. De Cand. Prodr. vol. x. p. 113.

HAB. Sunny rocks, with "*Bartonia parviflora.*" May. *n*. 348. On clay banks, trap-rocks, Kooskooskie and Spokan Rivers. June, July. *n*. 349.

2. Myosotis *glomerata*, Nutt.—Hook. Fl. Bor. Am. vol. ii. p. 82. tab. 162. Eritrichium? A. DC. Prodr. vol. x. p. 131. Gonospermum myosotoides, *Nutt*.

HAB. Moist saline calcareous clayey slopes of the Upper Platte hills. Flowers with the odour of *Heliotropium Peruvianum*. June, July. *n*. 70.

3. Myosotis *leucophæa*, Dougl. in Hook. Fl. Bor. Am. v. ii. p. 82. t. 163. Eritrichium, *A. DC. Prodr. v. 10. p.* 129.—Dasymorpha longiflora, *Nutt. MS.*

HAB. In the drift sand desert, between Upper Platte and Sweet-water Rivers, amongst *Opuntia Missourica*, growing with several species of *Hymenopappus*, and prostrate stemless species of *Erigeron*. July. *n*. 186.

4. Myosotis *Californica*, Fisch. et Mey.—Eritrichium Californicum, *A. DC. Prodr. v.* 10. *p.* 130. Myosotis Chorisiana, *Nutt. MS. Hook. et Arn. in Bot. Beech. Voy. p.* 152 (*not Cham. nor Hook. Fl. Bor. Am.*)

HAB. Stony places about springs, Missouri and Oregon territories, abundant about Tshimakeine. Prostrate. July. *n*. 548.

This differs somewhat from De Candolle's plant in the corollas being a little shorter than the calyces.

5. Myosotis *cymosæ* (Nutt.) affinis, sed carpellis lævibus.

HAB. Moist sunny rocky slopes of the high plains towards Cœur d'Alene River, also on the Kooskooskie. May. *n*. 344.

With only one specimen I dare not venture to describe this species.

6. Myosotis (Eritrichium) *Texanæ*, A. DC., proxima, sed pedicelli fructiferi plerisque erecto-patentes, nec horizontales aut subdeflexi, et imprimis carpella granulosa, nec lævia, trigona.

HAB. Gravelly banks of Upper Platte. July. *n.* 260.

7. Myosotis ——?

HAB. In the Kooskooskie valley, in rocky loamy places, growing with *Clarkia pulchella*. Flowers very conspicuous, bright orange. June. Annual or biennial. *n.* 339.

A large-growing species, from eight inches to a foot high, very hispid, with carpels which are solitary by abortion and ruguloso-tuberculate.

8. Myosotis (Dasymorpha) *tenella*, Nutt. MS.

HAB. Sunny rocky slopes of the mountains along the valley of Cœur d'Aleine River. April. Corolla white with an ochre-yellow belt around the faux. *n.* 290.

9. Myosotis *sericea*, Nutt. MS.

HAB. Calcareous clayey cliffs, on the hills of the Upper Platte River. Flowers white. Plant prostrate. June. *n.* 89.

Too near, I fear, to *Myosotis* (*Eritrichium*, A. DC.).

1. Echinospermum *floribundum*, Lehm.—Hook. Fl. Bor. Am. vol. ii. p. 84. t. 164.

HAB. Banks of rivulets, deep shady woods of the hills of Upper Platte River. July. *n.* 44.

1. Lithospermum *pilosum*, Nutt. De Cand. Prodr. vol. x. p. 79.

HAB. Grassy slopes of Cœur d'Aleine Mountains. Many stems from one thick ligneous caudex. Corolla pale yellow. May. *n.* 605.

Mr. Nuttall detected this on the Flat-head River of the Rocky Mountains, and Mr. Tolmie has sent it to me from about Fort Vancouver.

1. Mertensia *paniculata*, Ait. (*sub* Pulmonar.). Hook. Fl. Bor. Am. ii. p. 87.

HAB. Deep shady defiles, in mountains of the Nez Percez country. 3-5 feet high. Male fl. June. *n.* 485.

2. Mertensia *oblongifolia*, Nutt. (*sub* Pulmonar.). De Cand. Prodr. x. p. 92.

HAB. Wet rocks, Upper Oregon, abundant on the Columbia, with the small *Fritillarias* and *Claytonia*. March, April. *n.* 316.

3. Mertensia *alpina*? Torr. (*sub* Pulmonar.). De Cand. Prodr. vol. x. p. 91. Lithospermum strictum, *Gey. MS.*

HAB. Fertile meadows in the Lower Platte valley, in shady grassy spots, under *Populus candicans*. May, June. *n.* 24.

This has narrower leaves than any authentic specimen of Dr. Torrey's *Pulmonaria alpina*; but it does not appear to be otherwise specifically distinct.

Coldenia? (*Sect.* Stegnocarpus) *Nuttallii*; annua procumbens dichotoma, foliis fasciculatis longe graciliterque petiolatis rhombeo-ovatis acutis hispidis, floribus axillaribus glomeratis, calycibus hispidis 5-partitis segmentis hispidis subulatis.—Tiquilia *parvifolia, Nutt. MS.*

HAB. Rocky Mountains, *Nuttall*, in Herb. Hook. On decomposed calcareous rocks in the sandy desert of muddy rivers near the great salt-lake Timpanagos. Never seen before nor afterwards. August. Flowers red. *Geyer*. *n*. 80.

Flores in dichotomiis ramorum sessiles aggregati, foliis floralibus et bracteis suffulti. *Folia* floralia longe petiolata, ovata, margine revoluta, venosa, pilis rigidis canescentia. *Bracteæ* subulatæ, hispidæ, apice pilis rigidioribus 1-2-aristatæ. *Calyces* 5-partiti, foliolis subulatis hispidis. *Corolla* infundibuliformis, tubo intus basi 5-squamato, fauce nuda, limbo patente 5-lobo. *Stamina* 5 parum inæqualia, prope basin tubi supra squamas inserta, tubo inclusa. *Ovarium* 4-lobum. *Stylus* intra lobos ovarii insertus, profunde bifidus, ramis apice capitato-stigmatiféris. *Nuculæ* 4, non acuminatæ, ultra medium usque ad insertionem styli cohærentes, maturitate solutæ, nitidæ, glabræ. *Embryo* exalbuminosus, cotyledonibus bipartitis, radicula accumbente cotyledonibus paulo longiore. *Benth*.

I am indebted to Mr. Bentham for the above note, drawn up from Mr. Geyer's imperfect specimens. It is unquestionably the *Tiquilia* (Coldenia) *parvifolia* of Nuttall's MS., but as there is a species evidently of the same genus in my herbarium from Peru, with much smaller leaves, I venture to change the name. A more complete analysis of this curious plant would probably prove it to belong to Dr. Hooker's genus *Galapagoa*, published in Linn. Trans. vol. xx. p. 196, as does also the *Coldenia dichotoma* of Lehmann (Lithospermum dichotomum, *R. et P. t.* 111), and C.? canescens, *De Cand.*, and the small-leaved *Coldenia* from Peru above mentioned.

LABIATÆ.

1. Monarda *fistulosa*, L.—β. mollis, *Benth. in Hook. Fl. Bor. Am.* v. 2. p. 112. M. menthæfolia, *Graham, in Bot. Mag. t.* 2958.

HAB. Thickets and rough places in the fertile parts of the Platte valley. July, August. *n*. 9.

2. Monarda *aristata*, Nutt.—Benth. Lab. p. 318.
HAB. Plains of the Upper Platte, in denuded gravelly situations, with *Gaura coccinea*. July. *n*. 91.
1. Monardella *odoratissima*, Benth.—Hook. Fl. Bor. Am. vol. ii. p. 113.
HAB. Stony islands and cataracts of Kooskooskie. Odour like *Hedeoma pulegioides*. July. *n*. 468.
1. Hedeoma *Drummondii*, Benth. Lab. p. 368.
HAB. Gravelly slopes of the high banks of Upper Platte. Flowers pale blue. June. *n*. 124.
1. Scutellaria *galericulata*, L.—Hook. Fl. Bor. Am. vol. ii. p. 114, var. floribus majoribus.
HAB. Stony declivities and valley of Kooskooskie River. June. *n*. 381.
1. Lophanthus *urticæfolius*, Benth.—Hook. Fl. Bor. Am. vol. ii. p. 115.
HAB. Thickets in rocky places, Upper Oregon. July. *n*. 87.
1. Dracocephalum *parviflorum*, Nutt.—Hook. Fl. Bor. Am. vol. ii. p. 115.
HAB. Neglected fields of the Spokan Indians. September. *n*. 533.
1. Physostegia *Virginiana*, Benth.—Hook. Fl. Bor. Am. vol. ii. p. 116.
HAB. Fertile inundated meadows, Upper Oregon. August. *n*. 585.
1. Stachys *aspera*, Mich.—Hook. Fl. Bor. Am. vol. ii. p. 116.
HAB. Moist springy meadows and in thickets, plains of Tshimakeine, Spokan country. July. *n*. 433.

OROBANCHEÆ, *Juss*.

1. Orobanche *Ludoviciana*, Nutt.—Hook. Fl. Bor. Am. vol. ii. p. 92. *Phelipæa*, Reut. in De Cand.
HAB. On roots of "*Psoralea verrucosa*," in the high drift sand desert at the mouth of Lewis and Walla-Walla Rivers. September. *n*. 650. On the roots of "*Artemisia tridentata*," in the desert of Missouri. Whole plant violet-colour. August. *n*. 514. A stunted and ill-developed specimen.
2. Orobanche *Pinorum*, Gey. MS.
HAB. Top of the high mountains near St. Joseph, Cœur d'Aleine country, growing on the roots of *Abies balsaminea*. June. *n*. 445.
This very distinct-looking species has only undeveloped flowers, and it is impossible to form a specific character from them. In my specimens the stem is nearly a foot high, thick and scaly below, gradually tapering to a very slender extremity, beset with numerous branches or

racemes of flowers, from within a few inches of the base to the summit: the whole glanduloso-pubescent. The pedicels arise from a subulate bract, and there is a smaller bract of the same shape on each side the base of the calyx. Calyx ovate, with 5 linear-subulate teeth, shorter than the tube.

1. Anoplanthus *uniflorus*, Endl.—Orobanche uniflora, *L.*—O. biflora, *Nutt.*—Hook. Fl. Bor. Am. v. 2. p. 93.

HAB. High wet rocks, Cœur d'Aleine mountains; on the roots of *Sedum stenopetalum*. April. n. 372.

2. Anoplanthus *fasciculatus*, Walp.—Orobanche fasciculata, *Nutt.*—Hook. Fl. Bor. Am. v. 2. p. 23. t. 170.

HAB. Flat tops of the trap masses of the Kooskooskie, growing on *Eriogonum macrophyllum*. May. n. 369. High stony table-lands, Upper Platte River, on "*Artemisia frigida*." July. n. 443.

SCROPHULARINEÆ, *Juss.*

1. Scrophularia *Marilandica*, L.—Hook. Fl. Bor. Am. vol. ii. p. 94.

HAB. Waste places, Upper Oregon. June. n. 589.

1. Linaria *Canadensis*, L.—Hook. Fl. Bor. Am. vol. ii. p. 94.

HAB. High fertile plains of Upper Platte, near the Black Hills. July. n. 263.

1. Collinsia *grandiflora*, Lindl. Bot. Reg. t. 1107.—Hook. Fl. Bor. Am. vol. ii. p. 94.

HAB. Sunny places, rocky willow thickets, Kooskooskie valley. June. n. 354.

2. Collinsia *parviflora*, Lindl. Bot. Reg. t. 1082.—Hook. Fl. Bor. Am. vol. ii. p. 94.

HAB. Wet sunny rocks, very abundant at the Kettle Falls, Upper Columbia. March to May. n. 462.

PENTSTEMON, *L'Her.*

Sect. 1. Erianthera.

1. P. *Douglasii*, Hook. Fl. Bor. Am. vol. ii. p. 98.

HAB. Crevices of naked exposed granite masses, declivity of the high mountains of Tshimakeine, Spokan country. June. n. 438.

Sect. 2. Sepocosmus.

2. P. *laricifolius*, Hook. et Arn. Bot. Beech. Voy. p. 376. P. filifolium, *Nutt. MS. in Herb. Hook.*

HAB. Foot of a high granite mountain, Upper Sweet-water River. June. *n.* 239.

3. P. *grandiflorus,* Fras.—P.˚Bradburii, *Ph.*

HAB. Fertile valley of Upper Kansas, near Platte River, on the slopes of the hills. June. *n.* 248.

4. P. *speciosus,* Dougl. in Lindl. Bot. Reg. t. 1270.—Hook. Fl. Bor. Am. vol. ii. p. 98.

HAB. Sandy woods and protected situations along Spokan River. July. *n.* 641.

5. P. *glabra,* Ph.—P. Gordoni, *Hook. Bot. Mag. t.* 4319.—β. foliis paulo latioribus.—P. erianthera, *Fras.*—P. glabra, *Sims, Bot. Mag.* t. 1671.—P. alpinus, *Torr.*

HAB. Sunny sandy slopes of the argillaceous bituminous slate hills of Upper Platte, near the junction with Horse and Laramie Rivers. July. *n.* 117.—β. on a sunny sandy declivity of Spokan River mountains. July. *n.* 477.

6. P. *cæruleus,* Nutt.—P. angustifolius, *Fras.*—*Ph.*—β. floribus albidis roseo-tinctis.—P. albidus, *Nutt.*

HAB. High gravelly calcareous plains of Upper Platte, in the Black Hills. Corolla light azure-blue. Leaves glaucous. June, July. *n.* 154.—β. stony gravelly table-lands and ridges of Lower Platte; May (*n.* 662); and elevated fertile gravelly plains, Upper Kansas and Lower Platte. *n.* 199.—Excepting in colour, the plants which I here consider a var. of *P. cæruleus,* Nutt., quite agree with that species.

7. P. *cristatus,* Fras.—P. erianthera, *Ph.* (*non Fras.*)

HAB. Clayey slopes, hills of Upper Platte (and Colorado) towards the valley, with "*Erigeron hirsutum.*" June. *n.* 237.

8. P. *Cobæa,* Nutt.—Hook. Bot. Mag. t. 3465.

HAB. Fertile slopes of hills of the valley of Kansas River. May, June. *n.* 90.

Sect. 3. Eupentstemon.

9. P. *pubescens,* Sol.—P. lævigatus, *Sol.—Sims, Bot. Mag. t.* 1428.— Hook. Fl. Bor. Am. v. 2. *p.* 97.

HAB. Moist sunny fertile places, valley of Upper Platte. July; (*n.* 264); and foot of mountains, Nez Percez highlands. *n.* 407 or 418?

10. P. *confertus*, Dougl. in Bot. Reg. t. 1360.—Hook. Fl. Bor. Am. vol. ii. p. 96.—β. *nanus*; omnibus partibus triplo minoribus.

HAB. Stony plains, Upper Columbia. *July. *n.* 464.—β. Stony places and table-lands, Upper Sweet-water River. July. *n.* 238.

11. P. *deustus*, Dougl. in Bot. Reg. t. 1318.—Hook. Fl. Bor. Am. vol. ii. p. 95.

HAB. Basaltic declivities of St. Joseph's, Cœur d'Aleine River. June. *n.* 448.

12. P. *ovatus*, Dougl. in Bot. Mag. t. 2903.—Hook. Fl. Bor. Am. vol. ii. p. 96.

HAB. Shady alpine places, valley of Kooskooskie River. July. *n.* 642.

13. P. *procerus*, Dougl. in Bot. Mag. t. 2954.—Hook. Fl. Bor. Am. vol. ii. p. 97.—P. Tolmiei, *Hook. Fl. Bor. Am. v. 2. p.* 97.

HAB. Sunny rocky woods and banks of streams, Missouri and Oregon territory. July to September. *n.* 515.

Sect. 4. Saccanthera.

14. P. *Richardsoni*, Dougl. in Bot. Reg. t. 1121.—Hook. Bot. Mag. t. 2391.

HAB. Rocks at the Kettle Falls, Fort Colville, Upper Columbia. August. *n.* 582.

15. P. *venustus*, Dougl. in Bot. Reg. t. 1309.—Hook. Fl. Bor. Am. vol. ii. p. 95.

HAB. Slope of the high table-lands towards Kooskooskie. About fifty stems spring from a thick ligneous caudex. June. *n.* 487.

16. P. *glandulosus*, Dougl. in Bot. Reg. t. 1162.—Hook. Fl. Bor. Am. vol. ii. p. 95.

HAB. Dry shady rocky ravines of the trap mountains on the banks of the Kooskooskie, five miles above its junction with Lewis River. *n.* 362.

Sketch of the VEGETATION *of the Isthmus of* PANAMA; by M. BERTHOLD SEEMANN, Naturalist of H. M. S. Herald.

(*Continued from p.* 270.)

Nor is the flora destitute of plants which claim attention on account of their beauty, rarity, or singular configuration. The *Espiritu santo*, or Holy Ghost plant (*Peristeria elata*, Hooker), bears a flower resembling a dove, and is, like the *Flor de semana santa* (another *Orchidea*), almost held in religious veneration by the inhabitants, and eagerly sought for when in blossom. The *Biura* (*Petræa volubilis*, Jacq.) is a flower of whose beauty those who have only seen it in European conservatories can form but an inadequate idea. Nothing can be more charming during the dry season, than the sight of whole groves overspread with the long blue racemes of this creeper—it almost baffles description. The *Palo de buba* (*Jacaranda Bahamensis*, R. Brown) is another of those exquisite plants, on which poets delight to try their pen, and painters their brush. When this noble tree rises on the banks of the rivers, amidst the dark foliage of a luxuriant vegetation, and waves its large azure panicles in the air, the foot is involuntarily arrested, and we gaze for some time quite lost in wonder and admiration.

There are also a number of plants which exhale a delicious perfume. A long list of them could be cited, but it may suffice to enumerate the *Flor de Aroma* (*Acacia Farnesiana*, Willd.), *Buenas tardes* (*Mirabilis Jalapa*, Linn.), the different *Caracuchas* (*Plumieria* sp.), *Copecillo oloroso* (*Clusia* sp.), *Dama de noche* (*Cestrum paniculatum*, Willd.), *Guavito cansaboca* (*Pithecolobium ligustrinum*, Bth.), *Jasinto* (*Melia sempervirens*, Swartz), *Jasmin de monte* (*Tabernæmontana alba*, Mill.), *Nnorbo* (*Passiflora biflora*, Lam.), and *Manglillo* (*Ternströmia brevipes*, De Cand.) Some of them emit an odour almost too strong to be agreeable. I recollect, when ascending the Chagres in September 1846, all the trees of *Pithecolobium ligustrinum*, which adorn the banks, were in full flower; and so powerful was their smell, even in the middle of the river, that I became quite giddy, and was ultimately compelled to put cotton into my nostrils to exclude the perfume.

The most famous, however, of all the ornamental plants is the *Couroupita odoratissima*, Seem., combining a most delicious fragrance with a splendid flower. In the *Morro*, a forest near the village of Rio Jesus, are four of these trees, which are considered by the inhabitants as the only ones that exist in the country, and the greatest curiosities Veraguas can boast; and, indeed, I myself have never observed them in any other locality. They form a groupe, and are vernacularly termed *Palos de Paraiso* (i. e., *Paradise trees*), or *Granadillos*, deriving the former name from their beauty and the latter from the close resemblance which their flowers bear in shape and size to those of the *Granadilla* (*Passiflora quadrangularis*, Linn.). The trees are from sixty to eighty feet high, and up to an elevation of twenty feet; where the branches diverge, their stems are thickly covered with little sprouts, bearing, from February until May, blossoms, the odour of which is of so delightful and penetrating a nature, that in a favourable breeze it may be perceived at nearly a mile's distance. The flowers are one and a half to two inches in diameter, and their petals are of a beautiful flesh-colour with yellow stripes, contrasting charmingly with the golden stamens of the centre. The people of Veraguas, whose apathy is not easily roused by the beauties of nature, often repair to these trees during their flowering season, in order to behold the bright tints of the blossoms, and enjoy the delicious perfume which they exhale.

A production, less beautiful but equally singular, is the *Palo de velas*, or *Candle-tree* (*Parmentiera cereifera*, Seem.). This tree is confined to the valley of the Chagres, where it forms entire forests. In entering them, a person might almost fancy himself transported into a chandler's shop. From all the stems and lower branches hang long cylindrical fruits, of a yellow wax-colour, so much resembling a candle as to have given rise to the popular appellation. The fruit is generally from two to three, but not unfrequently four, feet long, and an inch in diameter. The tree itself is about twenty-four feet high, with opposite, trifoliolated leaves, and large white blossoms, which appear throughout the year, but are in greatest abundance during the rainy season. The *Palo de velas* belongs to the Natural Order *Crescentiaceæ*, and is a *Parmentiera*, of which genus, hitherto, only one species, the *P. edulis*, De Cand., was known to exist. The fruit of the latter, called *Quauhxilote*, is eaten by the Mexicans; while that of the former serves for food to numerous herds of cattle. Bullocks, especially, if fed with the fruit of this tree, Guinea grass, and *Batatilla* (*Ipomœa brachypoda*, Benth.), soon get fat. It is generally admitted, however, that the meat partakes in some degree of the peculiar apple-like smell of the fruit; but this is by no means disagreeable, and easily prevented, if, for a few days previous to the killing of the animal, the food is changed. The tree produces its principal harvest during the dry season, when all the herbaceous vegetation is burned up; and on that account its cultivation in tropical countries is especially to be recommended: a few acres of it would effectually prevent that want of fodder, which is always most severely felt after the periodical rains have ceased.

A tree, which has attained great celebrity, is that called *Cedron* (*Simaba Cedron*, Planch.). The most ancient record of it which I can find is in the 'History of the Buccaneers,' an old work published in London, in the year 1699. Its use, as an antidote for snakes, and place of growth, are there distinctly stated; but whether on the authority of the natives, or accidentally discovered by the pirates, does not appear. If the former was the case, they must have learned it while on some of their cruizes on the Magdalena, for in the Isthmus the very existence of the tree was unsuspected until about 1845, when Don Juan de Ansoatigui ascertained, by comparison, that the *Cedron* of Panamà and Darien was identical with that of Carthagena. The virtues of its seeds, however, were known, years ago, from those fruits imported from the Magdalena, where, according to Mr. William Purdie, the plant grows in profusion about the village of San Pablo. In the Isthmus it is generally found on the outskirts of forests in almost every part of the country, but in greater abundance in Darien and Veraguas, than in Panamà. The natives hold it in high esteem, and always carry a piece of the seed about with them. When a person is bitten, a little, mixed with water, is applied to the wound, and about two grains scraped into brandy, or, in the absence of it, into water, is administered internally. By following this treatment the bites of the most venomous snakes, scorpions, centipedes, and other noxious animals, have been unattended by dangerous consequences. Doses of it have also proved highly beneficial in cases of intermittent fever. The *Cedron* is a tree, from twelve to sixteen feet high, its simple trunk is about six inches in diameter, and clothed on the top with long pinnated leaves, which give it the appearance of a palm. Its flowers are greenish, and the fruit resembles very much an unripe peach. Each seed, or cotyledon I should rather say, is sold in the chemists' shops of Panamà for two or three reals (about 1*s*. or 1*s*. 6*d*. English), and sometimes a much larger price is given for them.

Highly interesting is the *Antà*, a species of *Vegetable Ivory* (*Phytelephas* sp.) distinct, probably, from that of the Magdalena. It grows in low damp localities, principally on the banks of rivers and rivulets, and is diffused over the southern parts of Darien, and the vicinity of Portobello, districts which are almost throughout the year deluged by torrents of rain, or enveloped in the thick vapour that is constantly arising from the humidity of the soil and the rankness of the vegetation. It is always found in separate groves, seldom or never intermixed with other trees or bushes, and where even herbs are rarely met with, the ground appearing as if it had been swept. In habit it resembles the *Corozo colorado,* or Oil Palm (*Elais melanococca,* Gærtn.); so much, indeed, that at first sight the two are easily mistaken for each other. Both affect similar localities, and have trunks which, after creeping along the ground a few yards, ascend, and attain about an equal height. Their leaves, also, resemble each other; and their fruit grows in a similar way, attached to short peduncles, and almost hidden in the axils. The habit, however, is nearly the only link that connects the *Antà* with the order of Palms: in flower, stamens, the organization of the fruit, in fine, in almost every essential character, it differs so widely from that family, that it cannot but be separated, and united with *Pandaneæ.* This species of *Phytelephas,* as has already been stated, is probably distinct from that growing on the banks of the Magdalena. The trunk creeps along the ground, and then ascends, seldom, however, higher than from four to six feet; it is always pulled down, partly by its own weight, partly by the aerial roots, and thus forms a creeping caudex which is not unfrequently more than twenty feet long. The top is crowned with from twelve to sixteen pinnatifid leaves, the entire length of which is from eighteen to twenty feet. The leaflets, or rather segments, are towards the base of the leaf alternate, towards the apex opposite: they are three feet long, two inches broad, and their entire number generally amounts to 160. All the plants which I saw were diœcious, the males always being more robust, and their trunks more erect and higher, than the females. The flowers of both emit a most penetrating almond-like smell, which attracts swarms of honey-bees, chiefly the stingless species inhabiting the forests. The male flowers are attached to fleshy spikes, which are from four to five feet long, and are hanging down. The female flowers appear in bundles, on short thick peduncles, and stand erect. The fruit, being a collection of drupes, forms large heads, and is at first erect, but when approaching maturity its weight increases, and the leaf-stalks, which so long supported the bulky mass, have rotted away, it hangs down. A plant bears at one time from six to eight of these heads, each of which contains on an average eighty seeds, and weighs, when ripe, about twenty-five pounds. The uses to which the *Antà* is applied by the Indians are nearly the same as elsewhere. With its leaves their huts are thatched, and the young liquid albumen is eaten. The "nuts," however, are turned to no useful purpose. The Spanish-

Isthmians did not know, before I visited the Isthmus, that *Vegetable Ivory*, or *Marfil vejetal* as they call it, existed in their country; and, although they have been told that with the produce of the groves of Darien whole ships might be loaded, no one has yet taken advantage of the discovery.

An indigenous production deserving especial notice is the *Jipijapa* (*Carludovica palmata*, R. et Pav.), a palm-like plant, of whose unexpanded leaves the far-famed "Panamà hats" are plaited. This species of *Carludovica* is distinguished from all others by being terrestrial, never climbing, and bearing fan-shaped leaves. The leaves are from six to fourteen feet high, and their lamina about four feet across. The spatha appears towards the end of the dry season, in February and March. In the Isthmus, the plant is called *Portorico*, and also *Jipijapa*, but the latter appellation is most common, and is diffused all along the coast as far as Peru and Chili; while in Ecuador a whole district derives its name from it. The *Jipijapa* is common in Panamà and Darien, especially in half-shady places; but its geographical range is by no means confined to them. It is found all along the western shores of New Granada and Ecuador; and I have noticed it even at Salango, where, however, it seems to reach its most southern limit, thus extending over twelve degrees of latitude, from the tenth N. to the second S. The Jipijapa, or Panamà hats, are principally manufactured in Veraguas and Western Panamà: not all, however, known in commerce by that name are plaited in the Isthmus; by far the greater proportion is made in Manta, Monte Christi, and other parts of Ecuador. The hats are worn almost in the whole American continent and the West Indies, and would probably be equally used in Europe, did not their high price, varying from 2 to 150 dollars, prevent their importation. They are distinguished from all others by consisting only of a single piece, and by their lightness and flexibility. They may be rolled up and put into the pocket without injury. In the rainy season they are apt to get black, but by washing them with soap and water, besmearing them with lime-juice or any other acid, and exposing them to the sun, their whiteness is easily restored. So little is known about these hats, that it may not be deemed out of place to insert here a notice of their manufacture. The "straw" (paja), previous to plaiting, has to go through several processes. The leaves are gathered before they unfold, all their ribs and coarser veins removed, and the rest, without being separated from the base of the leaf, is reduced to shreds. After having been put in the sun for a day, and tied into a knot, the straw is immersed into boiling water until it becomes white. It is then hung up in a shady place, and subsequently bleached for two or three days. The straw is now ready for use, and in this state sent to different places, especially to Peru, where the Indians manufacture from it those beautiful cigar-cases, which fetch sometimes more than 6*l*. apiece. The plaiting of the hats is very troublesome. It commences at the crown, and finishes at the brim. They are made on a block, which is placed upon the knees, and requires to be constantly pressed with the breast. According to their quality, more or less times is occupied in their completion: the coarser ones may be finished in two or three days, the finest take as many months. The best times for plaiting are the morning hours and the rainy season, when the air is moist: in the middle of the day and in dry clear weather, the straw is apt to break, which when the hat is finished is betrayed by knots, and much diminishes the value.

Here I must pause. Sufficient has been said to show that in the Isthmus of Panamà Nature has distributed her gifts with no sparing hand; and I have, hitherto, strictly confined myself to noticing these; in another article I shall endeavour to give a general view of the Agriculture, and to point out those productions which, by the agency of man, have been introduced from foreign countries.

(*To be continued.*)

FLORULA HONGKONGENSIS: *an Enumeration of the Plants collected in the Island of Hong-Kong, by* Capt. J. G. Champion, 95*th Reg., the determinations revised and the new species described by* GEORGE BENTHAM, ESQ.

(*Continued from p.* 264.)

TERNSTRŒMIACEÆ.

A more detailed account of the Hong-Kong *Ternstrœmiaceæ*, by Capt. Champion, was read before the Linnean Society in the autumn of 1850; but as a considerable time must elapse before that paper can appear in the Society's Transactions, the diagnoses of the new species are here repeated, with some additional observations, suggested by further examination and comparison, and the paper itself is referred to for the detailed descriptions.

1. Eurya *Chinensis*, R. Br. in Abel's Voy. Append. p. 379.

Dr. Gardner considered this to be the *Eurya Japonica*, Thunb., and supposed that the following species (*E. Macartneyi*) might be the *E. Chinensis*; but before I left Hong-Kong, having had access to Mr. Brown's figure and description, I had no hesitation in pronouncing the present species to be the true *E. Chinensis*, Br. In the British Museum I have, with Mr. Bennett, looked over the *Euryæ* brought home at that period, and found both my species. These specimens have confirmed me in the belief that the *E. Chinensis*, Br., is my plant, including two varieties, which, at first sight, would almost appear to be distinct species, but which are the mere results of situation, as to whether growing on barren hills, exposed to the wind, or under shelter in damp woods. I paid considerable attention to the plant in the autumn of 1849, from supposing at first the Happy Valley species to be distinct from the Victoria Peak specimens, gathered in 1847. The season of 1849 was cool, and favourable to the development of the flowers of *Ternstrœmiaceæ*, whilst 1847 was hot and dry. There are, also, intermediate forms, which prevent fixing any precise distinctions between the two varieties. The fruit of the more floribund variety is pea-shaped, purple, 3-celled, with about three seeds in each cell, attached at first to a placenta suspended from the apex of the axis; finally the fruit becomes filled with a purple mucilage, which nearly obliterates the cells. I believe the *E. Japonica* to be a mere variety of the same plant, slightly pubescent at the extremities of the branchlets. If this should prove to be the case, Thunberg's older name should be adopted. (*J. G. Champ.*)

2. Eurya *Macartneyi*, Champ., sp. n.; divine, frutescens, glabra, foliis majusculis coriaceis subellipticis obtuse serrulatis, floribus majusculis, staminibus circa 12-xx, stylis tribus, 3-4 distinctis revolutis, fructu purpureo circa 14-spermo.

In woods and on rocks, Hong-Kong, in flower and fruit from August to November. The ovary is 3-celled, with 9 to 10 ovules in each cell; the ovules, in place of being suspended, are attached nearly horizontally to the central axis. As a species this is quite distinct from the last. In the British Museum are specimens collected during Lord Macartney's voyage, but hitherto unnamed. (*J. G. Champ.*)

3. Cleyera *fragrans*, Champ. sp. n. arborea, tota glabra, ramulis

bi-trichotomis, foliis lanceolatis margine integriusculis leviter revolutis coriaceis, ramulis floribundis floribus axillaribus pallidis fragrantibus fugacibus solitariis bibracteolatis, sepalis fimbriatis petalisque glabris parvis, staminibus brevibus glabris, connectivo acuto, stylis 2 profunde divisis, stigmatibus reniformibus.

This tree, which I have described more in detail in the above-mentioned paper, constitutes much of the woods of Hong-Kong, flowering in May and October, and bearing fruit in October and November. It may require further comparison with the East Indian *C. gymnanthera*, from which, however, it appears to be sufficiently distinct. (*J. G. Champ.*)

4. Cleyera *dubia*, Champ., sp. n.; frutescens, tota glabra, foliis lanceolatis margine integriusculis leviter revolutis coriaceis, floribus majoribus (8¼ lin. diam.) axillaribus pallidis bibracteolatis, sepalis fimbriatis petalisque glabris, staminibus brevibus connectivo acuto, stylo saepius trifido, fructu globoso diametro supra-semiunciali.

Mount Victoria, flowering in February and March, fruit in June.—I have my doubts as to this being really different from the last. Their general resemblance is very striking; and the only specific differences I perceive are that the branchlets are more swollen, ash-coloured, and less compact; the flowers larger, nearly scentless, the petals less fugacious, and the fruit and seeds much larger,—differences which might almost be effected by situation. One of my specimens has the fruit nearly an inch in diameter. (*J. G. Champ.*)

5. Ixionanthus *Chinensis*, Champ., sp. n.; foliis petiolatis ovali-oblongis integerrimis reticulatis, pedunculis folio multo brevioribus, floribus decandris. Affinis *I. reticulatæ*.—*Folia* distinctius et longius petiolata, minora, basi angustata. *Pedunculi* multo breviores. *Calyx* profunde 5-fidus, laciniis orbiculatis lineam longis. *Petala* persistentia, orbiculata, calyce subduplo longiora. *Stamina* 10, exserta. *Discus* 10-crenulatus inter stamina et ovarium. *Ovarium* subglobosum, acutiusculum, 5-loculare, *ovulis* in quoque loculo 2, ex apice anguli centralis pendulis. *Stylus* longissimus, simplex, apice discoideo-dilatatus, supra stigmatosus. *Capsula* fere pollicaris, rigida, septicido-5-valvis. (*G. Bentham.*)

Grows in the Happy Valley to be a small tree, and is there rather common; flowers in May and June. See a more detailed description in the above-quoted paper.

✓ 6. Polyspora *axillaris*, Sweet.
Common all over the island, and constitutes much of its wood.
7. Schima *superba*, Gardn. et Champ. Kew Journ. Bot. vol. i. p. 246. ✓
A very rare tree in Hong-Kong, except in the woods on Little Hong-Kong, near the top of the slopes, where it grows abundantly. Its bunches of large white flowers, in May, resemble, at a distance, those of *Mesua*. Fruit from October to December.
8. Pentaphylax *euryoides*, Gard. et Champ. in Kew Journ. Bot. vol. i. p. 244.
An under-tree of great beauty when in flower, and exceedingly common in the Hong-Kong woods. The flowers are small and white, and the pseudo-racemes, growing in a pyramidal shape, owing to the lower flowers expanding first, have a very peculiar effect. The seed is dry, its coating membranaceous, embryo conduplicate, radicle terete, cotyledons elongate semicylindric. (*J. G. Champ.*)
9. Camellia *Japonica*, Linn.
Of this, two trees growing wild in Hong-Kong were discovered by Lieutenant-Colonel Eyre, of the Royal Artillery, and Mr. J. Bowring now mentions a third as having been found in the Happy Valley woods. It is a moderate-sized, smooth-barked tree, loaded in October with single pink flowers. The fruit is smooth and much smaller than in the *C. spectabilis*, being rather above an inch in diameter. The petals, about seven, adhere at the base in a ring, and soon fall off. The sepals are slightly silky, and the leaves more elongated than in most cultivated varieties. (*J. G. Champ.*)
10. Camellia *salicifolia*, Champ., sp. n.; arbuscula, ramulis pubescentibus flexuosis, foliis subsessilibus elongato-ovatis acuminatis serratis pubescentibus, floribus parvulis albis, sepalis acuminatis pubescentibus, capsulis glabris parvis rostratis 1-3-spermis.
Woods in Hong-Kong. This and the two following new species are more fully described in the above-mentioned paper. (*J. G. Champ.*)
11. Camellia *assimilis*, Champ., sp. n.; frutex, ramulis glabris, foliis subsessilibus lanceolatis acuminatis serratis glabris, floribus parvulis pendulis albis, sepalis sericeis obtusis, capsulis glabris parvis rostratis:
Mount Victoria and Mount Gough. I have seen this species growing almost alongside of the last, and the general resemblance is very striking. Its smooth habit, shorter and wider leaves, and more especially the difference of shape in the sepals, form the distinction.

Its form is more stunted, and it grows amongst rocks in ravines. Its pretty white pendulous flowers come out about January. The late Dr. Gardner referred it to *C. caudata*, Wall.; but a comparison of specimens shows it to be perfectly distinct. (*J. G. Champ.*)

12.? Camellia *Banksiana*, Lindl., is, I believe, a Hong-Kong species, but unknown to me. Some specimens, found on a hill near Mount Parker, and sent home by Mr. J. Bowring previous to my arrival, were considered as belonging to the *C. Banksiana*. Mr. Bowring mentioned its having sweetly-perfumed flowers, so that it could scarcely be the same as the *C. assimilis*, of which the flowers are scentless, notwithstanding a general affinity in habit and in the white pendulous flowers. The vegetation of the hill on which it grew having been burnt by the Chinese for agricultural purposes, we could not succeed in finding spceimens. (*J. G. Champ.*)

13. Camellia *spectabilis*, Champ., sp. n.; arborea, foliis lanceolatis acuminatis glabris crenatis subtus reticulatis, floribus solitariis magnis albis axillaribus et subterminalibus, sepalis coriaceis fructibusque pomi magnitudinis sericeis.

On Mount Victoria and more abundantly in the Happy Valley woods. A very handsome species, quite distinct from *C. oleifera*. It flowers sparingly when young, but abundantly when it has grown into a tree. (*J. G. Champ.*)

14. Thea *Bohea*, Linn., is cultivated in Hong-Kong, but is not indigenous. It frequently forms borders to garden-beds, just as we employ the Box. As a genus there seems no good distinction from *Camellia*. The *C. oleifera*, Abel, and *C. euryoides*, Lindl., have neither of them been found in Hong-Kong. (*J. G. Champ.*)

GUTTIFERÆ.

1. Garcinia *multiflora*, Champ., sp. n.; foliis ovatis obovatisve acuminatis, floribus (hermaphroditis) corymboso-paniculatis 4-sepalis 4-petalis, staminibus 4-adelphis.—*Frutex.* **Folia** breviter petiolata, $3-3\frac{1}{2}$ poll. longa, $1\frac{1}{2}-2$ poll. lata, apice rotundata et brevius longiusve acuminata, basi cuneata v. rotundata. **Panicula** corymbiformis, trichotoma, intra folia summa sessilis et ea subæquans. **Bracteæ** lanceolatæ, 2 lin. longæ, deciduæ. **Flores** numerosi, ad apices ramulorum terni, breviter pedicellati. **Sepala** orbicularia, 3 lin. lata. **Petala** obovalia, duplo longiora. **Staminum** phalanges breves, crasso-

carnosi, antheris numerosissimis. *Ovarium* quadratum, stigmate maximo peltato; *loculi* duo tantum in flore a me aperto. *Flores* masculos haud vidi.

A shrub common towards the Black Mountain, flowering in the heat of summer, which caused much difficulty in procuring specimens.

2. Garcinia *oblongifolia*, Champ., sp. n.; foliis oblongis basi longe angustatis breviter petiolatis, floribus terminalibus, fœmineis solitariis sessilibus, masculis 3-7 pedicellatis 4-sepalis 4-petalis, staminibus fere ad apicem monadelphis.—*Folia* in genere parva, 2½-3 poll. longa, raro pollice latiora, superiora ramorum fœmineorum subsessilia, cætera petiolo 2-4 lin. longo fulta. *Sepala* orbiculata, 2 lin. longa. *Petala* in flore masculo fere 5 lin. longa, in flore fœmineo minora. *Stamina* marium in massam tetragonam floris centrum occupantem connata. *Antheræ* subsessiles, circa 40.

Common in the Happy Valley woods, where it is arboreous. Capt. Champion states, that if he recollects right, the fruit is of the size of a small apple, and not lobed or grooved, but smooth as in the Mangosteen.

3. Calophyllum *membranaceum*, Gardn. et Champ. in Kew Journ. Bot. vol. i. p. 309.

In a ravine on Mount Victoria, where it is tolerably abundant, also in the Happy Valley woods.

HYPERICINEÆ.

1. Hypericum *Japonicum*, Thunb.
Common in marshes and fields.

2. Ancistrolobus *ligustrinus*, Spach, Ann. Sc. Nat. Par. ser. 2. vol. v. p. 352.
Very abundant in the low grounds of Hong-Kong.

MALPIGHIACEÆ.

1. Hiptage *Madablota*, Gærtn—Juss. Malpigh. p. 248.

In its leaves, rather less coriaceous than the common East Indian varieties, and in its full-sized flowers, this resembles the specimens gathered by Dr. Wallich at Prome. It is certainly not the *H. obtusifolia*, DC., described as a Chinese species. It flowers in May, festooning the trees in the Happy Valley woods, and is found also on rocks on Mount Gough, but rare.

ACERINEÆ.

1. Acer *reticulatum*, Champ., sp. n.; glabrum, foliis integerrimis ovatis oblongisve breviter acuminatis coriaceis reticulato-venosis utrinque viridibus, corymbis compositis glabris, alis fructus divaricatis.

Near *A. oblongum*, Wall., but the leaves are much firmer, more reticulate, and with a shorter point, the petioles shorter, the corymbs or panicles smooth (not hairy), the flowers rather larger, and the wings of the fruit rather longer and narrower and much more spreading. The *A. laurinum*, from Java, is at once distinguished by the leaves, white underneath.

On Mount Gough, in flower in December, in fruit in July and January, also in the Happy Valley in flower in June. Leaves pellucid on the midrib and margin, rose-coloured when young. Flowers 3-4 lines in diameter. Sepals oblong, rose-coloured. Petals spathulate, white. Style filiform, bifid, and acute at the apex. Ovary 2-celled, with two pendulous ovules from near the apex of the axis. Ovary and fruit rarely trimerous.

SAPINDACEÆ.

1. Nephelium *Litchi*, Camb. Mem. Sapind. p. 30.

Very common in Hong-Kong, but probably always cultivated.

(*To be continued.*)

Figures and descriptions of two species of BOEHMERIA, *of which the fibre is extensively used in making Cloth; by* SIR W. J. HOOKER, D.C.L., F.R.A. and L.S.

(With two plates, TAB. VII. & VIII.)

The *cloth* to which we here allude has been already noticed in this Journal, and mention made of the plants yielding the fibre. But the botanical appellations of plants can give little information to the merchant, or the manufacturer, or the consumer of the produce, or to any save a professed botanist, unless accompanied by a figure or full description; and we desire to convey information which should be useful to all classes of our readers. The scientific student of plants, upon being told that the species under consideration belong to the genus *Boehmeria*, will at once comprehend that the *Boehmerias* are of the NETTLE-FAMILY, and will call to mind many species of the same natural groupe, endowed with the same property, viz., that of yielding *textile fibre*, not even excepting our common *stinging Nettle** (Urtica urens). Indeed, so closely is the genus *Boehmeria* allied to the true *Nettles*, that most of the species were by the older botanists considered and were called *Urticæ*; for example, the very two species we now have under consideration, viz., *U. Puya*, Roxb., and *U. nivea*, Linn.

Jacquin, in his ' Selectarum Stirpium Americanarum Historia,' 1763, first established the genus *Boehmeria*, but so little did he understand the genus himself, and so minute are the characters, that in 1770 he published a figure of *Urtica nivea* as still an *Urtica*, which is now universally included in *Boehmeria*. The chief distinction consists in *Urtica* having a 2-valved perianth to the female flower ; while that of *Boehmeria* has a tubular perianth, more or less distinctly 4-lobed at the apex. Whenever the entire family of *Urticeæ* shall have received the attention it deserves from the studies of a competent botanist, the characters of these and allied genera will be reformed. Our object at present is chiefly with particular species.

At p. 25 of the first volume of the present series of the ' Journal of Botany,' or ' Kew Garden Miscellany,' we noticed a very valuable textile from China, recently known to the merchants of Europe by the name of " Chinese Grass," and the beautiful material manufactured from it as Chinese Grass cloth. By the assistance of Dr. Wallich and Sir George Staunton, we ascertained that this was no grass at all, but the produce of a kind of *Nettle* of the East Indies and China, known to botanists as the *Urtica nivea* of Linnæus, or *Boehmeria nivea*, Gaudichaud. Shortly after, at p. 159 of the same volume, we gave an extract from a periodical at Berlin, ' Naturforschende Freunde,' in which Dr. Münter had endeavoured to show that the *Chinese Grass cloth* was derived from the fibre of the Jute, *Corchorus capsularis*, a plant of which an account may be found at p. 25 of our same volume of this Journal, and at p. 91 and tab. 3 of the following volume (vol. ii.)

---

* Mr. Seemann speaks of the great value of *Boehmeria albida*, on account of its fibre, in the Society Islands, and Dr. Campbell of the " gigantic stinging Nettle " of Nepal and Sikkim (*Urtica heterophylla*, Willd. ?) used in making cloth, called *Bangra*.

there is a description and figure. That, however, is a fibre we wholly derive from India proper, and, however valuable commercially, it is very inferior in quality to the Chinese Grass. This error we endeavoured to correct; nevertheless statements are again issued, contrary to what is the fact, by merchants interested in the subject of the "Chinese Grass," stating that it is a "kind of *Cannabis*, or *Hemp*;" and we presume, because the Hemp is a plant whose cultivation is suited to our climate, so they recommend this as a fit object for cultivation with us, and seed has been imported and distributed accordingly. Long previous even to the time when the commercial importance of the fibre became known among us, we had raised this plant, and had it in cultivation in a hot-house or in a warm green-house. This present year we have planted it in the open ground, rather with a view of showing that it cannot succeed, than with a hope of its bearing our climate unharmed, save during the hottest of the summer months.

But this is no reason why the "Chinese Grass" should not be cultivated, and advantageously, in our colonies, that is, such of our colonies as possess a climate nearly analogous to that of Canton; and we cannot doubt that it would, with due care, prove a most valuable and important article of export. It is only a true and correct knowledge of such plants, and of the peculiarities of soil and climate necessary for their being successfully reared, that can enable us to grow them to good purpose.

The second plant, of which we here give a figure (Tab. VII.) and brief description, we at present know less about; it is the "Pooah," "Puya," so called in Nepal and in North-eastern India (*Boehmeria Puya*, Wallich). What we do know is given at p. 26 of the first volume of this Journal; but although all that we have yet seen of the cloth made from it by the natives of Sikkim be of a very inferior quality, yet, properly prepared, it is likely, judging from the close affinity of the two plants, that it may be found equal to that from the *B. nivea*. We trust it will be soon put to the test.

There can be no doubt that many mistakes about plants originate in the English vernacular names that are assigned to them, too often in such a way as wholly to mislead: as, for example, the *Prunus Lauro-Cerasus* is called "Common Laurel" (it being a kind of Plum or Cherry); and *Ilex*, which means a Holly, is applied a kind of Oak, &c. &c. Why the Chinese Nettle was called "Chinese Grass" I cannot tell, nor is it worth the time to consider or inquire; but assuredly, as far as can be done, since the Latin scientific names are not palatable to the uninitiated, it is very desirable to give such English names as may lead the mind to a familiar object, with which the one in question has some resemblance. In the present case the name of "Chinese Nettle," in lieu of Chinese Grass, would appear to be an improvement; but then, it may be asked, why, if a "Nettle," is not its proper Latin name "*Urtica*"? To this we have to answer, that the niceties of botanical discrimination require that this plant and the *Puya* should be separated from the true Nettles. We then propose the word Boehmer-Nettle for those plants which in botanical language are called *Boehmerias*.

1. BOEHMERIA NIVEA, *Gaud.*
Chinese Boehmer-Nettle, or Chinese Grass-Plant.
(TAB. VIII.)

Fruticosa erecta, caule petiolisque patenti-pilosis, foliis alternis longe petiolatis lato-cordatis basi 3-nerviis subito anguste acuminatis grosse serratis subtus dense albo-pannosis, paniculis axillaribus, masculis superioribus, floribus glomeratis, glomerulis subsessilibus, pericarpiis basi attenuatis.

Boehmeria nivea, *Gaudich. in Frey. Voy. Bot. p.* 499 (*excl. syn. Rumph.*) *Miq. in Plant. Jungh. p.* 33.

Urtica nivea, *Linn. Sp. Pl. p.* 1398 (*excl. syn. Rumphii*). *Lour. Fl. Cochin. v.* 2. *p.* 682. *Jacq. Hist. Schœnbr. v.* 1. *t.* 166. *Willd. Sp. Pl. p.* 565. *Wall. Cat. n.* 4606.

Urtica tenacissima, *Roxb. Fl. Ind. v.* 3. *p.* 590. *Wight, Ic. Pl. Ind. Or. v.* 2. *t.* 688.

HAB. China. Inhabiting walls, according to Linnæus; most extensively cultivated there. Native, also, of the island of Sumatra, according to Marsden, cultivated on account of its fibre, and called "Caloose." Pulo Penang, probably only cultivated, where it has the Malay name of "Rami." (*Roxb.*). Cultivated in the Botanic Garden of Calcutta, where it thrives exceedingly, and "strikes as readily from cuttings as the Willow." *Roxb.*

A *shrub*, 3–4 feet, or probably more, in height, erect, branched, the young *branches* green and hairy with short spreading hairs. *Leaves* alternate (characteristic of the division *Procris* of *Boehmeria*, according to Gaudichaud). *Leaves* on long petioles, broadly cordate but having no sinus or lobes at the base, the base rather truncate or tapering suddenly into the petiole, the apex suddenly acuminated into a slender point, almost caudate, the margin coarsely serrated, full green above, white with dense fine down beneath, 3-nerved at the base, the rest of the leaf penninerved, the principal nerves united by slender nervelets. *Petiole* terete, hairy. *Stipules* subulate, brown, deciduous, soon becoming brown. *Peduncles* about 2 together, axillary, slender, filiform, paniculate, bearing nearly sessile *clusters* or *glomerules* of *flowers* throughout their length, hairy. *Male panicles* below; *female* above. *Male flowers* few in each glomerule. *Perianth* deeply 4-partite, hairy externally; the segments ovate. *Stamens* 4, spreading: rudiment of an ovary. *Female flowers* several in a globose glomerule. *Perianth* of one piece, oblong-cylindrical, very hairy, 4-toothed. *Ovary* included: *style* thick-subulate, much exserted beyond the perianth, hairy. *Achænium*, when ripe, obovate, substipitate.

TAB. VIII. *Boehmeria nivea.* Fig. 1. Male flower; fig. 2, glomerule of female flowers; fig. 3, single female flower; fig. 4, pericarps (or seeds, as they are generally called), *nat. size*; fig. 5, pericarp :— *all but fig. 4 more or less magnified.*

2. BOEHMERIA PUYA.
Nepal Boehmer-Nettle, Pooah or Puya.
(TAB. VII.)

Fruticosa elata erecta, caule petiolisque appresso-hirsutis, foliis alternis sublonge petiolatis lato-lanceolatis e basi parallelo-trinerviis anguste acuminatis pergrosse serratis subtus albido-lanatis, paniculis axillaribus, floribus glomeratis, glomerulis subsessilibus, pericarpiis basi obtusissimis.

Urtica Puya. *Herb. Ham.—Wall. Cat. n.* 4605.—*Hook. in Kew Gard. Misc. v.* 1. *p.* 26 (*Boehmeria*).

Urtica frutescens, *Roxb. Fl. Ind. v.* 3. *p.* 589. (*excl. Syn. Thunb. et Willd.*)

Pooah, *Campbell in Trans. of Agric. Soc. of India,* 1847.

HAB. Mountains north of Bengal and Oude, *Roxburgh*. Nepal, *Hamilton, Wallich*. Mountains of Eastern Nepal and Sikkim, at the foot of the hills skirting the Terai to the elevation of 1,000 and 1,200 feet, and on the mountains up to 3,000 feet, in open hilly places,

Dr. Campbell. Dr. Wallich gives, doubtfully, in Taong-Dong, Ava. The north-eastern boundary of Bengal appears to be its native locality.

This grows to the height of 6–8 feet. The leaves, it will be seen, are very different from those of *B. nivea*; the male flowers very similar: the female we have seen only few and imperfect, and much advanced. The pericarp, or seed, is quite unlike that of the *B. nivea*. For the particulars and uses of this species we must refer to our first volume of the 'Kew Garden Miscellany,' and to Dr. Campbell's Memoir above quoted.

TAB. VII. *Boehmeria Puya.* Fig. 1. Male flower; fig. 2, pericarp, enclosed in its perianth; figs. 3, 4, pericarps separated from the perianth :—*magnified*.

BOTANICAL INFORMATION.

Plants of ALGERIA.

[The following circular has been recently issued from Paris.—ED.]

Nous avons l'honneur de vous prévenir que M. B. Balansa, encouragé par l'adhésion de plusieurs membres de la Société d'Exploration Botanique, vient d'explorer avec soin les environs de Mostaganem (province d'Oray), dont la végétation encore peu connue est bien différente de celle des points de l'Algérie visités par M. Jamin. M. Balansa, grâce à la connaissance qu'il avait du pays, a pu se borner à ne recueillir que des espèces rares ou propres à l'Algérie en négligeant les espèces répandues dans presque toute la région Méditerranéenne. Il a rapporté environ 200 espèces de choix en assez grand nombre pour pouvoir être offertes à tous les membres de la Société. Les échantillons qui composent la collection sont préparés avec grand soin et recueillis en fleurs et fruits, toutes les fois que cela a été utile ou possible. La détermination des espèces est due à l'obligeance de M. Durieu de Maisonneuve, auteur de la partie botanique de l'Exploration Scientifique de l'Algérie. Chaque plante sera munie d'une étiquette imprimée, portant un numéro d'ordre. Le prix de la collection est fixé à raison de 20 francs par centurie.

M. Balansa se propose de continuer incessamment ses voyages

botaniques en Algérie, et de visiter, en particulier, la portion du désert ou Sahara Algérien qui avoisine Biskara.

Nous espérons, Monsieur, que vous voudrez bien nous honorer de votre adhésion, et nous vous prions de vouloir bien faire parvenir votre réponse dans le plus bref délai possible à M. Balansa, Rue Suger, 1, à Paris, ou à M. Ernest Cosson.

Paris, le 22 Juillet, 1851.

P.S.—Nous profitons de cette occasion pour vous informer qu'il reste à la disposition des souscripteurs un certain nombre des collections de M. Jamin (environ 100 espèces), recueillies en 1850 aux environs d'Alger, ainsi qu'une centurie de plantes intéressantes recueillies en 1850, par M. Dœnen, dans le Val Lassina, Lombardie.

MICROSCOPES *of the late* PROFESSOR LINK.

Two excellent microscopes, used by Professor Link for preparing his representations of the microscopical anatomy of plants, are to be sold at Berlin. One of these instruments was constructed by the celebrated optician, Plössl, of Vienna. It has five ocular and six object-glasses, and magnifies by the first ocular from 25 to 210 times, and by the last 1145 times. A revolving micrometer, of excellent workmanship, and measuring with the greatest precision, a large prism and a large lens mounted on separate statives, both for illuminating, and various other implements, are added. The instrument cost altogether about £45. The lenses are in no way injured, nor the screws worn out. The other instrument is a catadioptric microscope, by Professor Amici, of Florence, the discoverer of the pollen-tubes going down to the ovules. It is excellent for its achromatism and its large field of view. There is a fine prism for refracting objects. It has four ocular and five object-lenses. Its magnifying power reaches from 21 to 1558 times, or with the first ocular to 286 times. The object-lenses, separately mounted, are also used either alone or all together as a simplex. The object-table slides up and down, and is furnished with a turning diaphragm and two revolving micrometers, the one moving from side to side, the other forward and backward. Two different kinds of camera lucida, and various other implements, especially for illuminating and excluding the false light, are added. This instrument has been but little used. Its price is said to have been about £45. Offers will be received by Dr. Pritzel, Berlin, Alexandrinen-strasse, 88.

On some facts tending to show the probability of the Conversion of ASCI *into* SPORES *in certain* FUNGI; *by the* Rev. M. J. BERKELEY, M.A., F.L.S., *and* C. E. BROOME, ESQ., M.A.

Some very interesting observations on the supposed spermatozoids of Lichens, and their relation to the spores of certain genera of Fungi, such as *Cytispora, Septoria,* and other sporophorous *Hypoxylaceæ,* were laid before the French Academy by our excellent friend Mons. L. R. Tulasne, on the 24th and 31st of March of the present year. He has not confirmed the observations of M. Itzigsohn as to the mode of growth or activity of the contents of the black specks so common on the fronds of Lichens, and which have been conjectured, for half a century, to contain the male organs, though without attracting the attention which they deserved. It may be observed, however, that in *Borrera ciliaris,* the species more immediately examined by Itzigsohn, the pustule when young is filled with cellular tissue, as indeed is the case with those *Sphæriæ,* such as *S. herbarum,* which we have examined at an early period of growth; that each cell contains a distinct nucleus, and that when the oblong bodies, endowed with a slight molecular action but by no means with any great mobility, are present in extreme profusion, not only is the cellular tissue still observable, but a very careful examination with a first-rate compound microscope, and also with deep doublets, has failed to detect anything like sporophores, and left a complete conviction that the bodies were indeed produced within the cells. Indeed it was believed that a sort of oscillating motion, as of a body attached at one extremity, was observed before the corpuscles had attained their freedom. It is possible, therefore, that both M. Itzigsohn and M. Tulasne may be correct, and that much still remains to be discovered with respect to the structural as well as the physiological relation of these bodies.

M. Tulasne, it should be observed, with a commendable caution, does not profess to come to any distinct conclusion respecting the nature of the bodies in question, or of the naked spores of many Hypoxylaceous Fungi, which resemble some of them so closely as to indicate something more than mere analogy, especially when the fact is taken into consideration that a vast portion of the supposed species of certain genera are so constantly connected with ascophorous *Sphæriæ* as to make it highly probable that, notwithstanding the immense difference in point of structure, they are really distinct forms of so many species. He does not seem, however, to believe that their spores are mere modifications of asci, though he is not prepared to assert that they indicate differences of sex.

It is not our intention to offer any more positive opinion as to their nature, though the observations now adduced tend rather to show the convertibility of organs at first apparently so different. The facts, indeed, have already been adverted to in Morton's 'Cyclopædia of Agriculture,' but they seem to us so interesting as not to be unworthy of attention, especially as the figures illustrative of them, with a single exception, have not been published.

The species which have afforded the materials for the following remarks are more especially three; viz., *Tympanis saligna*, Tode; *Sphæria inquinans*, Tode; and *Hendersonia mutabilis*, Berkeley and Broome.

Tympanis saligna scarcely differs from a Lichen, except in the total absence of a crust, and consequently of gonidia. Its apothecia are at first closed, but at a later period of growth the fructifying disc or hymenium is exposed. On the same twig, in this instance of the common Privet, some specimens exhibited all the characters of *Tympanis*, and others those of *Diplodia*. As long as the naked spores occurred only in specimens in which the disc was not expanded this caused no surprise, for nothing is more common than to find *Sphæriæ* and *Diplodiæ* on the same matrix, which cannot be distinguished externally; but further examination exhibited the proper fructification of *Diplodia* in specimens with an open disc, a character quite at variance with that of the genus; and then the same hymenium was detected, producing rather large uniseptate naked spores, and broad elongated asci, exceeding them many times in length, and containing a multitude of minute oblong sporidia. Professor Fries, in a letter lately received, informs us that he has observed a similar fact in *Hendersonia syringæ*.

The circumstance of the constant or occasional occurrence of *Uredo* and *Puccinia*, *Uredo* and *Aregma*, *Uredo* and *Xenodochus*, or two species of the same genus, in the same spot, may be adduced as analogous; but where there is no perithecium, the occurrence of two species on the same spot of the same matrix is not matter of so much surprise, though suggestive of further consideration.

The case, however, of *Sphæria inquinans*, which we have now to bring forward, is still stronger. This species and *Stilbospora macrosperma* were extremely abundant on an old elm at Batheaston in January last. Not only were they so intermixed as to make it difficult, from the close resemblance of the fruit, that of the one being merely a little more elongated, and in a very slight degree more attenuated at either extremity, with rather a browner tinge, to say at once which was the *Sphæria*, which the *Stilbospora*, but the same orifice in the bark gave egress both to the sporidia of the one and the spores of the other. At the base of the spores of the *Stilbospora*, where seated on their sporophores, from one to three short sheaths were observed, as if the spore had burst through one or more enveloping membranes. But not only was the *Stilbospora* produced in the same portion of the bark as the *Sphæria*, or perhaps, to speak more correctly, in the same stroma, but in one case it was actually developed on the external surface of a perithecium, the inner surface giving rise to perfect asci, with their proper sporidia. In a certain stage of growth the sporidia of the *Sphæria* are furnished at either extremity with a cirrhiform appendage, but this is not always visible in the ejected mass which surrounds the common orifice of the perithecia. Analogous appendages occur in some other species.

The third case to which we invite attention is *Hendersonia mutabilis*, a species which occurs on twigs of Plane. The main perithecium contains one or more cavities, more or less isolated, which produce far smaller hyaline bodies, and which do not accord with the genus *Hendersonia* but rather with *Phoma*. This is not indeed a case bearing upon the conversion of asci into spores, but is interesting as exhibiting one perithecium within another; and whether considered as a new cell developed within the old one, and consequently containing younger spores (a view at first adopted, but which, on maturer consideration, seems scarcely tenable), or as two forms of spores both belonging to the same species but produced in distinct cavities, or finally as two genera united within the same common receptacle, it is full of interest. We are not prepared, as in the last case, to say that the same wall from its two sides produces different forms of fruit, though in some sections the two fertile surfaces are so confluent above that it is very probable that the same fact will be found to obtain here also.

These instances certainly seem to indicate rather a transformation of organs than any totally distinct productions; a view, indeed, which is not at variance with the possibility of the transformed organs being an indication of sexual functions, if we may be allowed to form any inference from known analogies in the animal world. Dr. Hooker, when examining the fruit of *Laminariæ*, on his return from the Antarctic expedition, felt convinced of the possibility of the transformation of an ascus into a spore; a view entertained long since by Fries, and which is certainly supported by those instances in Fungi and Lichens where a single spore only is developed in an ascus. The difference in the analysis of the genus *Sphærophoron*, as given by Dr. Montagne in the 'Annales des Sciences Naturelles' and by Dr. Hooker in the 'Antarctic Flora,' is probably due to a similar change; the former exhibiting true asci containing sporidia, the latter moniliform threads breaking up into spores. In earlier times the analysis of one would have been pronounced erroneous, but the present age, with deeper knowledge of the apparent anomalies exhibited by nature, and, consequently, a greater measure of diffidence, regards such discrepancies as calls to further investigation, and as the possibly available keys for the solution of difficulties which have been hitherto insurmountable.

Explanation of the Figures.

TAB. IX.

1. *Tympanis saligna*, producing both sporidiferous asci and naked spores from the same hymenium :—*highly magnified.*
2. *Sphæria inquinans* and *Stilbospora macrosperma* growing together on the same matrix, and having a common orifice for the emission of their sporidia and spores :—*slightly magnified.*
    a. a. a. *Sphæria inquinans*.   b. b. *Stilbospora macrosperma*.
3. The same :—*more highly magnified.*
4. Portion of the perithecium of *Sphæria inquinans*, producing asci internally and naked spores externally.
9. Normal form of spores in *Stilbospora macrosperma*.
10. Ditto in *Sphæria inquinans*.

TAB. X.

5. Section showing the stroma of *Stilbospora macrosperma* beneath the cuticle with its spores :—*slightly magnified.*
6. Part of the same, *more highly magnified*, exhibiting the sporophores with their spores and the sheaths at the base of the latter.

b. Stroma.

c. c. c. Sporophores with young spores apparently bursting through an outer membrane.

d. Lateral and dorsal view of the mature spores.

7. Asci of *Sphæria inquinans* in different stages.
8. Sporidia of the same with the curled appendages; the left-hand spore is germinating.
11. Section of *Hendersonia mutabilis*, showing the two cells containing different bodies.

a. Colourless bodies from the upper cell, like the spores of a *Phoma*.

b. Brown bodies from the lower cell, like those of a *Hendersonia*.

c. Dark bodies which seem to be the young state of those marked (*b*).

ANGIOPTERIS LONGIFOLIA, *Grev. et Hook., of Sir William Jackson Hooker's Herbarium, and its synonyms; communicated by* DR. W. H. DE VRIESE, Professor of Botany in the Royal University at Leyden.

Fronde bipinnata, rachi secundaria terete, glabra; pinnulis valde remotis, petiolatis, inæqualiter cordatis, longe lineari-lanceolatis, rectis vel curvis subfalcatisve, obtuse dentatis vel crenatis, vix serrulatis, subrepandisque, apice angustato fere sinuatis, in dorso pulvere quasi farinaceo firmissime adhærente adspersis; costa valida, fusco-nigra, venis furcatis et simplicibus; soris a margine remotis, distinctis, ad extremum usque apicem continuatis.

A. longifolia, *Grev. et Hook. Enum. Fil. in Hook. Bot. Misc. v.* 3. *p.* 227. 1838 (*excl. omnibus synonymis !*).

HAB. Pitcairn's and Society Islands, *Mathews* (no. 2). Vidi in Herb. Hook. ipsum Mathewsii specimen, cui species Hookeriana innititur. Adscripta sunt: "*March to May*, 1830. *Fronds* 6–7 *feet. A. M.*" (manu Mathewsii).

Adumbratio. Duo adsunt specimina, quæ ad eandem pertinent frondem. In his hæc agnosco. Longitudo est 0,9; latitudo 0,42. Rachis secundaria propter desiccationem fusca, 0,005 crassa, sursum attenuata, recta. Pinnarum numerus æquat 26. Inferiores sunt fere 0,175 longæ, mediæ 0,22 æquant, apicales sensim decrescunt. Latitudo 0,015, in omnibus fere eadem est. Apex solus 0,03 longus et fere 0,002 latus est. Pleræque pinnulæ fere 0,028 vel magis etiam a se invicem remotæ, horizontales sunt, paucæ curvatæ vel et subfalcatæ, petiolulatæ, petiolis fere 0,002 vel 0,035 æquantibus, in aliis brevioribus, squamuloso-pilosis, lanuginosis, vel demum glabris. Structura pinnularum est membranacea, superne opacæ sunt atque glaberrimæ, costa proëminente atque venis primariis conspicuis. Dorsum pallidius est propter pulverem subtilissimum tenacissime adhærentem, quo tota fere illa superficies inter venulas obtegitur. Costa ibi prominet, badii coloris et hic illic fusca lanugine tecta. Venulæ sunt alternato-pinnatæ, badiæ, inde ab origine ex costa jam furcatæ, aliæ vero primum simplices demum furcatæ, aliæ simplices sunt; omnes in marginem, ut videtur, abeunt. Venulæ secundariæ ex dentis crenæve sinu aliæ ad costam, aliæ ad furcaturæ alam decurrunt (*Eu-angiopteris*, Presl, Suppl. Tent. Ptcrid. p. 19). Sunt hæ tenuissimæ, nudæ, badiæ. Basis utrinque rotundata est, latere superiore minore, inferiore majore. Sori ad 0,001 distantiam a margine sunt remoti, nunquam contigui, oblongi, fere 0,002 longi. Sporangia sunt ovata vel obovata, alterna 9–10–11 vel pauciora, in soris acuminis 3–5na.

Angiopteris longifolia, Grev. et Hook., was first described by those authors in 1833 (l. c.). They remark that, "if examined with a little attention, it can never be confounded with the *A. evecta*, Hoffm." Notwithstanding, the species has hitherto been confounded by all the botanists who have since made investigations on the subject.

It is not easy to say what *Angiopteris evecta*, Hoffm., is. I have had the very rare advantage of seeing all the specimens in the different herbaria of England and the Continent, in which I thought I might find specimens of plants of the small tribe *Marattiaceæ* (except only the Royal Herbarium at Leyden, under the direction of Dr. Blume). But I have not been fortunate enough to ascertain which plant must be considered as the species in question. My much-lamented friend, Professor Gust. Kunze, of Leipsic, informed me in the spring of this year, a few days before his death, that he had just received the type-specimen of Forster, from Sprengel's herbarium. But it is not improbable that there are among the Forsterian specimens very different forms belonging to different species. The Banksian specimens are probably those of Forster, and have not the least resemblance to Hoffmann's figure of *Angiopteris evecta* (Polypodium evectum, *Forst.*). It must be here remarked that a considerable number of specimens of *Angiopteris*, originally from the Society Islands, belonging to the most different species according to my researches, are to be found in the herbaria and gardens, under the erroneous name of *A. evecta*, Hoffm. It has been sufficient for an *Angiopteris* to come from Tahiti, to obtain the name of *A. evecta*.

I am very much obliged by the liberality of my esteemed friends, Sir William Jackson Hooker, Director of the Royal Gardens at Kew, of the President of the Linnean Society, Robert Brown, Esq., of Professor Fenzl, of Vienna, Professor Presl, of Prague, the directors of the Botanical Gardens and Museum at Paris, Professor De Jussieu, Ad. Brongniart and Decaisne, to that very noble patron of science, François Baron Delessert, and Professor Miquel of Amsterdam, for having afforded me the opportunity of examining, without any restriction, all specimens of *Marattiaceæ* in their collections. I am sure that there is not a single species among the very great number described and undescribed (nearly forty), to be compared with that of Mr. Mathews, the type-specimen for Greville's and Hooker's species.

For this reason, and on account of the great confusion which exists among the synonyms of authors, I have thought it not superfluous to give the above more extensive descriptions of *A. longifolia*, Grev. et Hook. Guillemin's *A. longifolia*, Grev. et Hook. (Zephyr. Tait. p. 15, Paris, 1837), is not Mathews's specimen, but Moerenhout's, sent by this gentleman from Tahiti to M. D'Orbigny. It is in Baron Delessert's herbarium, but without a name. It belongs to Presl's *A. commutata* (Tent. Pterid. Supplem. p. 15).

I can hardly think that the very acute Professor Presl has had an opportunity of seeing the *A. longifolia* of the English botanists. If so, it would have been impossible that he should have referred so very distinct a species to *A. evecta*, Hoffm., as a synonym (Suppl. Tent. Pterid. p. 19). On another occasion I shall enter more fully into the investigation of the subject.

In Kunze's 'Index Filicum in Hortis Europæ Cultarum' (Linn. vol. xxiii. 1850), a specimen from the Leyden garden, presented in 1847 to that of Leipsic, is considered to be *A. longifolia*, Grev. et Hook. At that time I was still less acquainted with the different forms of that very polymorphous genus. But I can say, without any doubt, that the plant in question was not *A. longifolia*, Grev. et Hook. Consequently the other specimens, from the Amsterdam, Berlin, Leipsic, and Schönbrunn gardens, mentioned by the author in the same part of his work, must be, like ours, of another species.

Professor Miquel, in the Index Sem. Horti Amst. 1849, and in his Anal. Bot. Ind. p. 49 (Verh. Kon. Ned. Inst. vol. iv. l. c.), has mentioned as *A. longifolia*, Grev. et Hook.? (but very prudently with an interrogation) an inhabitant of Java. I have received, through the kindness of that botanist, all the specimens of his herbarium; but I can now affirm that there is as great a difference between this species and the specimens of Mr. Mathews, as there can be between any two species of this varied genus. The apex in the Javanese specimen is sterile, in the Mathews specimens the apices of all the leaflets are, without exception, fertile. The bases are generally truncate, not unequally cordate. The colour is not brown, but pale yellow. The under surface is not pulverulent, but smooth. The costa, the veins, and the veinlets are not brown, but yellow. The sori are not separated, but contiguous. The whole structure differs, being not membranaceous but coriaceous.

If perhaps I have been successful in tracing out distinctions in some species of a genus of plants, of which other botanists have formed a different opinion, I must in this case only thank those gentlemen for the opportunity afforded me of thoroughly examining the objects of their own investigation.

FLORULA HONGKONGENSIS : *an Enumeration of the Plants collected in the Island of Hong-Kong*, by Capt. J. G. Champion, 95th Reg., *the determinations revised and the new species described by* GEORGE BENTHAM, ESQ.

(*Continued from p.* 312.)

AURANTIACEÆ.

1. Murraya *exotica*, Linn.
Scarcely indigenous to Hong-Kong, but generally naturalized.
2. Cookia *punctata*, Retz.—Wight et Arn. Prodr. vol. i. p. 95.
Naturalized like the last.
3. Glycosmis *citrifolia*, Lindl. Trans. Hort. Soc. vol. vi. p. 72.— *Limonia citrifolia*, Willd.—*Limonia parviflora*, Bot. Mag. t. 2416.
Rather scarce in Hong-Kong.
4. Sclerostylis *buxifolia*, Benth.; glabra v. ramulis minute puberulis, foliis obovali-oblongis emarginatis crebre parallele venosis, floribus axillaribus sessilibus subfasciculatis pentameris, staminibus liberis,

ovario biloculari, loculis uniovulatis.—*Severinia buxifolia*, Ten. Cat. Hort. Nap. p. 96.

Abundant on the sides of hills near the race-course.

Frutex humilis v. in hortis nostris arbuscula nana, divaricato-ramosa, ramulis junioribus obscure angulatis vel compressiusculis sæpe puberulis, mox glabratis teretibusque. *Spinæ* axillares validæ, foliis sæpius multo breviores, nunc rarius longiores. *Folia* 1-1½ poll. longa, 6-9 lin. lata, obtusissima, basi in petiolum brevem teretem angustata, coriacea, rigida, crebre punctata, glabra, venis a costa media divergentibus numerosis parallelis, utrinque multo magis quam in plerisque speciebus conspicuis. *Flores* intra bracteas minutas ad axillas foliorum omnino sessiles v. pedicello brevissimo rarius fulti, nunc solitarii, nunc sæpius per 2-3 fasciculati. *Calyx* late campanulatus, linea brevior, glaber, laciniis 5 rotundatis obtusissimis. *Petala* 5, vix 2 lin. longa, oblonga, erecta, glabra. *Stamina* 10, corolla breviora. *Filamenta* dilatata, apice attenuata. *Antheræ* ovatæ, loculis parallelis. *Ovarium* intra discum cupulatum sessile, subglobosum, carnosum, loculis in centro minimis, ovulis paulo infra apicem axis centralis affixis. *Stylus* brevissimus, stigmate crasso ovoideo pulvinato. *Bacca* depresso-globosa, matura nigrescens, seminibus magnis ovoideis.

Specimens communicated to me by Professor Tenore, from the Royal Botanical Garden of Naples, have enabled me to identify his *Severinia* as a not uncommon Chinese plant. It was gathered also at Macao by Mr. Millett, and referred, in the Botany of Beechey's Voyage, to the *Atalantia monophylla*, DC. The latter plant appears to be confined to India; and all the more Eastern species referred to it by myself, as well as by other botanists, prove to be species of *Sclerostylis*.

5. Sclerostylis *venosa*, Champ., sp. n.; glabra, foliis subsessilibus ovatis v. ovali-oblongis obtusis subemarginatisque coriaceis crebre parallele venosis reticulatisque, floribus axillaribus breviter racemosis 4-5-meris, ovario biloculari loculis biovulatis, stylo stigmate longiore.

Fever Hospital Hill, and other localities.

Habitus sequentis (*S. Hindsii*), folia ramulorum floriferorum fere *S. buxifoliæ*, sed basi obtusissima et brevius petiolata, inferiora multo majora, 3 poll. longa, 1¼ poll. lata. *Venatio* fere *S. buxifoliæ*. *Spinæ* validæ. *Racemus* unicus suppetit axillaris, rachi semipollicari, floribus 4 breviter pedicellatis. *Calyx S. Hindsii*, sed laciniæ nunc 4 tantum, nunc

5. *Petala* et *Stamina* jam delapsa non vidi. *Ovarium* ovoideum, in stylum brevem attenuatum, stigmate multo minore quam in *S. Hindsii*. *Ovula* ex apice loculorum gemina pendula. *Bacca* aurantiaca, ea *S. Hindsii* minor.

6. Sclerostylis *Hindsii*, Champ., sp. n.; glabra, foliis ovali-ellipticis oblongisve obtusis tenuiter venosis, floribus axillaribus subfasciculatis pedicellatis pentameris, staminibus subconnatis, ovario biloculari, loculis biovulatis, stylo stigmate breviore.—*Atalantia monophylla*, Benth. in Lond. Journ. Bot. vol. i. p. 483.

Common on wooded hills in Hong-Kong.

Ramuli novelli angulati v. compressiusculi, adulti teretes. *Spinæ* validæ, pleræque semipollicares. *Folia* 1½–3-pollicaria, basi obtusa, coriacea, crebre punctata, venis primariis quam in præcedentibus multo paucioribus, rete venularum parum prominulo. *Petiolus* brevis, nunc teres, nunc marginatus v. in eodem ramo late alatus. *Pedicelli* per anthesin lineam longi, fructiferi paulo longiores. *Calyx* pedicello brevior, laciniis triangularibus acutis. *Petala* 3 lin. longa. *Stamina* paulo breviora; *filamenta* dilatata, ante anthesin connata, per anthesin plus minus soluta, apice attenuata; *antheræ* ovatæ, loculis parallelis. *Ovarium* disco carnoso concavo impositum, ovoideo-globosum; *ovula* ex apice loculorum geminatim pendula. *Stylus* brevissimus, stigmate crasso ovoideo. *Bacca* globosa, majuscula, aurantiaca.

OLACINEÆ.

1. Schœpfia *Chinensis*, Gardn. et Champ. in Kew Journ. Bot. vol. i. p. 308.

At first considered to be rare, but found afterwards abundantly in the Happy Valley woods, growing to a tall straggling under-tree, with brittle branches, and flowering profusely. The flowers are usually light pink, but they vary, being sometimes nearly white: they have a delicious perfume of almond paste.

2. Cansjera *lanceolata*, Benth. in Lond. Journ. Bot. vol. i. p. 491.

A climber, by no means common, flowering in November. Found at East Point, in the Happy Valley, and Little Hong-Kong.

VITACEÆ.

1. Cissus *diversifolia*, Walp. Pl. Meyen. p. 314.

Very frequent in ravines and on barren hills.

RUTACEÆ.

1. **Xanthoxylum** *nitidum*, DC. Prod. vol. i. p. 727.
Rather abundant in ravines.

2. **Xanthoxylum** *cuspidatum*, Champ., sp. n.; aculeis sparsis, foliolis 15-25 petiolulatis ovatis longe obtuseque cuspidatis coriaceis nitidis, paniculis axillaribus terminalibusque floribundis folio pluries brevioribus, floribus (masculis) tetrameris.

Less common than the *X. nitidum*, and rather local.

Species *X. nitido* affinis, at pluribus notis distincta. *Ramuli* petiolique parce aculeati, cinerascentes sed glabri. *Petioli* communes 8 poll. et ultra longi. *Foliola* opposita vel sæpius alterna, tria ultima ad apicem petioli more generis digitata, omnia longiuscule petiolulata, 1-2-pollicaria, vix inæquilatera, apice in acumen longum abrupte contracta, margine recurva et minute crenulata v. subintegerrima, basi angustata v. cuneata, crassiuscula et rigida, supra nitidissima, subtus pallida et fusco-punctata, punctis non pellucidis. *Paniculæ* florum masculorum a basi ramosissimæ, 1½-2-pollicares, cinerascentes. *Sepala* minuta. *Petala* 4, linea paulo longiora, glabra, æstivatione imbricata. *Stamina* 8, glabra, petalis duplo longiora. *Ovarii* rudimentum parvum, disco immersum. *Flores fœmineos* non vidi. *Fructus carpella* sæpius 4, reticulato-rugosa, 3 lin. diametro.

3. **Xanthoxylum** *lentiscifolium*, Champ., sp. n.; aculeatum, glabrum, foliolis 7-11 oblique obovatis oblongisve obtusis basi acutis, petiolo angulato, paniculis fructiferis longe pedunculatis laxe corymbosis folia subæquantibus, floribus (fœmineis) pentameris.

Common on Mount Gough and other localities, flowering in August. *Frutex* erectus. *Aculei* ramulorum breves, pauci, sursum incurvi. *Petiolus* communis 3-5-pollicaris. *Foliola* pleraque opposita, 1-2-pollicaria, vix obtusissime acuminata, basi sæpius in petiolulum brevem longe et inæqualiter angustata, tenuiter coriacea, nitidula, pellucido-punctata. *Paniculæ fœmineæ* bis terve trichotomæ v. umbellatim ramosæ, glabræ. *Sepala* 5, parva, ovata, æstivatione imbricata. *Petala* lineam longa, æstivatione imbricata. *Stamina* nulla. *Ovaria* 2, semiobovoidea, singula apice in stylum brevem stigmate magno peltato coronatum attenuata. *Ovula* in quoque ovario 2, collateralia. *Carpella* 2 lin. diametro, tuberculata. *Flores masculos* non vidi.

4. **Xanthoxylum *pteleæfolium*,** Champ., sp. n.; inerme, glabrum, foliis oppositis, foliolis ternis oblongis acuminatis basi longe angustatis, paniculis axillaribus laxifloris petiolum paulo superantibus, floribus (masculis) 3-4-meris.

On Mount Gough, rare, flowers in August (Champ.). Also in the Philippines (Cuming, *n.* 1819).

Frutex erectus, speciminibus siccitate pallescentibus. *Ramuli* ad nodos compressi. *Petioli* communes circa 1½ poll. longi. *Foliola* majora 3-4-pollicaria, tenuia, integerrima v. obsolete crenata, creberrime pellucido-punctata. *Inflorescentiæ* siccitate flavicantes, graciles, floribus numerosis parvis. *Petala* ¾ lin. longa, æstivatione imbricata. *Stamina* glabra, duplo longiora. *Flores fœmineos* non vidi. *Carpella fructus* sæpius 4, haud rugosa, 2-2¼ lin. lata.

5. **Boymia *glabrifolia*,** Champ., sp. n.; foliis pinnatis glabris, foliolis 3-7 ovatis acuminatis supra nitidis, corymbi ramis puberulis.

Scarce in Hong-Kong, but abundant on the Chinese coast. It flowers in September, and is often blighted when coming into flower by the typhoons of that season (Champ.). It was also gathered in China by Mr. Parkes, collector to the Horticultural Society.

Arbor, ramis teretiusculis glabris. *Folia* opposita, inferiora sæpe trifoliolata, pleraque 5- rarius 7-foliolata. *Foliola* opposita, longiuscule petiolulata, terminalia 2-3-pollicaria, inferiora sæpius minora, omnia longiuscule acuminata, basi acuta, lateralia sæpe basi inæqualia, penninervia, parce pellucido-punctata, subtus pallida. *Panicula* trichotoma, corymbosa, ampla, floribunda, ad ramificationes articulata et minute bracteata, ramulis compressis adpresse puberulis. *Flores* in specimine omnes masculi, pedicello lineam longo fulti. *Calyx* minutus, apertus, 5-dentatus. *Corolla* 1¼ lin. longa, glabra, petalis 5 oblongis æstivatione valvatis. *Stamina* 5, hypogyna, petalis alternantia et iis paulo longiora. *Filamenta* basi incrassata et longe ciliata. *Antheræ* magnæ, loculis longitudinaliter dehiscentibus. *Ovarii* rudimentum disco tenui impositum, carnosum, oblongum, apice in lobos 5 breves lineares parce ciliatos v. glabros divisum. *Ovula* nulla detexi. Adest etiam pars paniculæ fructiferæ absque foliis separatim lectæ. Huic inflorescentia omnino maris, carpella 5, juniora approximata, demum radiatim divaricata, fere 2 lin. lata, dorso biangulata convexa et rugulosa, endocarpio cartilagineo demum soluto. *Semina* solitaria cum vestigiis ovuli alterius abortivi, subglobosa, ventre affixa, hilo oblongo, omnia in specimine adhuc immatura.

6. Toddallia *floribunda*, Wall. Pl. As. Rar. vol. iii. p. 17, t. 232.
Rather scarce in Hong-Kong.

7. Cyminosma *resinosa*, DC. Prod. vol. i. p. 722.
A common tree in the Happy Valley woods. The fruit, when ripe, is white, that of *C. pedunculata* is black. Flowers and fruits in autumn.

SIMARUBEÆ.

1. Brucea *Sumatrana*, Roxb.—Planch. in Lond. Journ. Bot. vol. v. p. 575.
Low grounds and roadsides.

STAPHYLEACEÆ.

1. Eyrea *vernalis*, Champ.*— *Staphylea simplicifolia*, Gardn. et Champ. in Kew Journ. Bot. vol. i. p. 309.
Not uncommon in ravines on Mount Victoria and Mount Gough, forming a shrub of three to four feet in height, in flower in March and April, and in fruit in October. There are two, if not three varieties of it, with broader or narrower leaves, and flowers a dirty white or purplish, especially when in bud.

Notwithstanding the similarity of the flowers of this plant to those of *Staphylea*, yet the general habit of the plant, its leaves always simple, its erect trichotomous inflorescence, and the fruit (observed by Captain Champion since Dr. Gardner's description was written), are so distinct, that we cannot any longer retain it in that genus. Dr. Gardner's description of the ovary is also rather a theoretical one than that of its actual appearance, for the carpels are completely combined into a single three-celled ovary, with a simple style, of which even the stigmatic apex is scarcely lobed. We, therefore, now propose it as a separate genus, under the above name of *Eyrea*, with the following character:—
Flores hermaphroditi. *Sepala* 5, decidua, æstivatione valde imbricata, 2 exteriora minora. *Petala* 5, sub disco hypogyno integro v. sub-

---

* This new genus is named by Captain Champion in honour of his friend Colonel Eyre, lately returned from Hong-Kong, with a considerable herbarium collected there, containing several additional species, which will be duly recorded in the present florula. He has also brought over a valuable collection of Chinese seeds, which he has presented to the Royal Gardens at Kew, and a large and interesting set of rice-paper drawings of plants.

crenato inserta, oblonga, sessilia, subæquilonga, æstivatione imbricata. *Stamina* 5, petalis alterna et cum iis inserta; *filamenta* complanata; *antheræ* ovatæ, versatiles, loculis parallelis longitudinaliter dehiscentibus. *Ovarium* supra discum subsessile, obovoideum, triloculare, ovulis in quoque loculo 6–8 axi centrali affixis horizontalibus anatropis. *Stylus* columnaris, triqueter, apice truncatus, stigmate leviter dilatato obscure trilobo. *Fructus* subglobosus, haud inflatus, indehiscens, abortu bilocularis, loculis monospermis. *Semina* planoconvexa, pulpo farinaceo involuta. *Embryo* (exalbuminosus?) cavitatem implens, cotyledonibus carnosis, radicula conica.

The affinities of the genus, as well as of the whole of this small Order, are evidently much closer with *Sapindaceæ* and their allies than with *Celastrineæ*.

2. Turpinia *Napalensis*, Wall.—Wight et Arn. Prod. vol. i. p. 156. Wight. Ic. t. 972.

Common in ravines.

HIPPOCRATEACEÆ.

This Order, or rather suborder, only differs from the exalbuminous genera of *Celastrineæ* in that the number of stamens corresponds with that of the carpels, and not with that of the petals or sepals.

1. Hippocratea *obtusifolia*, Roxb.—Wight et Arn. Prod. vol. i. p. 104, var. ? pauciflora; cymis folio brevioribus paucifloris, pedicellis corolla glabra duplo longioribus.

Frequent in ravines on Victoria Peak; flowers early in summer. The fruit, which ripens in summer, varies in shape, the carpels being sometimes elliptical and entire, sometimes broadly obovate and more or less emarginate.

CELASTRINEÆ.

1. Evonymus *nitidus*, Benth. in Lond. Journ. Bot. vol. i. p. 483.

Common in ravines. The flowers are pea-green and the capsules reddish-coloured.

2. Evonymus *longifolius*, Champ., sp. n.; glaberrimus, foliis oppositis petiolatis elongato-oblongis remote pauciserratis subcoriaceis nitidis, cymis paniculatis paucifloris petiolo paulo longioribus, floribus pentameris, petalis margine leviter crenulato-crispulis, ovulis erectis.—*Folia* 4–6 poll. longa, 1–1$\frac{1}{3}$ poll. lata, laurina, nitida, paucivenia.

Cymarum pedunculi communes 3–6 lin. longi. *Corolla* circa 8 lin. diametro.

Very rare in the Happy Valley woods. The capsule of this species is much smaller and slighter than in *E. nitidus*. It is reddish-coloured, and its segments are narrow obcordate. The flowers are light green.

3. Evonymus *laxiflorus*, Champ., sp. n.; glaberrimus, foliis petiolatis ovali-ellipticis obtuse acuminatis integerrimis paucicrenatisque basi acutis nitidulis pauciveniis, cymis laxis paucifloris folio paulo brevioribus, floribus majusculis pentameris, petalis margine crenulato-crispulis, ovulis erectis.—*Rami* subteretes. *Folia* consistentia laurina, 2½–3-pollicaria. *Cymæ* circa 7-floræ, ramulis pedicellisque tenuibus. *Flores* roseo-purpurei, 5 lin. diametro.

Happy Valley woods, rare.

4. Evonymus *hederaceus*, Champ., sp. n.; scandens vel radicans, glaberrimus, foliis petiolatis ovatis acuminatis rotundatisve paucicrenatis basi acutis nitidis subcoriaceis, cymis folio multo brevioribus paucifloris, floribus majusculis tetrameris, petalis integerrimis, ovulis reverso-pendulis, capsula subglobosa leviter quadrisulca non muricata.—*Ramuli* tetragoni, virides, siccitate minute rugulosi. *Folia* forma valde variabilia, pleraque latiuscule ovata, 2–3 poll. longa, 1½ poll. lata, longiuscule acuminata et basi in petiolum 3–4-linearem angustata; variant tamen nunc anguste ovali-elliptica v. fere lanceolata, nunc obtusissima fere orbicularia, omnia paucivenia et consistentia laurina. *Cymæ* 3–7-floræ, pedunculo communi semipollicari leviter compresso, pedicellis ultimis 2 lin. longis. *Flores* 5 lin. diametro, albo-virescentes, inodori. *Calycis* laciniæ 4, breves, latissimæ, margine brevissime crenulato-laceræ. *Petala* 4, sessilia, orbiculata, demum reflexa, integerrima, glabra. *Filamenta* ad marginem disci lati integri inserta, petalis dimidio breviora, basi dilatata, demum recurva; *antheræ* primum erectæ, mox reflexæ, basifixæ, didymæ, biloculares. *Ovarium* disco semi-immersum, depresso-hemisphæricum, 4-loculare. *Ovula* in loculis gemina, collateralia, ex apice anguli interni reverso-pendula. *Capsula* fere 5 lin. diametro, læviuscula v. leviter corrugata, glabra, 4-locularis, 4-valvis. *Semina* arilla coccinea involuta.

Abundant in a ravine of Victoria Peak, climbing over the rocks, and rooting. Flowers in May. In habit it comes near to the American *E. obovatum* (now considered as a variety of *E. Americanum*) so admirably illustrated in Gray and Sprague's Illustrations of the North American Genera, vol. ii. t. 171; and the general structure of the flowers, especially of the ovary and ovules, is the same. It differs in its tetramerous not pentamerous flowers, the smooth fruit, and some other minor particulars.

4. Celastrus *Hindsii*, Benth.; scandens, glaberrima, foliis oblongis obtuse acuminatis calloso-serratis rigide coriaceis, cymis subsessilibus laxe paucifloris axillaribus v. ad apices ramorum anguste subpaniculatis, capsula ovato-globosa basi obtusa trisulca monosperma.— *Catha monosperma*, Benth. in Lond. Journ. Bot. vol. i. p. 483, excl. syn. Roxb.

A trailing plant, gathered in Hong-Kong, both in flower and fruit, Mr. Hinds's specimens previously described having been in fruit only. Captain Champion's notes on the precise locality are, however, unfortunately lost. The species bears considerable resemblance to the *C. monosperma*, Roxb. (a Silhet plant of which I have now good specimens both in flower and fruit), but is quite distinct in foliage, inflorescence, and fruit. The leaves are stiffer, more narrowed at the base, and the crenatures much fewer and more distant. The peduncles are mostly axillary, branching nearly from the base, and bearing from three to ten flowers of the size of those of *C. paniculata*, each having a pedicel of from $1\frac{1}{2}$ to 2 lines long; whilst those of *C. monosperma* are very much smaller and more numerous, crowded together at the top of a peduncle near half an inch long. The form, size, and structure of the capsule are like those of *C. paniculata*, without any narrowing of the base, as in *C. monosperma* and *C. Championi*, although it is always monospermous (by abortion) as in the two latter, not three-seeded as it usually is in *C. paniculata*. All these species belong to the true *Celastri*, not to *Catha*, to which I had erroneously referred them in my former paper.

5. Celastrus *Championi*, Benth.—*Catha Benthamii*, Gardn. et Champ. in Kew Journ. Bot. vol. i. p. 310, excl. syn. Benth.

The species described by Gardner is different both from Roxburgh's *C. monosperma* and from the above *C. Hindsii*. I have not myself seen the flowers, but the leaves are broader than in both those species, and the capsule, which is remarkably narrowed at the base, as in *C. monosperma*, is, as mentioned by Gardner, very considerably larger. It was found trailing over the bare rocks, in a ravine of Victoria Peak.

(*To be continued.*)

Journal of a Voyage from SANTAREM *to the* BARRA DO RIO NEGRO;
by RICHARD SPRUCE, ESQ.

(*Continued from p.* 278.)

After passing the mouth of the Maué, there was no perceptible current in the Ramos: the water was very warm, and so thick with the slime of decomposing *confervæ*, as to be very unwholesome to drink. We were told by parties of Indians whom we met, that the upper mouth was still closed, and consequently that we should be unable to get out into the Amazon; but on the 18th the water, though still unchanged in colour, began to run a little, and several small Ilhas de Capim and branches of trees passed us, indicating that some force was in action above. When day broke on the following morning, the water had assumed a yellow tinge, and as we proceeded on our voyage several masses of scum floated by us, and the current was decidedly strong. There was now no doubt that the waters of the Amazon had entered the *Ramú-orômoçáua*, as the upper mouth of the Ramos is called by the Indians; and towards night of the same day we had fuller proof of it, in occasional sudden influxes of water, making the whole river a series of whirlpools. On one occasion, when one of these irruptions caught us near the middle of the river, the canoe became quite ungovernable: it whirled round I suppose a hundred times, and all our exertions did not suffice to bring it into smoother water. We were drifting rapidly downwards, and in continual danger of thumping against the side or on some sandbank, when fortunately a breath of wind sprang up, and though it did not last more than ten minutes, it sufficed to put us nearly across the river, and into comparatively still water. The meeting of the cold waters of the Amazon and the heated waters of the Ramos had an extraordinary effect on the fish in the latter, they floated on the surface quite benumbed and stupefied, and we caught as many of them as we liked with our hands. On the 19th we had fresh fish in superabundance, and we salted down as many *pescados* (a delicate fish, the size of a large trout) as served us for ten days afterwards. This phenomenon takes place every year, not only in the Ramos, but in many other furos of the Amazon; but I had not been previously informed of it, and therefore had not ascertained the temperature of the water of the Ramos before it was mixed with that of the Amazon, as I ought to have done.

A little after noon on the 21st we reached a group of three sitios, called "As Pedras," on account of several large blocks of volcanic rock lying close by the river. Here we learnt that the current was so strong in the Ramú-orômoçáua that unless we were content to wait several days for the river to fill, or could procure the assistance of three or four men, there was no possibility of our passing it. It was on the 18th that the Amazon burst into the Ramos, with a noise which was distinctly heard here, although nearly a day's journey distant, and a montaria attempting to pass on the 20th was split by the force of the current. We determined on the latter alternative, and until the men could be found I occupied myself in examining the surrounding vegetation, which, however, presents great sameness throughout the Ramos, and in its general features does not differ from that of the Amazon. Since leaving Villa Nova, I had constantly made inquiries respecting the *Victoria*, but could not hear of it, excepting that Capitaô Pedro described it as existing abundantly in the Lago de Sarucú; but this was very far out of our way. At the "Pedras" I was glad to learn that it grew in a small lake on the opposite side of the Ramos, and within a short distance; but I had no montaria to enable me to reach it, for one of our men—a Júma Indian—had run away a few nights previously with our montaria and all our fishing-tackle, nor was there any montaria at the sitio where we were staying, but I was told I might borrow one at a sitio a little higher up. To this sitio I accordingly proceeded, and found at it an old man and his three sons, men of middle age, with their children. Two of the sons had just come in from a fishing expedition: the third had his arm in a sling, and on inquiring the cause, I learnt that seven weeks ago he and his father had been fishing in the very lake I wished to visit, in a small montaria, which remains constantly in the lake (the outlet being dried up), and that having reached the middle and laid aside their paddles, they were waiting for the fish with their bows and arrows, when, unseen by them, a large jacaré glided under the montaria, gave them a jerk which threw them both into the water, and seizing the son by the right shoulder, dived with him at once to the bottom, the depth being, as they supposed, about four fathoms. In this fearful peril he had presence of mind to thrust the fingers of his left hand into the monster's eyes, and after rolling over three or four times, the jacaré let go his hold, and the man rose to the surface, but mangled, bleeding, and helpless. His father immediately swam to his assistance, and providentially the two reached the shore without being further attacked. I was shown the wounds—*every tooth had told*; and some idea may be formed of this one terrible gripe, when I state that the wounds inflicted by it extended from the collar-bone downwards to the elbow and the hip. All were now healed, except one very bad one in the armpit, where at least one sinew was completely severed. Even this seemed to me in a fair way to heal soon; but though this should be the case, the deep scars and the useless arm (for it seems improbable that he will ever again be able to move his elbow or his shoulder) will ever remain to tell the tale. The only remedy applied was the milk of a tree called Acaráipu-rana, which I much wished to see, but was told it was a long distance by water to the place where it grew.

The sight of the wounded man was no encouragement to prosecute my enterprise, but I was very desirous to procure the fruit of the *Victoria*, and as three of the little fellows who were running about "ao fresco," offered to row me over, and their grandfather made no objection, I did not hesitate to avail myself of their services. The outlet of the lake was speedily reached, when we disembarked and followed the dried bed of the igarapé, in which my guides were not slow to detect the recent footsteps of a jacaré. In five minutes more we reached the lake, and embarked in the frail montaria, in which it was necessary so to place ourselves before starting as to preserve an exact balance, and then coasted along towards the *Victoria*, which appeared at a distance of some 150 yards. We had made but a few strokes, when we perceived by the muddy water ahead of us that some animal had just dived. As we passed cautiously over the troubled water a large jacaré came to the surface a few yards from the off-side of our montaria, and then swam along parallel to our course, apparently watching our motions very closely. Although the little fellows were frightened at the proximity of the jacaré, their piscatorial instincts were so strong, that at sight of a passing shoal of fish they threw down their paddles and seized their mimic bows and arrows (the latter being merely strips of the leaf-stalk of the Bacúba, with a few notches cut near the point), and one of them actually succeeded in piercing and securing an Aruaná, of about eighteen inches long. Our scaly friend still stuck to us and took no notice of our shouting and splashing in the water. At length the eldest lad bethought him of a large harpoon which was lying in the bottom of the montaria; he held this up and poised it in his hand, and the jacaré seemed at once to comprehend its use, for he retreated to the middle and there remained stationary until we left the lake.

There were three plants of the *Victoria*, only the largest of which was in flower and fruit. I suppose this plant covered a surface of six hundred square feet, although none of its leaves exceeded six feet in diameter. The fruits were all under water, and some of them partially decomposed, with the ripe seeds falling out. One of the flowers was the most magnificent I have seen: its petals of the deepest rose, perfectly expanded, and exhaling a delicious odour.

On the morning of the 23rd we left the Pedras, having obtained a promise of assistance on the following day to pass the mouth of the Ramos from a brother of the wounded man. There was no wind to aid us, and our progress was very slow against the strong current, for our crew was now reduced to two, having originally numbered only three, the captain included. It was night when we reached a place where the water began to run furiously, being about a mile distant from the real mouth, which we could see very plainly. We anchored on the right bank, adjacent to a broad sandy delta, extending to the Amazon, and in winter deep under water. After supper I started with the captain to explore the passage by starlight, and after stumbling into sundry holes, and rounding a good many *pocinhos* ("little wells," as the lagoons left in the sand are called), we reached the mouth. Here we found the waters of the Amazon entering with a force and a noise truly formidable, and ploughing through the sand in such a manner as to make a wall in it of fifteen feet high, from which the increasing torrent was every moment tearing large masses and thus widening its bed. On the following morning, after waiting for some hours in vain for the promised aid, we resolved to attempt the perilous passage. It is impossible for any one to travel much on these rivers without acquiring something of the practice of navigation, and Mr. King and I had constantly taken the helm for more than half the day; we were indeed heartily sick of the protracted voyage, and glad to do anything in our power to accelerate it. On this occasion the strong cable of the anchor was secured to the fore-mast, and carried on shore to serve as a hauling-rope, Mr. King and our man yoking themselves to it, while I took the helm, and the captain placed himself in the prow with a pole.

As long as there was water deep enough to float our canoe within five or six yards of the side we got on well enough, but when we were obliged to put out a little farther the current was too strong for our united force, and the canoe was in great danger of being carried away. We toiled on until noon, making very little headway, and as it began to be excessively hot, we allowed the canoe to take the ground, and resolved to wait until the air became cooler. In the interval we occupied ourselves in cooking our dinner, and were just about to fall on our boiled pirarucú, when a canoe appeared behind us, containing our friend of the Pedras, with two stout Indians and two boys. We made a hasty meal and by two o'clock were again under way. This was the disposition of our force: I was voted to the helm, the captain placed himself in the prow with a long pole to ascertain continually the depth of water, while the remaining hands tugged at the rope on land. We could now stand out more into the middle of the stream, where the current ran fast and furious, making a deep roaring against the prow as we ploughed through it; and my principal object was to keep the head of the vessel well out, as the force applied to the rope tended continually to draw her in shore, and had she turned in that direction, the current would have borne her violently on the bank, and either have swamped her or at least have brought on us a mountain of sand. The exertion required was so great, that the perspiration ran down my arms and legs, and my hands were quite sore; but most happily we succeeded in getting clear out into the Amazon without once grounding, though we had rarely so much as a fathom of water. Those on shore could not have suffered less than myself, for the sun and the white sand were scorching hot. The rope pressing on the edge of the cliff brought down, every few seconds, large masses of sand, but we stood far enough out to avoid them. It would be difficult to express what a load was taken off our minds when we found ourselves once more on the broad Amazon, and our previous silent anxiety was changed into noisy expressions of joy. The wind was blowing fair, and lasted until near sunset, sufficing to put us over to the opposite shore of the Amazon, along which our course now lay.

The Ramú-orômoçáua, and the dangers of its passage, are well known to the Indians on the Amazon. Last year a canoe, larger than ours, attempting to pass it, was wrecked, the captain rashly scorning to seek the advice and assistance of any of the neighbouring settlers. The last reach of the Ramos is nearly due north.

The inhabitants of the numerous sitios on the Ramos are either Tapuyas or Mestiços, of various shades of colour. The only white man we met with was the Capitaô Pedro, and he could not be reckoned as more than a visitor. Notwithstanding that the land is exceedingly fertile, and the lakes abound in fish and waterfowl, the people live in a state of comparative destitution; their only care being to eat up all their provisions to-day and reserve nothing for the morrow. Money they rarely see, and when they have it they are unable to count it. Their sole article of commerce is pirarucú, and even this is generally sold before it is caught. When we visited them there was great lack of farinha, their custom being to make it almost from day to day; and they levied frequent contributions on my biscuits, coffee, salt, &c.

Throughout the Amazon and its branches, I have found that the indigenous inhabitants have no idea of a country, save as of land bordering a great river. I am often asked, "Is the river of your land large?" "Is there much campo?" "Are the matos very extensive?" They are filled with astonishment when I say that nearly all our forests are planted. "Why, here," say they, "when one wants to plant a tree, one must first cut down a dozen to make room for it!" I have often had occasion to remark that people not born in, or not accustomed to, a romantic country, are slow to appreciate the picturesque. A Paraënse's idea of beautiful scenery supposes a land perfectly flat, with broad rivers, the stiller the better. The idea of mountains always suggests cachoeiras and rapids, impossible or dangerous to be traversed by canoes. When I make inquiries respecting an unvisited region, hoping to hear of "antres vast and deserts wild," they on their part expect to give me pleasure by describing it as a "terra bonita, *plaina* —lá naô ha lugares feios, nem serras nem cachoeiras." One essential of a fine country to them, and not an object of indifference to any traveller, is that it contains " muita caça, muito peixe."

On the 29th we passed Serpa, on the north shore. It is the exact counterpart of Villa Nova, but rather more pleasantly situated. On the morning of the 2nd of December a montaria came up with us, in which was an old man, who was bound for a sugar-engenho, which an Englishman, Mr. M'Culloch, is forming on a Paraná-mirí, about a day's journey below the river and lake of Paraquecoára. I had made

Mr. M'Culloch's acquaintance last year at Pará, and I gladly seized the opportunity of going forward to visit him. We reached his engenho about two o'clock, and I remained with him until our canoe came up, about noon on the following day. There is no manufactory of sugar in the province, except near Pará, and at this distance in the interior the difficulties to be overcome in commencing such an undertaking are immense. Mr. M. has already been a year employed in clearing away forest, planting cane, arranging his water-power, &c., and he has yet a great deal to do ere the engenho is completed. The only workmen on whom he can depend are a few slaves of Senhor Henrique of the Barra. At daybreak on the 3rd I found him occupied with a number of Múra Indians, of all sorts and sizes, who had come to work for the day. There are several small colonies of these people on the lakes hereabouts, and when they take it into their heads to work, this is the way they do.—They come to Mr. M'Culloch early in the morning, when he gives each of them a *pinga* of cashaça. Afterwards any one who is so rich as to possess a palm-leaf hat—and, if not, he is provided with a fragment of cloth of some kind—holds it out, and Mr. M. dispenses into it a cuya-full of farinha, and as much dried fish as will serve for the day. Every one takes his rations separately—even the father and child—the husband and wife. They now work until sunset, when they again come to their employer for a parting *pinga*, and then betake themselves to their forest-homes. This lasts for two or three days, when they begin to feel fatigue or *ennui* at this mode of life, and Mr. M'Culloch sees no more of them for perhaps a week. You will easily perceive that the sole inducement of these people to labour is the cashaça; and I was amused to see the ease and the gusto with which little naked urchins tipped off their *pinga*—a quantity assuredly sufficient to choke me twice over.

At Mr. M'Culloch's I saw some blocks of wood lying about, resembling Itaüba in colour and texture, but much harder and heavier. He was so kind as to cut off for me a portion of one of them, which I send you, and I should be much obliged if you would ascertain its specific gravity, for I have never seen so heavy a wood, and I doubt if there be a heavier in existence. You will remember that when Edwards visited the Barra, Mr. M'Culloch had a saw-mill near: this has been since burnt down, but Mr. M. informed me that the main-shaft in it was made of this heavy wood, its dimensions being fourteen feet long by fifteen inches in diameter, and that he estimated its weight at a ton. If this be correct, its weight, in proportion to the same bulk of water, is as 3584 to 1718, or *more than twice as heavy* ; but this is scarcely credible. Mr. M. took me up the igarapé which supplies his engenho to show me the tree, which is called *Piranha-uba*. It is one of the largest forest-trees, with slight sapopemas at the base, and from its leaves I do not hesitate to refer it to *Rhizoboleæ*; possibly it may be a *Caryocar*, but it seems a different species from any I have hitherto gathered. The timber is excellent for mill-work, or for anything which requires to be kept under water.

I will not further weary you with details of a wearisome voyage. The rains had lately been almost incessant, the river began to run furiously, and the winds were generally from above. With these obstructions combined against our slow-sailing craft, we did not enter the mouth of the Rio Negro until the morning of the 10th. The change from the yellow waters of the Amazon to the black waters of the Rio Negro is very perceptible, and indeed abrupt. The latter are black as ink when viewed from above, and not the deep blue of the waters of the Tapajoz: in shallow places they appear purple, and stones or sticks at the bottom seem red: when taken up into a glass they are of a pale amber-colour, and quite free from any admixture of mud.

The Solimoês enters from the left, while the Amazon seems a more direct continuation of the Rio Negro. The last is broader than the Solimoês, but it is less deep, and its waters are placid almost as a lake.

It was dusk when we reached the Barra, which lies two miles within the mouth of the Rio Negro. I went on shore immediately, and waited on Senhor Henrique Antonio, an Italian merchant, to whom my letters of credit were addressed. From him we received a most kind and cordial reception, and he immediately gave up to us three rooms in one of his houses (for he owns half the Barra), where we are still residing: his reputation for princely hospitality is indeed almost world-wide.

We are gradually recovering from the weakness and exhaustion produced by our tedious, disagreeable voyage, and from the day of our landing have been hard at work. But the weather continues excessively rainy, insomuch that we have been for three days together without being able to stir out, while at Santarem there never passed an entire day on which we could not have got into the woods. I am, however, delighted to find myself in the midst of a novel vegetation, and with excellent forest-tracks made to my hands; though the field of our operations is becoming from day to day more contracted by the rapid rise of the rivers and igarapés. The *Myrtaceæ, Melastomaceæ, Solanaceæ*, and *Rubiaceæ*—families which give a character to the flora of the valley of the Amazon—are here quite different in species, and partly in genera, from those of Santarem and Pará. I have been glad to meet with abundance of a fine *Cattleya*, which, unless I am mistaken, is unknown to our English collections; it is the first of the genus I have seen in the province of Pará, where *Orchidaceæ* are few and far between.

(*To be continued.*)

Contributions to the Botany of WESTERN INDIA; *by* N. A. DALZELL, Esq., M.A.
(*Continued from p.* 282.)

Nat. Ord. ORCHIDEÆ-VANDEÆ.

SARCANTHUS.

S. *peninsularis*; caule simplici tereti flexuoso folioso pendulo, foliis linearibus acuminatis crassis coriaceis subtriquetris, racemis oppositi-foliis folio dimidio brevioribus, labelli calcare perianthii longitudine cornuto obtuso pendulo *complete biloculari*, fauce callis duobus clausa, limbo brevi integro ovato obtuso crasso carnoso erecto, facie interiore sulcato perianthio breviore. *Dalz. Ic. ined.*

Perianthium subpatens, diametro 5-lin. *Sepala* et *petala* conformia, obovata, flava, purpureo-marginata; *sepalum* supremum in columnam incumbens. *Pollinia* bipartibilia, glandula in apice rostelli horizontalis *sphærica*. *Dentes* in labelli utroque margine parvi, truncati. *Labellum* basi albo et violaceo pictum.—Crescit in arboribus prope vicum Virdee in regno Warreensi; fl. Jul. et Aug.

EULOPHIA.

E. *bicolor*; foliis 2–3 lineari-lanceolatis acutis multiplicatis *serotinis*, sepalis lineari-oblongis acutis 7-nerviis, petalis oblongis obtusis sepalis brevioribus medio trinerviis, nervis exterioribus unilateraliter penninerviis, labio obtuse saccato trilobo, lobis lateralibus abbreviatis planis erectis intermedio elongato recurvo, marginibus crispis, disco *venis* 10 *cristatis* prædito, scapo sesquipedali stricto pauci-(9–10)-floro foliis longiore, floribus distantibus (*haud secundis nec pendulis*) purpureis vel flavo-virentibus, bracteis floralibus subulatis ovario unciali dimidio brevioribus, capsula lineari tereti deflexa bipollicari.

Folia 9–12 poll. longa, 2–2¼ lata.—" Amberkund " Indigenorum.—Crescit in jugo Syhadrensi; fl. Jun.

Tribe OPHRYDEÆ.

HABENARIA.

Erostres; petalorum lacinia anterior angustior *sed non elongata*.

H. *uniflora*; foliis 2–3 oblongo-lanceolatis basi angustatis apice mucronulatis planis inæqualibus, scapo gracillimo unibracteato unifloro, sepalis lateralibus supremo duplo majoribus obtusis patentibus, petalorum laciniis æquilongis, anterioribus linearibus acutis patentibus, posticis *obtusis* cum sepalo supremo ovato obtuso galeam efficientibus, labelli porrecti ultra medium trifidi laciniis lateralibus linearibus acutis intermedia lanceolata paulo longioribus, calcare filiformi pendulo curvato, ovario pedunculato triplo longiore.

Herba semipedalis. *Folia* 5-nervia, majora 4 poll. longa, 1 poll. lata. *Sepala* semiorbiculata, obtusa, petalorum *lacinia anteriore longiora*. *Columna* apice mucrone longiusculo instructa. *Glandulæ* magnæ, hemisphæricæ, subtus excavatæ. *Processus* carnosi, obtusissimi. *Flores* albi.—*H. rariflora*, Rich., proxima; differt petalorum laciniis æquilongis, posteriore obtuso.—Crescit rarissime in Concano australiore; fl. temp. pluviali.

PERISTYLUS.

P. *elatus*; caule basi vaginato medio folioso, foliis patentibus paucis (4–5) ellipticis, inferioribus obtusis amplexicaulibus, superioribus longioribus acutis calloso-mucronatis *scapo brevioribus* abrupte in squamas acuminatas transeuntibus, sepalo supremo rotundato, lateralibus oblongis apice cucullatis dorso infra apicem mucrone instructis, petalis longioribus labelloque *subintegro* rotundatis subæqualibus albis, calcare sphæroideo-scrotiformi, bracteis lanceolatis acuminatis flore longioribus.

Caulis sesquipedalis, strictus, pennæ cygneæ crassitie. *Spica* cylindrica, multiflora, folia longe superans. *Flores* conferti, parvi. *Folia* 5–7 poll. longa, 2½–3 lata, 5–7-nervia.—Crescit prov. Malwan; fl. Julio.

Tribe MALAXIDEÆ.

DENDROBIUM, § 1.

D. *microchilos*; pseudo-bulbis profunde bilobis, lobis orbiculatis valde depressis reticulatis epidermide albida, foliis 3–4 linearibus obtusiusculis superne subplanis basi angustatis ibique scapum floriferum vaginantibus, floribus alternis secundis spicatis, bracteis ovato-subulatis ovario longioribus, capsula sessili ovata glabra, sepalis e basi lata acuminatis, petalis conformibus æquilongis angustioribus labello ovato indiviso subduplo longioribus.

Herba pusilla, 2–3-uncialis. *Scapus* filiformis, multiflorus. *Flores* straminei, minuti. *Folia* ½–2 poll. longa, 2½–3 lin. lata, *pseudobulbi folia non gerentes*.—Crescit supra arbores, præcipue in ramis *Mangiferæ Indicæ*, in regno Warreensi; fl. Aug.

Nat. Ord. RUBIACEÆ.

ARGOSTEMMA.

1. A. *glaberrimum*; erectum, 4–6-unciale, foliis 4 verticillatis lanceolatis acuminatis inæqualibus basi inæquilateris, inflorescentia trichotoma umbellata foliis breviore, umbellis paucifloris, floribus 5-meris, corollæ fauce nuda, filamentis apice valde incrassatis per anthesin adhærentibus, stylo apice muricato.

Flores diametro 3–4 lin.—Crescit supra arbores in regno Warreensi.— Si ex descriptione Doniana (Syst. Gard. et Bot.) judicare liceat, ab *A. verticillato*, Wall., satis diversa; foliis margine non ciliatis, corolla intus eglandulosa. An *A. inæquilatera*, Benn., differt?

2. A. *connatum*; caule pubescente prope basin bisquamato, foliis 2 vel 4 verticillatis subsessilibus ovatis inæqualibus utrinque parce puberulis, pedunculo simplici glaberrimo, umbella brevi multiflora (12–15) bracteis foliaceis *connatis* suffulta, floribus 4-meris, calyce pedicellisque pubescentibus, antheris erostratis obtusissimis.

Calycis corollæque laciniæ subconformes, late ovatæ, fere rotundatæ. *Corolla* vix ¼ 4-fida (in specie præcedente ad basin 5-partita).— Crescit in rupibus prope Chorla-ghât, lat. 15° 30'; fl. Aug.

Nat. Ord. AROIDEÆ.

TAPINOCARPUS, genus novum.

GEN. CHAR. *Spatha* basi convoluta, limbo angusto elongato, acuminato, plano, patente. *Spadix* inferne interrupte androgynus (genitalibus rudimentariis supra et infra stamina confertis filiformibus), apice cylindricus, gracilis, spatham subæquans. *Antheræ* distinctæ, sessiles, loculis cylindricis parallelis, connectivo multo majoribus, apice poro dehiscentibus. *Ovaria* plura, libera, obconica, circa basin spadicis verticillata, biserialia, unilocularia. *Ovula in loculis* 6, funiculis longiusculis spongiosis, 2–3 *basilaria erecta, reliqua ex apice loculi pendula*. *Stigma* sessile, depresso-hemisphæricum. *Dalz. Ic. ined.*

Herba acaulis, rhizomate parvo tuberoso, perennans. *Folia* simultanea, longe petiolata, cordato-hastata, integra, lobis baseos obtusis. *Scapus* e foliorum basibus vaginantibus longe exsertus, *fructiferus contortus, apice solum osculans*. *Fructus* angulatus, 4–5-spermus.— *Aro* et *Dracunculo* affine; differt ovulis ditropis, a posteriore præcipue foliis integris, nam ovulorum positio, ut videtur, dubia.—Crescit in graminosis humidis Concani australioris.

(*To be continued.*)

BOTANICAL INFORMATION.

Letter on the successful cultivation of the VICTORIA REGIA *in Philadelphia, U. S. A., addressed to* Sir W. J. Hooker, *by* CALEB COPE, ESQ.*

Philadelphia, September 27, 1851.

As you so kindly responded to my wishes in supplying me with seeds of the far-famed *Victoria*, it is due to you that I should render some account of my stewardship.

On the 21st of March last I planted, in small seed-pans, four out of the twelve seeds you sent to me. Three of these germinated as follows: —the first on the 10th of April, the second on the 14th of April, and the third on the 22nd of May. On the 24th of May one of the plants was transferred to the house constructed for it, and placed in a tank, of an octagon form, about 24 feet in diameter. The largest leaf was then $4\frac{1}{2}$ inches in diameter. Fire-heat was applied till the 21st of

---

* An eminent cultivator and patron of horticulture in that city.

June, and then wholly dispensed with till within a few days. The solar heat, with the house kept very close, was sufficient to raise the temperature of the water to 85°, whilst the atmosphere was about 10° higher. The temperature of the house and water, however, was materially below these points at times, there being occasionally as great a difference as 15°. Notwithstanding these great variations, and the low temperature at intervals, the plant flourished in the highest degree, and on the 21st of August produced a flower 15½ inches in diameter. Another flower succeeded it in a week, which was pronounced by the Committee of the Pennsylvanian Horticultural Society to be 17 inches in diameter; the petals being 7 inches, and the disc or crown of the flower 3 inches. For the past three weeks the plant has contributed two flowers a week, the ninth flower blooming this evening. The leaves have reached a diameter of 6½ feet. The silver edge made its appearance with the twenty-fourth leaf, and every successive leaf has been thus formed. This twenty-fourth leaf attained at maturity a diameter of 5 feet 3½ inches. It was the twenty-seventh leaf that measured 6 feet 6 inches. Since the plant has begun to bloom, the leaves are not so large; still they have reached 5 feet 2 inches in diameter. That we have succeeded in producing larger leaves and flowers than you have in England, I ascribe to the more favourable character of our climate. It is certainly more natural to the plant that the temperature of the air should be higher than that of the water. These conditions, you are aware, are reversed in England : an effect produced by artificial heat, which is of course less favourable than solar heat.

I have in my kitchen-garden a small basin for catching the rain-water and overflow from the aquarium and other houses. It is about 8 or 10 feet in diameter. Into this basin I planted one of the *Victorias*, on the 25th of June last. The plant has grown remarkably well, the largest leaves attaining a diameter of more than 4 feet. It has not yet, however, bloomed, and may not, as cold weather is near at hand. Sash, blocked by whiting, has remained over the basin during the whole time. I am satisfied that we can flower the plant next season in this position.* The third plant is still in the small seed-pan in which the seed was sown.

The excitement produced by the successful cultivation of the

---

* In our less-favoured summer climate of England, Mr. Weeks, at Chelsea, has successfully flowered this plant in the open air, by a few heating pipes in the water.

Victoria on this side of the Atlantic has been very great, and I am happy to say that no one has affirmed that the glowing accounts of the plant were at all exaggerated. Indeed the universal sentiment is, that no tongue or pen *can* exaggerate it. Our worthy friend and accomplished botanist, Dr. Darlington, who spent a night with me recently, has enjoyed the sight as much as anybody. The pages of our favourite periodical, the 'Horticulturist,' will record what has been done, and to whom the honour is due of sending so valuable an exotic to our shores.

Since my last letter to you, in which I made some inquiries touching the winter treatment of the *Victoria*, I see the fact stated of all the plants in England having died, and the opinion expressed that it is an annual. What the plant may do here under milder treatment remains to be seen.*

C. C.

NOTICES OF BOOKS.

Papers and Proceedings of the Royal Society of Van Diemen's Land. Vol. I. 1850. 8vo. Hobart Town, V. D. L.

In this work, so creditable to the rising colony of Van Diemen's Land, and so powerfully fostered by the present Governor, Sir William Denison, there is more than one interesting paper relating to botany, on the uses and properties of plants. We may particularly mention the late lamented Mr. Bicheno's, on "the Potato as an article of food, and on the Potato-disease;" J. Britchell, Esq., on "the export and consumption of Wattle Bark, and the process of Tanning;" Captain Collinson on "Timber Trees of New Zealand;" Dr. Thomas Anderson, on "a new species of Manna, of New South Wales;" Sir William Denison, "on the manufacture of Potash from Tasmanian woods" (one of many communications from his Excellency).

---

* Whether annual or not (though it is, in its native country, considered to be perennial) is of little consequence. Those that have survived the winter in England, are in a measure dormant during the coldest season; but seeds are produced so abundantly, and the growth of the plant is so rapid, that we can never be at a loss to have good plants at a good season.—ED.

Under the head of "*Proceedings*" we find the following interesting letter, addressed to Joseph Milligan, Esq., the very zealous and intelligent secretary, on some gigantic native trees.

"New Town Parsonage, near Hobart Town, March 19, 1849.

" I went last week to see a very large tree, or rather two very large ones, that I had heard of since 1841, but which were not re-discovered until Monday last. As they are two of the largest, if not *the* largest trees ever measured, I have determined to send you an account of them, in order that a record may be preserved in any future publication of the Royal Society. They are within three-quarters of a mile of each other, on a small stream, tributary to the North-west Bay River, pretty far up on the ridge which separates its waters from those of Brown's River. They are easily reached from the Huon footpath, and are in a beautiful vale of Sassafras and Tree-ferns, and not in an inaccessible gully, like most of our gigantic trees. I have never before seen the Tree-ferns growing in such luxuriance, bending over the stream like enormous cornucopias. The fire has never reached them, as they and the forest around them plainly show; and every here and there you are puzzled on seeing a Sassafras-tree with a root on either side; one in particular forming a natural arch, underneath which you can walk. And it was some time before I could tell how it was ever possible for the tree to have grown there, until, on looking further, I perceived that the Sassafras must have originally sprung from seed lodged in the bark of some *Swamp-gum* that had fallen across the brook, and as it grew it gradually sent out roots along the trunk, until they met *terra firma*. The trunk having in the course of ages decayed, has left the Sassafras-tree in the odd position in which we now see it. I say so much before I give you the measurements. I am sure the whole scene would amply repay you for the trouble of a ride. In addition to the giants below, there are, I feel confident, within a mile, at least a hundred trees of 40 feet in circumference. One, about forty yards from the biggest, was 60 feet at 4 feet from the ground, and at 130 feet must have been fully 40 feet in circumference: it was without buttresses, but went up one solid massive column, without the least symptom of decay. A Silver Wattle was 120 feet high and 6 feet round. In fact, we named it the *Vale of Giants*, for puny indeed did men appear alongside these vegetable wonders. The largest we measured was, at 3 feet from the ground, 102 feet in circumference, and at the ground 130 feet. We had no means of estimating its height, so dense was the neighbouring forest, above which, however, it towered in majestic grandeur. This noble Swamp-gum is still growing, and shows no signs of decay; it should be held sacred as the *largest* growing tree. The largest Oak on record is the Cowthorpe, in Yorkshire, which is 48 feet in circumference at 3 feet from the ground. Some hollow pollard Oaks are larger, such as the Winfarthing, in Norfolk, which is 70 feet at the ground.—The second tree, also a Swamp-gum, is prostrate. It measures, from the root to the first branch, 220 feet, and the top measures 64 feet, in all 284 feet! without including the small top, decayed and gone, which would carry it much beyond 300 feet. The circumference at the base is 36 feet, and at the first branch 12 feet, giving an average of 24 feet. This would allow for the solid bole 10,120 feet of timber, without including any of the branches. Altogether, as green timber, it must have weighed more than 400 tons. The Oak that gave the most timber was the Gelonos Oak, in Monmouthshire, which, with its branches, turned out 2,426 feet, but the body alone only 450 feet.

"THOMAS J. EWING."

LEHMANN : *Novarum et minus cognitarum stirpium Pugillus nonus addita nova Recensione nec non enumeratione specierum omnium generis* POTENTILLARUM *earumque Synonyma locupletissima; auctore* CHRISTIANO LEHMANN. Hamburgi, 1851.

This is a quarto pamphlet of seventy-two pages which is very acceptable to every botanist. The first twenty-three pages are devoted to descriptions and observations of twenty new species of the genus *Potentilla*, from various parts of the world, known to the indefatigable and talented author, and the rest of the brochure is occupied by an enumeration, with synonyms and references to works in which are described all the known species, amounting to no less than 193. The whole are well arranged; primarily into sections, "FRUTICULOSÆ et SUFFRUTICOSÆ;" and "HERBACEÆ." The first contains twelve species. The second section is divided into "TERMINALES" and "AXILLIFLORÆ;" and these again into Series; and again into "Tribes" and "Subtribes."

Catalogue of CRYPTOGAMIC PLANTS *collected by* PROFESSOR W. JAMESON *in the vicinity of Quito; by* WILLIAM MITTEN.

(*Continued from* p. 57.)

51. Macromitrium *longifolium*, Brid. Fruit immature.
52. Zygodon *fasciculatus*, Mitten, MSS.; dioicus dense cæspitosus gracilis humilis fastigiatim ramosus viridis inferne ferrugineus, foliis caulinis lanceolatis nervo sub apice evanido carinatis e cellulis subrotundis minutissime papillosis areolatis divergentibus, perichætialibus erectis longioribus superne angustioribus, theca in pedunculo breviusculo ovato-pyriformi deoperculata sex-striata, operculo subulato, peristomio duplici, externo dentibus 8 bigeminatis interno ciliis 8 linearibus pallidis.—Flos masculus *Z. intermedio similis.*

A small species near to *Z. Brehmianii* and to *Z. Brownii*, judging from the figure of that species (Schwægr. Suppl. vol. iv. t. 317), but the leaves are longer and narrower, and the inflorescence is dioicous; the peristome resembles that of *Z. Brehmianii*, but is not so Arenicum.

53. Fabronia *Jamesoni*, Engl. Lond. Journ. Bot. 1848. p. 197.
54. Phyllogonium *fulgens*, Brid.
55. Neckera *leptomyra*, Hook. *in Wils. Lond. Journ. Bot.* xxx. p. xx. t. 16. L. Pterogonium *Jamesoni*, Tayl. *l. c.* p. 264. p. 99.
56. *Eadem.*
57. Neckera *longiseta*, Hook. Musc. Exot. t. 94.
58. N. *viridula*, Mitten, MSS.; *caespitose ramosa, ramulis pinnatim ramulosis, foliis e basi dilatata amplexicauli lanceolato-acuminatis integerrimis vel minutissime arguto-denticulatis e cellulis angustissimis interstitiis grossis minutissimis papilloso-punctulatis, nervo supra medium evanescente.*

Closely allied to *N. nigrescens*, but more robust and apparently not pendulous; the younger portions have the leaves of a rather deep green colour, the older and lower of a pale brown and all destitute of gloss. In size and general appearance it has some resemblance to *Pilotrichum hypnoides.*

59. Neckera *illecebra*, C. Muller, Synops. p. 137.
60. Pilotrichum *densum* (Sw.). C. Muller. Hypnum densum, *Swartz*,

Hedw. Sp. Musc. t. 74. Neckera densa, *Wils. Lond. Journ. Bot.* v. 4. t. 10. N. luteovirens, *Tayl. l. c.* 1846. *p.* 59.

b. *Idem.* A state with the leaves broken.

61. Pilotrichum *Jamesoni* (Tayl.), C. Muller. Cryphea Jamesoni, *Tayl. Lond. Journ. Bot.* 1848. *p.* 192.

b. P. *ramosum* (Wils.), Mitten; monoicum, caule longiusculo rigido, ramis divergentibus et recurvis, foliis ovatis breviter acuminatis concavis, marginibus reflexis, nervo ultra medium evanido, perichætialibus oblongis superne subulato-attenuatis, nervo ex apice ad medium folii descendente, theca immersa cylindrica subsessili, operculo conico breviter oblique rostrato, peristomio externo e dentibus angustis albidis, interno angustioribus et paulo brevioribus, calyptra glabra.—Cryphea ramosa, *Wils. in schedula.*

A fine species, differing from *P. Jamesoni*, Tayl. in its shorter and wider leaves, with the nerve vanishing below the point, and in the rostrate operculum. In *P. Jamesoni* the nerve is excurrent, the leaves are serrate, particularly towards the point, and the operculum is shortly conical.

62. Neckera *Jamesoni*, Tayl. Lond. Journ. Bot. 1846. p. 59. The perichætial leaves of this species are almost subulate, those of *N. Douglasii* are oblong, obtuse, cuspidate.

63. N. *obtusifolia*, Tayl. Lond. Journ. Bot. 1848. p. 193.

64. N. *debilis* (Wils.), Mitten; monoica, caule flexuoso gracillimo pendulo ramoso, foliis patentibus laxis longe lanceolato-subulatis basi decurrentibus superne denticulatis, nervo infra ¼ evanido, e cellulis laxis teneris lævibus alaribus inconspicuis areolatis, perichætialibus latioribus enerviis, theca in pedunculo scaberrimo oblongocylindrica curvula et paululum inclinata, operculo breviter conico, peristomio externo dentibus flavidis, interno processibus pertusis longioribus in membranam exsertam positis, calyptra diminuta lævi.—Leskea debilis, *Wils. in schedula.*

Very closely resembling *N. aciculata*, but readily distinguished by its more shortly-nerved leaves, more cylindric capsules, and shortly conical opercula. The leaves of both are of the same structure, and, except under a very high power, appear to be smooth; but there exist here and there very minute papillæ, which are, however, too small to be rendered visible by merely bending the leaf.—The place of these species appears to be near to *N. nigrescens*, from which they differ in their smooth and loosely areolate leaves, but agree in general habits.

b. N. *aciculata* (Tayl.), Mitten. Leskea aciculata, *Tayl. Lond. Journ. Bot.* 1847. *p.* 339.

65. Pilotrichum *tenuissimum* (Wils.), C. Muller. Cryphea tenuissima, *Wils. Lond. Journ. Bot. v.* 5. *p.* 6. *t.* 15. *E.*

66. Hypnum *expansum*, Tayl. Lond. Journ. Bot. 1846. p. 64.

67. H. *superbum* (Tayl.), C. Muller. Leskea superba, *Tayl. Lond. Journ. Bot.* 1846. *p.* 61. *Wils. l. c. v.* 5. *t.* 15. *M.*

68. H. *porotrichum*, C. Muller. Neckera longirostris, *Hook. Musc. Exot. t.* 1. Hypnum floridum, *Tayl. Lond. Journ. Bot.* 1847. *p.* 339.

b. *Idem.* An older state. Leskea gymnopoda, *Tayl. Lond. Journ. Bot.* 1846. *p.* 62.

69. H. *gracillimum* (Tayl.), Mitten. Neckera gracillima, *Tayl. Lond. Journ. Bot.* 1848. *p.* 192.

Closely resembling *H. polycarpum* and its allies.

70. H. *Hedwigii* (Wils.), Mitten. Leskea Hedwigii, *Wils. in schedula.* L. cæspitosa, *Hedw. non Swartzii, fide Wils.*

71. H. *denticulatum*, Linn.

72. H. *scariosum*, Tayl. Lond. Journ. Bot. 1846. p. 65.

73. H. *Jamesoni*, Tayl. l. c. p. 63.

74. H. *conostomum*, Tayl. l. c. 1848. p. 194.

75. H. *pseudo-plumosum*, Brid.

76. H. *rusciforme*, Weis.

77. H. *rutabulum*, Linn.

78. H. *intermedium*, Mitten, MSS.; monoicum, viridissimum et lutescens, caule gracili rigidulo bipinnato, foliis cordatis acuminatis opacis papillosis apicem versus denticulatis, nervo sub apice evanido, perichætialibus lanceolatis subulato-longe-acuminatis elongate laxius reticulatis minute denticulatis, theca in pedunculo longissimo rubro lævi parum inclinata oblonga fusca, operculo longe rostrato, peristomio interno e processibus pallidis imperforatis ciliis nullis.

Intermediate between *H. minutulum* and *H. pauperum*, agreeing with the first in size and habit, but differing in its shorter cauline leaves, which are toothed about the apex and not recurved at the margins; with the second it corresponds in some degree in the form of its cauline leaves and capsules, but differs in its bipinnate stems and larger size. The peristome is different from those of both species, having the processes imperforate, and being destitute of cilia.

79. H. *molluscum*, Hedw.
80. H. *Schlimii*, C. Muller, Synops. p. 310. H. decumbens, *Wils. in schedula*.

This species was also gathered in Brazil by Gardner.

81. Hypopterygium *scutellatum* (Tayl.), C. Muller. Hypnum scutellatum, *Tayl. Lond. Journ. Bot.* 1847. *p.* 338.
82. Neckera *tetragona* (Sw.), C. Muller. Var. ?
83. Andreæa *rupestris*, Hedw.
84. Entosthodon *acidotus* (Tayl.), Mitten. Gymnostomum acidotum, *Tayl. Lond. Journ. Bot.*
85. E. *Jamesoni* (Tayl.), Mitten. Physcomitrium Jamesoni, *Tayl. Lond. Journ. Bot.* 1847. *p.* 329.

This differs from *E. acidotus* in having narrower leaves; the capsule is horizontal.

86. Barbula *fusca*, C. Muller?

A different species, apparently, from that named by Dr. Taylor *B. campylocarpa*, which may be identical with *B. fallax*.

87. B. *limbata*, Mitten; monoica dense cæspitosa humilis densissime et fastigiatim ramosa, foliis erectis siccitate appressis, basi laxe elongate reticulatis pallidis lanceolatis acuminatis, superne e cellulis minutis rotundatis subopacis areolatis, margine ubique cellulis elongatis incrassatis angustis flavidis late limbato e medio ad apicem serrulato, nervo crasso rubente excurrente, perichætialibus conformibus, theca in pedunculo stricto aurantiaco mediocri cylindrica, operculo subulato, peristomio nullo.—Gymnostomum Jamesoni, *Tayl. Lond. Journ. Bot.*

This species very closely resembles *B. densifolia*, but differs in the monoicous inflorescence, and the absence of peristome; the capsule is also longer, and the whole plant a little larger.

88. B. *Jamesoni*, Tayl. Lond. Journ. Bot. 1846, p. 48.

The peristome of this moss is truly that of *Barbula*, but it has no resemblance to *B. vinealis*, i. e., *Zygotrichia cylindrica*, Fl. Hib., to which it is compared by Dr. Taylor.

89. B. *elongata*, Wils.
90. Tayloria *Jamesoni*, C. Muller, Synops. p. 135. Brachymitrion

Jamesoni, *Tayl. Lond. Journ. Bot.* 1846, p. 44. Eremodon Jamesoni, *Wils. l. c. v.* 5. *t.* 15. A.

The male flower is minute, gemmiform in the axils of the upper leaves.

91. Encalypta *ciliata*, Hedw.
92. Grimmia *apocarpa*, Hedw.
93. G. *longirostris*, Hook. Musc. Exot. t. 62.
94. G. *crispipila* (Tayl.), C. Muller. Trichostomum crispipilum, *Tayl. Lond. Journ. Bot.* 1846, p. 47.
95. Polyporus *virgineus*, Fries.
96. Blindia *fastigiata* (Tayl.), Mitten. Weissia fastigiata, *Tayl.*
97. Trichostomum *mutabile*, Bruch. Syrrhopodon crispulus, *Tayl. Lond. Journ. Bot.*

This corresponds so nearly with British specimens of *T. mutabile*, that I have not considered it worth while to describe it as distinct; the principal differences exist in the rather wider cells at the base of the leaves and the narrower nerves.

98. Trichostomum *Jamesoni* (Tayl.), Mitten. Syrrhopodon Jamesoni, *Tayl. Lond. Journ. Bot.* 1847, p. 331.

This curious species appears to belong to the section *Leptodontium* of the genus *Trichostomum*, but it wants the yellow colour prevalent amongst those species; the present is of a red hue, and the capsule is arcuate.

99. Leptotrichum *gracile*, Mitten; dioicum? cæspitosum humile ramosum gracile sordide virens inferne fuscescens, foliis caulinis homomallis e basi lanceolata convoluta subito in subulam longam attenuatis summo apice parce denticulatis, nervo lato totam subulam occupante, perichætialibus basi latioribus subulatis, theca in pedunculo mediocri oblonga.

A small species with much of the appearance of a poor state of *Distichium capillaceum*; the leaves are very slightly inclined on one side at the apices of the stems.

It is possible this may be the *Didymodon complicatus* of Dr. Taylor, but his specimens led me to consider that species really a small state of *Distichium capillaceum*.

100. Trichostomum *densifolium*, Mitten.
101. Ceratodon *stenocarpus*, Bruch et Schimp.
102. Bartramia *patens*, Brid. Wils. in schedula.

103. *Eadem.* No specimens of this species are in any of the sets I have seen.
104. Bartramia *longifolia,* Hook. Musc. Exot. t. 68.
105. B. *intertexta,* Schimp. C. Muller, Synops. p. 503. B. aristata, *Tayl. MSS.* B. capillifolia, *Wils. in schedula.* Gymnostomum setifolium, *Hook. Icon. Plant. Rar.* t. 135.
106. B. *aciphylla,* Wils. in schedula; dioica rigida robusta erecta elongata subsimplex, foliis caulinis reflexis squarrosis rigidis e basi latissima colorata breviter erecta utrinque ad margines cellulis laxis præditis in subulam latissimam lanceolatam elongatam productis, siccitate spiraliter tortis margine erecto serrulato e basi ad apicem pluries plicatis, nervo tenero excurrente, e cellulis angustissimis areolatis.—Flos masculus terminalis disciformis, antheridiis et paraphysibus numerosissimis.

Very sparing specimens have been sent of this noble species, which appears to exceed a foot in height: it belongs to the same group of species as *B. gigantea,* and in general appearance has some resemblance to the larger species of *Dicranum.*

107. Bryum *obconicum,* Hsch.

There appears to be no appreciable difference between these specimens and European.

108. B. *candicans,* Tayl.

Probably a form of *B. argenteum,* but much smaller.

109. B. *speciosum* (Hook. et Wils.), Mitten.
110. Zygodon *Pichinchensis* (Tayl.), Mitten. Z. hispidulus, *Wils. in schedula.* Didymodon? Pichinchensis, *Tayl. Lond. Journ. Bot.* 1848, p. 280.
111. Z. *squarrosus* (Tayl.), C. Muller. Leptostomum squarrosum, *Tayl. Lond. Journ. Bot.* 1846, p. 43. Aulocamnion squarrosum, *Hook. et Wils. l. c.* 1846, p. 448. t. 15.
112. Z. *pusillus,* C. Muller, Synops. p. 684. Gymnostomum euchloron, *Schwægr.* t. 176.

b. Z. *papillatus,* Mont. Ann. des Sc. Nat. 1845, p. 106. Z. stenocarpus, *Tayl. Lond. Journ. Bot.* 1847, p. 330.

113. Neckera *tomentosa* (Hook.), C. Muller. Leucodon tomentosus, *Hook. Musc. Exot.* t. 37.

The *Leucodon scabrisetus* of Dr. Taylor is a small moss with much the general appearance of *Hypnum albicans,* and in my opinion a doubtful *Leucodon.*

CATALOGUE OF CRYPTOGAMIC PLANTS. 357

114. N. *nigricans* (Hook.), Nees. Hypnum nigricans, *Hook.*
115. N. *nigrescens*, Schwægr. t. 244.
116. Pilotrichum *imponderosum* (Tayl.), C. Muller. Leskea imponderosa, *Tayl. Lond. Journ. Bot.* 1846, p. 62. Cryphea helictophylla, *Wils. l. c.* p. 453.
117. Hookeria *Mulleri*, Hampe. C. Muller, Synops. p. 195.
118. Daltonia *Jamesoni*, Tayl. Lond. Journ. Bot. 1848, p. 283.
119. D. *ovalis*, Tayl. l. c. 1846, p. 66.
120. Hookeria *venusta*, Tayl. MSS.; dioica pusilla, caule adscendente subsimplici, foliis symmetricis ovato-lanceolatis acuminatis integerrimis concavis margine plano apicem versus hic illic recurvo nervis ad medium productis flavidis, cellulis elongatis densis minutis haud chlorophyllosis, perichætialibus minoribus enerviis apice parum dentatis, theca in pedunculo asperrimo oblonga erecta vel paululum inclinata, operculo conico-subulato, calyptra late multifida sparsim pilosa, peristomio dentibus externis lamina exteriore angusta flava lamina interna latiore rugulosa, internis æquilongis luteis linea pallidiore exaratis densissime punctulatis rugulosis.

A small species, less than *H. adscendens* (Actinodontium adscendens, *Schwægr.*, t. 164), but agreeing in structure with those of the section *Lepidopilum* of the genus *Hookeria*.

121. H. *pilifera*, Hook. et Wils. Lond. Journ. Bot. 1844, p. 160. H. papillata, *Tayl. l. c.* 1848, p. 283.
122. H. *falcata*, Hook. Musc. Exot. t. 54.
123. Hypnum *scariosum*, Tayl.
124. H. *leptocladum* (Tayl.), C. Muller. Leskea leptoclada, *Tayl. Lond. Journ. Bot.* 1847, p. 283.
125. Hypopterygium tomentosum (Sw.), C. Muller.
126. Andreæa *subenervis*, Hook. et Wils. Lond. Journ. Bot. vol. vi. t. 10. Andreææ nova species, *Jameson, MSS.*
127. Plagiochila *Jamesoni*, Tayl. Lond. Journ. Bot. 1847, p. 340.
128. P. *pachyloma*, Tayl.
129. P. *bursata*, Ldbg. var. *ærea*. P. ærea, *Tayl.*
130. P. *fragilis*, Tayl. Lond. Journ. Bot. 1848, p. 198.
131. P. *centrifuga*, Tayl. Probably a form of the preceding.
132. P. *Raddiana*, Ldbg.
133. P. *bursata*, Ldbg.
134. Jungermannia *grandiflora*, Ldbg. et Gottsche.

135. Plagiochila *revolvens*, Mitten, MSS.; repens, ramis erectis prolifero-ramosis apicibus circinato-incurvis, foliis imbricatis appressis orbiculatis margine dorsali reflexis basi longe decurrentibus integerrimis, perianthio brevi obconico compresso ore integerrimo.

Allied to *P. ansata*, and in the form of its leaves to *Alicularia occlusa*, but different from both in their reflexed dorsal margin and long decurrent bases.

136. Jungermannia *nigrescens*, Mitten, MSS.; adscendens subramosus, foliis ovato-oblongis remotis subsecundis concavis apice brevi emarginatis angulis acutis vel obtusiusculis, basi dorsali longe decurrentibus et versus basin lacinia parva dentiformi instructis.

In the curious remote tooth at the base of the decurrent dorsal margin, this species resembles *J. leucostoma*, but the present is a much larger plant, the leaves are more concave, the sinus is more shallow, and the laciniæ more obtuse.

137. Gymnanthe *Bustillosii*, Mont.

138. LEPTOSCYPHUS, *Mitten*, nov. gen.—*Perianthium* terminale, compressum, ore truncato. *Involucri folia* caulinis similia. *Caulis* horizontalis vel adscendens. *Folia* horizontalia. *Amphigastria* bifida.

L. *Liebmannianus* (Ldbg. et Gottsche), Mitten. Jungermannia Liebmanniana, *Ldbg. et Gottsche*. Chiloscyphus fragilifolius, *Tayl.* Lond. Journ. Bot. 1848. p. 284.

This is one of a small group of *Hepaticæ* on which I propose to found the genus *Leptoscyphus*, and I have little doubt that the above reference of *Chiloscyphus fragilifolius* to the *Jungermannia Liebmanniana* of Lindenberg and Gottsche is correct, for I possess the same species from Mexico as well as from Cotopaxi. Other species of this genus are *L. succulentus*, (L. et L.), Mitten; *L. Taylori* (Hook.), Mitten; *L. pallens*, Mitten, MSS.; *L. æquatus* (Tayl.), Mitten; *L. turgescens* (Hook. fil. et Tayl.), Mitten; *L. strongylophyllus* (Tayl.), Mitten; *L. Chamissonis* (Ldbg. et Gottsche), Mitten; *L. nigricans*, Mitten, MSS.; *L. gibbosus* (Tayl.), Mitten; *L. juliformis*, Mitten, MSS.; *L. decipiens*, ejusdem; *L. reclinans* (Tayl.), Mitten; and probably *L. cuneifolius* (Hook.), Mitten. The species are remarkable for their conspicuous intercalary spaces.

139. L. *Chamissonis* (Ldbg. et Gottsche), Mitten.

140. Trichocolea *tomentella*, Nees. β. *tomentosa*.

141. *Eadem*, var. ε. *subsimplex*? Synops. Hepat. p. 237.

142. Sendtnera *juniperina*, Nees.
143. S. *æquabilis*, Tayl.
144. Radula *Jamesoni*, Tayl.
145. Madotheca *subciliata*, L. et Ldbg. M. arborea, *Tayl.*
146. M. *squamulifera*, Tayl.
147. Bryopteris *filicina* (Sw.), Nees. B. tenuicaulis, *Tayl.*
There does not appear to be any essential difference between *B. tenuicaulis*, Tayl., and *B. filicina*.
148. Phragmicoma *laxifolium*, Tayl.
149. Omphalanthus *filiformis*, Nees.
150. Lejeunia *robusta*, Mitten, MSS.; pinnatim ramosa, foliis subimbricatis horizontalibus patentibus ovatis apice acute denticulatis et acutis basi in lobulo sinuato-complicatis, amphigastriis orbiculatis margine superiore reflexis denticulatis basi decurrentibus angulis undulatis, foliis involucralibus angustioribus erectioribus lobulis acutis, amphigastrio erecto argute dentato, perianthio obovato oblongo truncato depresso marginibus superioribus tuberculis paucis erosis.

Near to *L. languida*, Nees et Mont., but the stipules are decurrent and dentate as well as recurved, the perianth is much depressed, and the margins almost smooth.

 b. Omphalanthus *filiformis,* Nees. Var. *ζ. tenuis*; gracillimus, foliis et amphigastriis angustioribus.

151. Lejeunia *rotundifolia*, Mitten, MSS.; repens inordinate pinnatimque ramosus, foliis imbricatis orbiculatis integerrimis basi subtus sinuato-complicatis lobulo parvo decurrente, amphigastriis folia æquantibus imbricatis orbiculatis acute bifidis, fructu ad basin ramulorum sessili, involucri foliis conformibus, amphigastrio magno, perianthio oblongo-obovato obtuso medio parum constricto quinquangulari compresso.

This species agrees in many respects with *L. contigua* and *L. sordida*, but is much larger; some of the stems are pinnatedly branched, others almost simple: the whole plant is of a pale yellowish colour.

 b. Omphalanthus *sulphureus*, L. et Ldbg.
152. Lejeunia *xanthocarpa*, L. et Ldbg.
 b. L. *cyathophora*, Mitten, MSS.; vage ramosa gracilis, foliis ovatis basi subsinuato-complicatis lobulo parvo plicæformi, amphigastriis foliis duplo minoribus subrotundis bifidis laciniis acutis, perianthio

terminali et laterali obovato inflato dorso convexo ventre obtusissime bicarinato mucronato, mucrone in appendiculo cyathiformi expanso.

This small species approaches very near to *L. serpyllifolia* in ɩs ze and appearance, but the leaves are thinner although the cells are of the same size, the lobule is narrower; the perianth is inflated, and with only two very obtuse angles at the top on the ventral side, the apiculus is expanded into a curious cup-shaped appendage.

153. L. *pallescens*, Mitten, MSS.; procumbens vel ut videtur pendula longissima hic illic ramis brevibus subpinnatim ramosa, foliis remotis cordatis acuminatis basi decurrente in lobulum saccatum complicatis, amphigastriis magnis cordatis bifidis laciniis acutis conniventibus, foliis amphigastrioque florali dentatis, fructu in ramulo brevi laterali in caulibus primariis singulatim dispositis, perianthio parvo obovato-cylindrico obtuso superne obtuse quadrangulato.

The stems of this species are sometimes six inches long and of a pale whitish-yellow colour. In some respects it comes near to *L. cerina*, particularly in the form of its leaves and stipules, which are however more distant and their texture is more lax: the perianth is very minute and scarcely protrudes beyond the leaves, by the emission of the capsule it is burst into four acute laciniæ; the pedicel is so short that the capsule remains barely out of the mouth of the perianth.

154. Frullania *hians*, L. et Ldbg.
155. F. *Ternatensis*, Gottsche.
156. F. *Brasiliensis*, Raddi.
157. F. *atrata*, Nees. Var. δ. *Mexicana*, Synops. Hepat. p. 463.
158. F. *cylindrica*, Gottsche.
159. Sarcomitrium *multifidum* (Linn.), Mitten. Aneura dactyla, *Tayl*.
160. S. *algoides* (Tayl.), Mitten. Metzgeria algoides, *Tayl. Lond. Journ. Bot.* 1846. p. 410.

Except in size, this species scarcely differs from *S. bipinnatum* (Sw.), Mitten. The name *Sarcomitrium* of Corda has been adopted for this genus, in preference to *Aneura* which implies that the plants are nerveless, when in reality they are almost entirely composed of nerve. Besides the species described in the 'Synopsis Hepaticarum' under *Aneura*, the genus *Sarcomitrium* must contain those comprised in the second section of *Metzgeria*, which are precisely of the same structure, and differ but little in habit.

161. Metzgeria *furcata*, Nees.   Var. β. *major*, Synops. Hepat. p. 502.
162. M. *dichotoma*, Nees.
    More than twice the size of original specimens from Swartz in Hb. Hooker, but doubtfully distinct from *M. furcata*.
163. M. *furcata*, Nees.   Var. δ. *gemmifera*, Synops. Hepat. p. 503.
164. M. *filicina*, Mitten, MSS.; fronde pinnatim divisa lineari glabra margine setulosa, calyptris ut plurimum axillaribus.
    This species has all the habit and appearance of *M. furcata*, except that it is branched in a pinnate manner, and not irregularly dichotomous: the stems are about two inches high, and the ramuli about half an inch long: the calyptra resembles that of *M. furcata*.
165. Dendroceros *Jamesoni*, Tayl. Lond. Journ. Bot. 1848. p. 285.
166. Marchantia *Berteroana*, L. et Ldbg.
167. Fimbriaria *elegans*, Spreng.
168. Usnea *barbata*, Ach.   a. *florida*, Fries.
    b. U. *barbata*, Ach.
169. U. *melaxantha*, Ach.
170. Evernia *flavicans*, Fries.
171. Cladonia *gracilis*, var. *C. vermicularis*, Schærer.
172. Peltidea *rufescens*, Ach.
173. Sticta *macrophylla*, Fee.
    b. *Eadem.*
    c. *Eadem ?*
174. S. *dissecta*, Ach. ?
175. Parmelia *crenulata*, Eschweiler.
176. P. *sinuosa.*
177. Ramalina *calicaris*, Fries.
178. Parmelia *leucomela*, Fries, var.
179. Stereocaulon *ramulosum*, Ach.
180. Cladonia *Jamesoni* (Tayl.), Mitten.   Bæomyces Jamesoni, *Tayl.*
    The structure of the apothecia of this species is that of *Cladonia*, and not that of *Bæomyces*.
181. Collema *tremelloides*, Ach.

Sketch of the VEGETATION *of the Isthmus of* PANAMA ; by M. BERTHOLD SEEMANN, Naturalist of H. M. S. Herald.

(*Continued from p.* 306.)

In a country where nature has supplied nearly every want of life, and where the consumption of a limited population is little felt, agriculture, deprived of its proper stimulus, cannot make much progress. It is, therefore, in the Isthmus, in the most primitive state : our first parents hardly could have carried it on more rudely. A spade is a curiosity, the plough has never been heard of,* and the only implement used for converting forests into fields, are the axe and the machete (or chopping-knife). A piece of ground intended for cultivation is selected in the forests, cleared of the trees by felling and burning them, and surrounded with a fence. In the beginning of the wet season the field is set with plants by simply making a hole with the machete, and placing the seed or root in it. The extreme heat and moisture soon call them into activity, the fertility of a virgin soil affords them ample nourishment, and without the further aid of man a rich harvest is produced. The same ground is occupied two or three years in succession ; after that time the soil is so hard and the old stumps have thriven with so much energy, that a new spot has to be chosen. In most countries this mode of cultivation would be impossible to practise; but in New Granada all the unoccupied land is common property, of which anybody may appropriate as much as he pleases, provided he encloses it either artificially or by taking advantage of rivers, the sea, or high mountains. As long as the land is enclosed it remains in his possession ; whenever the fence is decayed the land again becomes the property of the republic.

Colonial produce, such as sugar, coffee, cacao, tamarinds, &c., which require more attention than the inhabitants are wont to bestow, are merely raised for home consumption. Although the provincial government has tried to encourage this branch of industry by offering premiums for growing a certain number of plants, and the soil and

* In 1846, an English gentleman, residing at Panamá, often used to boast jokingly that he was one of the few, if not the only one, in the city, possessing a spade. Since the commencement of the railway, and a more active traffic, both spades and wheelbarrows, the latter formerly an unknown vehicle, have become better known ; still they are far from being familiar sights in the country.

climate are favourable, yet none, except a few enterprising foreigners, have taken a prominent part in the cultivation; and there is reason to believe that while the country remains so thinly populated as at present, that natural consequence of such a state of society, the high price of labour, will be a lasting impediment to the establishing of plantations on a large scale.

Most of the cultivated plants have been brought over from foreign countries. The cerealia grown are Rice and Indian Corn. The former was introduced by the Spaniards; the latter was known, before the conquest, to the Aborigines, who raised it extensively, and used to prepare from it their bread, and also *chicha*, a kind of beer. At present there are several varieties of Indian Corn, chiefly distinguished from each other by the colour and size of their respective grains. Some successful experiments with Wheat have lately been made on the mountains of Veraguas, which will probably lead to an extensive cultivation of that grain.

Of dessert fruit probably no country can exhibit a greater variety. Besides many indigenous ones, there are to be found the *Aguacate* (*Persea gratissima*, Gærtn.), *Anona* (*Anona laurifolia*, Dunal), *Aqui* (*Cupania Akeesia*, Cambess.), *Chirimoya* (*Anona Cherimolia*, Mill.), *Granadilla* (*Passiflora quadrangularis*, Linn.), *Jobo* (*Spondias lutea*, Linn.), *Lima* (*Citrus Limetto*, Risso), *Limon* (*Citrus Limonum*, Risso), *Mammey de Cartagena* (*Lucuma mammosum*, Gærtn.), Mango (*Mangifera Indica*, Linn.), Melon (*Cucumis Melo*, Linn.), *Naranja agria* (*Citrus vulgaris*, Risso), *Naranja dulce* (*Citrus Aurantium*, Risso), *Palo de pan* (*Artocarpus incisa*, Linn.), *Papaya* (*Carica Papaya*, Linn.), *Pinna* (*Ananassa vulgaris*, Lindl.), *Pomarosa* (*Jambosa vulgaris*, DC.), different species of *Ciruelas* (*Spondias*, sp. pl.), and *Toronjil* (*Citrus Decumana*, Linn.). The Mangosteen was introduced in 1848, plants having been obtained from the Royal Botanic Gardens at Kew.

The Plantain is most extensively cultivated, and furnishes the inhabitants with the chief portion of their food. The question whether the Plantain and its kindred are indigenous to the New World, or whether they have been introduced, has hitherto formed a topic for historians rather than for naturalists, and no satisfactory conclusion has as yet been arrived at. Some incline to the former, others to the latter opinion; and again a third party thinks that while some species are indigenous, others have been brought from foreign countries. Robertson, following Wafer and Gumilla, classes the Plantain among the native productions of America. It was found by the latter two authors far in the interior, and in the hands of Indian tribes who had little or no communication with the Creoles. But as both Wafer and Gumilla travelled a number of years after Columbus's discovery, and as we know that many plants, even some less useful than the different *Musas*, were disseminated with great rapidity over the territories of the New World, the proofs appear insufficient. Prescott seems to look upon the Plantain as introduced, but thinks it is not mentioned in the works of Hernandez. Yet Hernandez does mention the Plantain; he even informs us that it was brought to Mexico from foreign parts, and in his Hist. Plant. Nov. Hisp. Libr. vol. iii. p. 172, has the following account:—" Arbor est mediocris, familiaris calidis regionibus hujus Novæ Hispaniæ, vocatur a quibusdam recentiorum *Musa*. Folia sunt valde longa et lata, adeo ut hominis superent sæpenumero magnitudinem : fructus racematim dependent incredibili numero et magnitudine, cucumerum crassorum et brevium forma, dulces, molles atque temperiei proximi, nec ingrati nutrimenti. Eduntur hi crudi, assive ex vino, atque ita sunt gustui jucundigris. Differt fructus magnitudine, et quo minores sunt, eo salubriores et suaviores. *Advenam esse aiunt huic Novæ Hispaniæ atque translatam ab Æthiopibus aut Orientalibus Indiis, quorum est alumna.* Caulis et radix, quæ fibrata est, multis constant membranis, saporis expertibus et odoris, lubricis et frigescentibus, ex quo facile quis conjiciat, quibus morbis possint esse utiles." Conclusive as is this statement, both as regards the identity of the plant, and its native country, still some may yet entertain doubts, as Hernandez wrote not at the time of the discovery of America, but towards the end of the sixteenth century. There is, however, another proof that the Plantain was introduced. Neither the Quichua nor the Aztec, the two most refined and widely diffused of all American languages, nor indeed any other indigenous tongue of the New World, possesses a vernacular name for this plant. Even Hernandez, who collected the Aztec names with the utmost care, could find none, and was compelled to place the Plantain near the *Quauhxilotl* (*Parmentiera edulis*, DC.), and call it *Quauhxilotl altera*; the cucumber-like fruit of the *Parmentiera* appearing to him to form the closest approach to that of the Plantain.

The esculent roots under cultivation are : *Nname* (*Dioscorea alata*,

Linn.), *Yuca* (*Manihot utilissima*, Pohl), *Batata* or *Camote* (*Batatas edulis*, Chois.), *Otò* (*Arum esculentum*, Linn.), and *Papas* (*Solanum tuberosum*, Linn.). All these plants, except the Potato, are propagated merely by cutting off the top of the roots (tubers, corms, &c.). The vitality of these cuttings is very great; they may be left for weeks on the field, exposed to sun and rain, without receiving any injury. Many other vegetables are treated in a similar manner in tropical America. In Ecuador the tops of the Aracacha remain three or four months out of the ground, and when at last they are planted, thrive as well as if they had been subjected to the most attentive treatment.

Other vegetables grown are the *Challote* (*Sechium edule*, Swartz), *Guineo* (*Musa sapientum*, Linn.), *Guandu* (*Cajanus bicolor*, DC.), *Mani* (*Arachis hypogæa*, Linn.), *Pepino* (*Cucumis sativus*, Linn.), *Sapallo* (*Cucurbita Melopepo*, Linn.), *Lechuga* (*Lactuca sativa*, Linn.), and *Col* (*Brassica oleracea*, Linn.). The Lettuce and Cabbage are raised with difficulty in the lower region; they never form any heads, and are not much liked. Tomatos (*Lycopersicum esculentum*, Mill.) and different kinds of *Aji* (*Capsicum*, sp. pl.) are cultivated in considerable quantities, and used as condiments for culinary purposes.

With this enumeration I conclude. It would be foreign to the subject to touch upon the other branches of agriculture, branches practised as rudely as that relating to the cultivation of the fields. Perhaps, in a few years the old system will be overturned, and a new one be established. The great impulse given to every kind of industry in the Pacific by the discovery of gold in California, and the constant demand for eatables which that event has occasioned, will do their work. In the Isthmus it must produce a progress in agriculture, which, besides increasing the opulence of the country, will have a most beneficial effect upon the climate. According to Dr. G. Gardner, the seasons of Rio de Janeiro were formerly similar to those of Portobello, Chirambirá, and other parts of the Isthmus,—they could hardly be divided into wet and dry. But since the axe was laid on the dense forests surrounding the city, the climate has become dry; in fact, so much has the quantity of rain diminished, that the Brazilian Government was obliged to pass a law prohibiting the felling of trees in the Corcovado range. The same effect will probably be produced in the Isthmus.. When the immense forests, which at present cover the greater portion of the country, have been reduced, and a free circulation of air from sea to sea has been established, the rainy season will be considerably shorter, and the climate cooler and more healthy; but to what extent that change may be accomplished is a problem which industry will have to solve.

Second Report on MR. SPRUCE'S *Collections of Dried Plants from* NORTH BRAZIL; *by* GEORGE BENTHAM, ESQ.

(*Continued from p.* 200.)

Since the last portion of this report was printed, a further collection of between three and four hundred species has arrived from the neighbourhood of Barra do Rio Negro, at the junction of the Rio Negro with the Amazon. These plants are henceforth included in our report.

The *Olacineæ* consist of *Heisteria cyanocarpa*, Pöpp. et Endl., a new genus from Barra, described below under the name of *Diplocrater acuminatus*, and a *Liriosma*, also from Barra, with the foliage and inflorescence of *L. Vellosiana*, DC., but with the buds too young to ascertain with certainty whether it belongs to that species, my specimens from Rio Janeiro being already past flowering. The Barra plant is described by Mr. Spruce as a slender tree of about eighteen feet, and much branched, growing in the thick forest.

The *Heisteria cyanocarpa* was from a weak slender tree, of about twenty-five feet, found on the steep cliffs called Barreras, near Santarem. The foliage is not unlike that of *H. Raddiana*, Benth., from Rio Janeiro, with petioles three to four lines long, but the pedicels are much longer, whilst in *H. Raddiana* they are always shorter, than the petioles, the fruit-bearing calyx very much smaller and less deeply divided, and the fruit itself about four lines long, oblong, not globose.

Gardner's Brazilian collection contains, besides the *H. Raddiana* (n. 5378, described in the Niger Flora, p. 258), and some specimens which, for want of fruit, I am unable to determine with certainty (n. 2516, 5974, and 5379, the two former probably *H. cauliflora*, and the third *H. Raddiana*), two new species, which it may not be out of place here to mention.

1. Heisteria *ovata*; foliis ovatis brevissime acuminatis basi rotundato-truncatis, pedicellis fructiferis petiolum (3–4-linearem) superantibus,

calyce fructifero patente ad medium diviso quam fructus oblongus longiore, lobis latis acutiusculis.—Pernambuco. (Gardner, n. 2787.)

2. H. *subsessilis*; foliis ovatis basi cuneatis rotundatisve, pedunculo fructifero petiolum brevissimum (1 lin.) duplo triplove superante, calyce fructifero profunde 5-fido, lobis erectis acutiusculis fructu oblongo paulo brevioribus.—Prov. Goyaz. (Gardner, n. 3040.)— Flores in hac specie alii uti in cæteris ad axillas foliorum fasciculati, alii ad apices ramorum breviter fasciculato-racemosi.

Our new genus, *Diplocrater*, has the small flowers and fasciculate axillary inflorescence of *Heisteria*, but with membranous leaves; the ovary is divided only up to the insertion of the ovules, they, as well as the axile placenta, being entirely free from the summit of the cavity, as in *Ximenia*, *Olax*, *Liriosma*, and *Schœpffia*; and the stamens are equal to and opposite the petals, as in *Schœpffia*. The general habit and foliage remind one strongly of the figure of *Rhaptostylum acuminatum* in Humboldt and Kunth's Nova Genera et Species, vol. vii. t. 621, but the structure of the flowers is very different, as will appear from the following character.

DIPLOCRATER, gen. nov.—*Involucrum* cupulatum, integerrimum, calycem mentiens. *Calyx* cupulatus, integerrimus. *Discus* petalifer calycem intus fere omnino vestiens, calycis margine angustissimo vix prominulo. *Petala* 5, margini disci imposita, æstivatione valvata, superne intus incrassata. *Stamina* 5, petalis opposita, ad basin eorum inserta et intra cavitatem petali quasi recepta, filamento brevissimo crasso, anthera adnata loculis rima brevi dehiscentibus. *Ovarium* sessile, orbiculare, medio carnoso-pyramidato in stylum simplicem abiens, apice ovoideo-stigmatosum. *Ovula* 2, ex apice placentæ centralis ad summitatem cavitatis non attingentis pendula, dissepimento separata ultra ovula non producto.—Species unica *D. acuminatus*. *Arbor* 15–20-pedalis, ramulis foliaceis elongatis tenuibus, ex omni parte glaber. *Folia* alterna, exstipulata, breviter petiolata, ovato-oblonga, acuminata, basi rotundata, membranacea, penninervia, reticulato-venosa, 3–5-pollicaria. *Flores* minimi, ad axillas foliorum fasciculati, pedicellis semilineam longis. *Cupula* exterior (e bracteis 2 coalitis formata?) explanata, fere lineam diametro, basi carnosula, margine tenui; interior (calyx cum disco adnato) paulo longior. *Petala* triangulari-lanceolata, semilineam longa, acuta, apice carnosula et subinflexa, a basi paulo ultra medium intus excavata, concavitate intus superne costa prominente divisa. *Stamina* ad tertiam partem petali attingentia, latiora quam longa, connectivo antherarum a filamento lato crassiusculo haud distincto. *Stylus* petalis brevior. *Fructus* non visus.

There are two *Humiriaceæ*, the *Umirì* (*Humirium floribundum*, Mart.) mentioned in our first report, and again found abundantly in the neighbourhood of Santarem, and the *Saccoglottis Amazonica*, Mart., a much more local species, although frequent on the margin of forests about Santarem. It there forms a spreading tree of about thirty feet, with the young leaves of a deep red. Its fruit is called *Uaxuà* by the Brazilians, and said to be very good eating.

Among *Meliaceæ*, the common *Azedarach* was found, to all appearance wild, as a tall shrub or low tree, in the vicinity of Santarem: a *Guarea* from the vicinity of Obidos may possibly be the same as the one previously gathered near Parà, both forming lofty trees; the specimens were very few, and could only be supplied to the first two or three sets. The following new *Trichiliæ* are more generally distributed.

1. Trichilia *excelsa*; foliis amplis, foliolis 5-7 alternis oppositisque ad venas petiolis ramulisque pubescentibus, paniculis brevibus, floribus vulgo tetrameris, filamentis ultra medium connatis apice bidentatis dentibus antheram glabram superantibus, ovulis collateralibus. —*Arbor* excelsa, ramulis teretibus pube molli haud densa. *Petioli* communes 4-8 poll. longi. *Foliola* 3 ultima ad apicem petioli digitata, 5-6-pollicaria, cætera minora, opposita v. alterna, omnia ex ovata v. oblonga forma in obovatam transeuntia, apice rotundata cum acumine brevi et obtuso, margine sæpe undulata v. leviter sinuata, basi plus minus angustata et petiolulo brevi fulta, membranacea, penninervia et reticulato-venosa, supra glabra nisi ad venas parce puberula et demum lucidula, subtus tactu mollia, ad costas venasque puberula, pallide virentia. *Paniculæ* vix pollicares, a basi ramosissimæ. *Pedicelli* ultimi ¾ lin. longi. *Calyces* late et obtuse 4-lobi, ¼ lin. longi. *Petala* 2 lin. longa, alba, extus adpresse puberula. *Stamina* undique glabra. *Filamenta* complanata, glabra, paulo ultra medium monadelpha, tubo basi disco carnosulo adnato. *Antheræ* ovatæ. *Ovarium* hirsutum, triloculare. *Ovula* in loculis gemina, minutissima, collateralia, medio lateraliter affixa. *Stylus* brevissimus, glaber.—Ovarium in omnibus quos examinavi flores semiabortivum videtur et planta verisimiliter polygamo-subdioica.— From the virgin forests near Santarem.

2. **Trichilia *macrophylla*,** sp. n.; foliis amplis, foliolis 3–7 alternis obovatis ellipticisve acuminatis membranaceis ad venas petiolis ramulisque minute puberulis, paniculis brevibus, floribus pentameris, filamentis infra medium connatis apice bidentatis, dentibus anthera hirsuta brevioribus, ovulis collateralibus, capsulis echinato-hispidis monospermis.—*Frutex* alte scandens, trunco crassitie brachii humani, ramulis tenuibus. *Folia* fere *T. excelsæ*, sed majora, glabriora. *Foliola* terminalia semipedalia vel interdum fere pedalia, cætera vulgo minora, apice plus minus acuminata, basi angustata et petiolulata, secus petiolum communem valde irregulariter disposita. *Paniculæ* breviores et flores multo minores quam in *T. excelsa*. Paucos tamen vidi *flores*, antheris ut videtur sterilibus subfœmineos, dum in speciminibus meis *T. excelsæ* omnes submasculi erant. *Calyx* puberulus, apertus vix ⅓ lin. longus, lobis acutis. *Petala* linea paulo longiora, glabra, æstivatione valde imbricata. *Stamina* 8, 9, v. 10, petalis paulo breviora, filamentis basi connatis glabrisque, dein liberis complanatis extus glabris intus villosis. *Antheræ* angustæ, villosæ. *Ovarium* disco tenui impositum, hispidum, triloculare. *Ovula* ex apice attenuata pendula, collateralia. *Capsula* ovato-triquetra, 4–5 lin. longa, extus tuberculis squamulisve breviter rigideque hispidis dense obtecta, trivalvis valvulis crassiusculis, intus abortu unilocularis et monosperma. *Semen* pendulum, testa fusca, arilla parva (exsiccatione partim destructa). *Cotyledones* crassæ, carnosæ, radiculam superam includentes.—From the Paranà-mirì dos Ranos, between Santarem and Barra.

3. **Trichilia? *micrantha*,** sp. n.; foliolis 5–7 ovali-ellipticis obtuse acuminatis coriaceis glabris v. subtus parce puberulis, paniculis laxis multifloris folio sublongioribus, floribus parvis subglobosis, filamentis ad apicem connatis tubo intus apice villoso, antheris erectis exsertis dentibus nullis interjectis, ovulis solitariis (v. geminis adglutinatis?). —*Arbor* 30-pedalis, ramulis teretibus glabris. *Foliola* ultima 4–5-pollicaria, inferiora minora. *Paniculæ* in axillis superioribus et ad apices ramorum semipedales et longiores, cinerascentes, ramis divaricatis, floribus subfasciculatis brevissime pedicellatis albis vix lineam diametro. *Calyx* parvus, laciniis 5 orbiculatis. *Petala* 5, orbiculata, calyce duplo longiora. *Staminum* tubus subinteger, brevis, extus glaber, intus basi glaber, apice dense villosus. *Antheræ* 10, erectæ, ovatæ, acutæ, glabræ. *Ovarium* disco orbiculari impositum, triloculare, stylo brevi.

This tree, from the Capoeiras, near Barra do Rio Negro, is very unlike all other *Trichiliæ* known to me, and the ovules appeared to be solitary in each cell, even under a high magnifying power; but as Jussieu observes that they are often so closely combined as to appear like a single ovule, it is possible such may be the case with this plant, or that, in the specimens before me, the flowers may be sterile and the ovules imperfectly developed. In the absence of fruit I have, therefore, left it in *Trichilia*, to which it approaches nearer than to any other genus.

The only plant belonging to *Vitaceæ* is a variety of *Cissus sicyoides*, Linn., from Santarem, with the leaves broadly but not deeply cordate, and sometimes scarcely more than truncate at the base, thus coming near to *C. ovata*, Lam., and showing that the two plants may very probably be mere varieties of one species, as already suggested by De Candolle.

There are four *Simarubeæ*,—a variety of *Simaruba versicolor*, St. Hil., forming a bushy tree of about twenty feet in the campos near Santarem, having flowers rather smaller and more closely clustered than in South Brazilian specimens, but precisely the same form as Blanchet's n. 2727, referred by Planchon without hesitation to this species; and three *Simabæ*, all belonging to the small-flowered group, which come much nearer in general aspect to *Simaruba* than to the large-flowered *Simabæ*, although the structure of the flowers renders it necessary to associate them with the latter genus. One of these, a small irregularly-branched tree, from the south shore of the Rio Negro, near Barra, is the *Simaba Guianensis*, Aubl., or at any rate the common Guiana species, which Planchon has, in all probability correctly, referred to the one so rudely figured by Aublet. The two following are new :—

1. Simaba *fœtida*, sp. n.; caule arboreo, ramulis glabris, foliis glaberrimis, foliolis 5–7 obovato-cuneatis obtusissimis emarginatisque, panicula laxa foliis longiore, floribus pro genere parvis, staminum squamis abbreviatis, ovario puberulo.—*Arbor* parva, ramulis leviter angulatis glaucescentibus. *Foliola* opposita, tria ultima digitata, $1\frac{1}{2}$–2 poll. longa, 1–$1\frac{1}{4}$ poll. lata, basi in petiolulum brevissimum longe angustata, integerrima, coriacea, nervis lateralibus subtus impressis. *Paniculæ* et flores fere *S. Guianensis*. *Petala* vix 2 lineas excedentes. *Drupæ* obovoideæ, rubræ, styli cicatrice versus apicem laterali, pericarpio crasso-carnoso, putamine lignoso oblique obovoideo

valde compresso rugoso pollicem longo semipollicem lato. *Semen* oblongum, compressum, infra apicem lateraliter affixum, testa membranacea. *Cotyledones* carnosæ. *Radicula* supera vix conspicua. *Plumula* semilineam longa.—From the beach of the Rio Tapajoz, near Santarem, the greenish flowers emitting a vile stercoraceous odour.

2. Simaba *angustifolia*, sp. n. ; caule arboreo, ramulis glabris, foliis glabris, foliolis 5-9 anguste oblongis apice obtuse angustatis basi longe acutatis, panicula laxa foliis sublongiore, floribus pro genere parvis, staminum squamis brevissimis, ovario glabro.—Affinis *S. Guianensi.*—*Foliola* 2-3 poll. longa, pleraque vix semipollicem lata, maxima rarius 4 poll. longa et fere pollicem lata. *Paniculæ* pube minuta cinerascentes. *Pedicelli* iis *S. Guianensis* longiores et tenuiores, flores ejusdem magnitudine vel vix majores.—From the south shore of the Rio Negro, near Barra.

In Planchon's enumeration of *Simarubeæ*, he appears to have overlooked the *Simaba Bahiensis*, Moric. Pl. Nouv. Amer. p. 11, t. 9, which is Blanchet's no. 3143, and perhaps also Gardner's n. 2514 from Piauhy. In the generic character of *Simaba* in the same paper, the " squama staminum longa sæpe bifida " applies only to the larger-flowered species, for it is very short in *S. Guianensis, Orinocensis, fœtida,* and *angustifolia.*

The *Ochnaceæ*, lately worked up with great care by Dr. Planchon (Lond. Journ. Bot. vol. v. and vi.), are nevertheless excessively difficult to determine from descriptions. Of the four *Gomphiæ* in the present collection, two from Santarem and one from Barra do Rio Negro, all in single specimens or nearly so, appear different from any I possess, yet I am unable to satisfy myself that they are really unpublished ; the fourth, from the Rio Tapajoz near Santarem, is however more evidently distinct, and has been more generally distributed. I therefore here describe it :—

Gomphia *microdonta*, sp. n.; glaberrima, foliis brevepetiolatis ovalibus apice obtusis margine subrecurvis minute calloso-dentatis, venulis utrinque subobsoletis, racemis simplicibus terminalibus brevibus, pedicellis calyce glabro 2-3-plo longioribus.—Affinis quodammodo *G. Blanchetianæ.*—*Folia* crebra, erecto-patentia, $1\frac{1}{2}$-$2\frac{1}{2}$ poll. longa, 1-$1\frac{1}{4}$ poll. lata, basi rotundata v. leviter cuneata, petiolo 1-$1\frac{1}{2}$ lin. longo, costa media prominente, venis tenuibus v. omnino inconspicuis. *Bracteæ* ad basin inflorescentiæ convolutæ, acutissime acuminatæ, 3 lin. longæ. *Racemi* 2-3-pollicares, omnes in speciminibus suppetentibus simplices, pedicellis unifloris 7-9 lin. longis, prope basin cicatrice bracteolæ notatis. *Sepala* obtusa, 2¼ lin. longa. *Petala* et stamina desunt. In fructu juniore, sepalis jam delapsis, discus crassus globosus, carpella gerens 5 obovoideo-globosa glaberrima.

The following very distinct *Chailletia*, from the vicinity of Santarem, is remarkable from the petals, the inside of the calyx, and the filaments being perfectly black, with white anthers. All the flowers I have examined have proved to be male by abortion.

Chailletia *vestita*, sp. n.; ramulis inflorescentiaque rufo-velutinis, foliis ovatis rugosis supra villosulis subtus molliter tomentoso-villosis, pedunculis dichotome corymbosis, inferioribus axillaribus petiolo adnatis, summis paniculatis aphyllis, floribus polygamis, petalis calyce subbrevioribus, staminibus marium breviter exsertis, (fructu oblique obovoideo compressiusculo velutino).—*Arbor* parva, ramis irregularibus fragilibus, ramulis teretibus. *Folia* inferiora vix brevissime petiolata, 3-6 poll. longa, 2-4 poll. lata, obtusa v. breviter obtuseque acuminata, basi rotundata v. subcordata, floralia similia nisi petiolo 2-6 lin. longo fulta. *Stipulæ* lanceolatæ, crassæ, 1-2 lin. longæ, rufo-tomentosæ. *Cymæ* dichotome corymbosæ, tomento rufo villosæ, inferiores pedunculatæ pedunculo petiolo æquilongo et ei adnato, superiores ad axillam bracteæ parvæ sessiles, thyrsum terminalem constituentes; omnes a basi repetite dichotomæ, 1½-2 poll. latæ, floribundæ. *Bracteæ* parvæ, caducæ. *Flores* quos vidi omnes masculi. *Calyx* profunde 5-lobus; *laciniæ* ovatæ, 1 lin. longæ, æstivatione imbricatæ, extus dense cano-tomentosæ, intus glabræ et nigræ. *Petala* nigra, calyce paulo breviora, semibifida, lobis longitudinaliter plicatis. *Stamina* 5, calyce paulo longiora, glabra, antheris ovatis, loculis longitudinaliter dehiscentibus. *Glandulæ* 5, bifidæ, petalis oppositæ. *Ovarii* rudimentum in centro floris disciforme, stylo brevissimo intra lanam abscondito.

A set of specimens gathered in young fruit in the neighbourhood of Barra appear to belong to the same species, although, according to Mr. Spruce's label, they are from a stout *sipò* or woody climber, not from a tree. The leaves are often larger, less downy, and often more decidedly and somewhat sharply acuminate; many of them, however, are precisely the same as in the flowering specimens above described. The fruit, not yet fully formed, is about half an inch long, somewhat compressed, very woolly, either roundly obcordate and two-celled, or by the abortion of one cell obliquely obcordate.

(*To be continued.*)

BOTANICAL INFORMATION.

Death of Dr. WILLIAM ARNOLD BROMFIELD.

Seldom has it been our duty to record the death of a naturalist more devoted to the cause of science, and more esteemed as a man, than Dr. Bromfield. The mournful event took place at Damascus, on the 9th of October of the present year, 1851.

William Arnold Bromfield, M.D., F.R.S., &c., was born in 1801, at Boldre, in the New Forest, Hants; a county so beautiful in its scenery, that it is no wonder that one whose infant years were passed in such a region should be ever after alive to the charms of nature. He was cradled, as it were, among woods. His father was the Rev. John Arnold Bromfield, M.A., formerly Fellow of New College, Oxford, and was, on account of his health, several years ago, obliged to quit his living at Market Weston, in Suffolk, for the warmer climate of Hampshire, where he died in the year which gave birth to his only son. In his childhood Dr. Bromfield was intelligent and remarkable for sweetness of temper, which rather increased than otherwise as years advanced, and he always took great delight in all representations of natural history, in mechanics, &c. One who was much his companion at that time, well remembers when at the age of five or six he stole away from the child's sport of the day to watch some workmen mending a pump: their movements were eagerly scrutinized, and he was ready even then, as on other similar occasions afterwards, to lose no opportunity of gaining information.

At about eleven years of age he was placed under the charge of Dr. Knox, of Tonbridge, for two years, and during that period received much notice and kindness from the Rev. Dr. Cartwright, who lived in the neighbourhood, and who frequently had him at his house, where he loved to encourage the boy's taste for the mechanical arts, in which he was himself no mean proficient. The rest of his school life was passed under Dr. Nicholas, of Ealing, and he was afterwards, for a short time, placed with a clergyman, the Rev. Mr. Phipps, in Warwickshire. The pleasure that he now took in chemistry and his acquirements in that science were such, that when it was thought necessary he should enter some profession, he was urgent that this pursuit should be made available for the purpose, and he was therefore sent to Glasgow, and became, in 1821, an inmate, and we have reason to know a great favourite too, in the family of Dr. Thomson, the distinguished professor of chemistry in the University of that city. There he not only attended the medical curriculum in the college, but was an ardent pupil in Dr. Thomson's public and private classes. At the end of two years he took his degree in medicine. It does not appear that he showed any particular bias for the study of botany till his visit to Glasgow; and then botanical excursions, more, perhaps, than the botanical lectures,—especially a short tour made in the Highlands of Scotland, where the zeal he evinced in pursuit of plants, and the inquiring turn of his mind, are still vividly recollected by the writer of this brief notice,—induced him to make botany the chief object of the remainder of his life. His surviving sister well remembers the pleasure with which he described the botanical trip which first attached him to the study.

He left Scotland finally in 1826, and, not feeling the necessity of engaging in medical practice, he commenced his travels in Germany, France, and Italy, returning to England in 1830. His private letters to his family, now before us, show how profitably he spent that time, and how much he must have enriched his mind in all kinds of subjects in any way connected with natural history. Soon after his return he lost his excellent and affectionate mother, and from that period he lived in the society of his sister, first at Hastings, then at Southampton; and finally, since 1836, at Eastmount, Ryde, Isle of Wight. The two latter places are situated in Hampshire, which beautiful county has been the principal and very successful field of his botanical researches, and gave rise to his 'Notes and occasional Observations on some of the rarer British Plants growing wild in Hampshire,' published in a series of papers in the third volume of Mr. Newman's 'Phytologist.' These 'Notes' evince great accuracy, diligence, and uncommon research, and an extensive reading of all works, foreign as well as English, which could in any way illustrate the subject. He says, with great truth, at the commencement of his communication: "In presenting the readers of the 'Phytologist' with the following list of Hampshire plants, my object has been to promote our knowledge of the geographical distribution of the species in Britain, which important branch of philosophical botany is now, through the impulse happily given to it by the labours of Mr. H. C. Watson, beginning to receive its due share of attention in this country; and the time is gone by when such catalogues are to be viewed, and their utility measured by their fitness, as vehicles for the communication of mere *rarities* to the collector." The whole essay would be well worth printing as a separate work, and be invaluable to every student of the botany of Hants.

But if the botany of Hampshire claimed a great share of Dr. Bromfield's attention, that of the Isle of Wight, his head-quarters for the last fourteen years of his life, was especially interesting to him, and his chief energies were directed to the preparation of a flora of that island. So anxious was he to make this flora perfect, that while other botanists might have deemed two years sufficient to investigate and publish the vegetation of so limited a spot, our valued and lamented friend did not think that at the end of fourteen years he had accomplished all that ought to be performed; and thus, advanced as we know the MSS. to be, and much as had been done by him for a botanico-geographical and geological map of the island with all the accuracy of that of the Ordnance survey which was the groundwork of it, and under the full persuasion that "another summer's" work would complete his flora, the "Flora Vectis" or "Flora Vectensis" (the exact title was not decided) still only exists in manuscript.* It was not given him to pass another summer in his favoured and favourite island.

He had indeed been unwearied in his researches into every nook and corner of the Isle of Wight, in the interior, and on the coast, "isle and islet, creek and bay," and scarcely any summer elapsed without his being rewarded with some discovery. Wild and cultivated localities were alike explored by him, and his visits, at first not

---

* Since the above was in type we are informed, on good authority, that the MS. is all but complete, and that it includes the Flora of Hampshire.—ED.

allowed in a very friendly spirit by the landlord or the tenant of the several properties, were latterly welcomed by the same individuals, and his person and pursuits became familiar to the residents in all parts of the island. It has been our privilege to see him going out at early morn and returning at dewy eve, laden with his collections, which were, for more convenient examination, transferred from the ordinary vascula in which they were packed in the field, to another, so large, that we should hardly be credited if we gave its exact dimensions, but which he used to call his "*witness-box.*" Nothing was left undone which could tend to the completing and the perfecting of his "*Flora.*"

In 1842 he made an autumnal tour, of some weeks, in Ireland, visiting Killarney, Limerick, Cork, &c.

But we should render little justice to Dr. Bromfield's varied knowledge, if we were to speak of him only as a British, or even as an European botanist or naturalist and traveller. In January 1844 he embarked for a six months' tour in the West India Islands, visiting several of them, and spending his chief time in Trinidad and Jamaica. At the former island he received much attention from the botanist Lockhart, one of the few survivors of the Congo expedition under Captain Tuckey, the Curator of the Botanic Garden there, and in the latter (Jamaica) from Dr. M'Fadyen, a physician of great practice, and author of the 'Flora of Jamaica,'—both since dead. In the same island, too, he joined Mr. Purdie, then collecting for the Royal Gardens of Kew, and they made several excursions together, in one of which our friend was bitten by a "*black snake*," which he had incautiously seized by the middle. The creature held to the back of Dr. Bromfield's hand by its fangs for nearly a minute, and could only be removed by a negro forcing the end of a stick into its mouth. Happily the species was not a poisonous one: a swelling of the hand, with slight pain, only ensued.

In the summer of 1846 he resolved to visit North America, and there to make a very extensive and, to his inquiring mind, profitable tour, embarking in July of that year, and not returning till the autumn of the following. How advantageously his time was spent on this occasion may to some extent be seen in the 'London Journal of Botany,' where, at the request of the Editor, he was with difficulty * induced to

---

* Dr. Bromfield was not anxious to appear in print at an early period. His first specimen of authorship, as far as we know, is his valuable account of his newly-discovered grass, in Britain, the *Spartina alterniflora*, Lois., in 1836, in the 'Companion to the Botanical Magazine,' vol. ii. p. 254.

lay before the public some 'Notes and Observations on the Botany, Weather, &c., of the United States of America, made during a tour in that country in 1846 and 1847.' "An indifferent state of health," he tells us, "rendering a change of scene, climate, and occupation absolutely necessary, I determined, towards the middle of 1846, on visiting the United States of America; a country I had long wished to see, as well on account of the great moral and political experiments of which it is the theatre, as of the analogy its vegetation bears to that of Europe, our own island of Great Britain included." This interesting journey extended from Canada north, to New Orleans south, and westward to St. Louis on the Missouri.

Happy would it have been for his friends if this had been the last of his wanderings. He had returned in greatly improved health, and somewhat more than half a Yankee, for a time, in appearance and dress; but his energetic mind longed to occupy itself with oriental scenery and productions, and his carefully-kept Journal, now before us, written in the form of letters to his beloved sister, shows how deeply he was interested in the wonders of Egypt and Nubia. He accordingly embarked from Alexandria in the Sultan steamer, in September 1850. He climbed the Pyramids and penetrated their interior recesses. He ascended the Nile, to visit Thebes, in company with two other English gentlemen, Lieut. Pengilly and Mr. Lakes, a young naval officer. Above Ekhmeon in Upper Egypt, they came at one and the same time on crocodiles and on the curious and famous Doum Palm, the latter remarkable especially for the forking (so unusual in the Palms) of its repeatedly dichotomous stem; each branch terminating in a tuft of fan-shaped leaves, and bearing a noble cluster of fruits, of which our valued friend did not fail to secure a fine specimen for the Museum of the Royal Gardens of Kew, where it is already deposited. Each fruit is, when fresh, the size of a large apple, and the outer eatable coat has a flavour always (and not inaptly) compared to gingerbread. "Yesterday (Dec. 14), whilst taking an evening ramble, we noticed the Doum Palms, growing in plenty along the eastern bank of the river, between Girgeh and Farshoot. The trees bore plenty of fruit, but still unripe." "Nearly coequal with the limits of the Doum Palm is the line that bounds the distribution of the crocodile northwards, at the present day, for in ancient times it would appear to have ranged lower down the Nile, and is said to have even inhabited the

Delta and Lower Egypt properly so called. In our days, the crocodile begins to make its appearance at or near Assiut, yet we saw none of them during our short stay at that city; but, on Dec. 14, on arriving within about a quarter of a mile of a sand-bank, which we learned from our boatmen was a favourite resort of these reptiles, and which is a little beyond Girgeh, between that town and Farshoot, we had the great satisfaction of beholding a whole herd, if I may use the term, of these river-monsters, emerge one by one from the stream as the sun gained power, and assemble on the sand-bank, where we counted no less than sixteen of various sizes, huddled together, and evidently enjoying the warmth of the bright unclouded morning-ray. The smallest of those we saw, as we watched them through our telescopes, seemed to be at least eight or nine feet in length; and several were absolutely leviathan monsters, as hideous and terrific as can well be imagined, not measuring less, certainly, than sixteen or eighteen feet long. Their bodies were as thick as that of a horse, and the huge jaws of some were gaping wide apart, as they basked listless and motionless on the sand, or, occasionally, dragged themselves forth from the water, to lie along like huge logs or trunks of Palm-trees, to which they bear no inconsiderable general resemblance in the rough scaly covering of their unwieldy forms, knotted with crested protuberances. We were so near them that, by aid of our telescopes, we could perfectly watch their motions, and discern their minutest characters, longing all the while to be amongst them with our guns, and planning an attack which we intend making on their stronghold when we return down the river. We purpose to throw up a masked battery of sand the day previous, and, landing on the bank before day-break the following morning, to open fire on them from behind our temporary fort as they emerge from the river to bask in the sun. We have furnished ourselves with balls of hardened lead expressly for the purpose, and trust to be able to achieve the feat of shooting a crocodile, and carrying off his jaws and his skull as a trophy of our campaign against the monster deities of Egypt's river. The young specimens of the crocodile of the Nile that are occasionally brought alive to England, give no idea whatever of the hideous deformity and ferocious aspect of the full-grown animal. A more revolting creature does not exist. Yet I believe the crocodile is seldom or ever dangerous to man, being an extremely watchful and timid animal, waddling slowly down and sliding into the water on the near approach of any person. We also observed the sand-bank occupied by numbers of aquatic birds, as geese, cranes, pelicans, &c., walking among the outstretched monsters as if convinced that they were in no peril of their lives in the society of these ugly reptiles. A boat, in rounding the bank, fired a gun at the crocodiles, but not within range, which had the effect of sending them all off, pell-mell, into the water; but, in a few minutes afterwards, the snouts of one or two might be seen emerging, and soon the sand-bank became again peopled by the discomfited, but now re-assured, fugitives."

Our travellers were now on their way to Wady Hafeh, 2,000 miles up the Nile. About Henneh, or Genneh, they were struck with the great fertility of the soil, and immense abundance of cultivated produce. "Both the valley and the river are here of great breadth, and the latter is richly adorned with lofty groves of Date Palms, interspersed with Doum Palms, which are now abundant in all the fields, and of which I have to-day (Dec. 17) seen very fine specimens in full fruit. The country is everywhere beautifully green with the tender springing wheat and barley, which are as far advanced as in England they commonly are in April or May, and will be ready for harvesting about the former of these months, or even at the end of March. At this time the Guinea-grass, of which vast quantities are grown in Egypt, is being gathered in, and the sugar-harvest will succeed it, a week or two later. The quantity of garden and vegetable produce raised in Egypt is prodigious. The whole valley of the Nile may be regarded as one vast kitchen-garden, and all the ancient plant-deities of Egypt still find favour in the sight of her modern inhabitants."

The ruins of Thebes engaged much of the attention of the travellers, both in the ascent of the river and on their return, as well as the lofty obelisk at Luxor and the avenue of sphynxes at Karnak. On the 20th of March, Dr. Bromfield wrote to his sister from Khartoum, at the junction of the White and Blue Niles, in lat. 15° 11′ and long. 34° 10′, almost in the heart of tropical Africa;—" a region of dust, dirt, and barbarism." Their land journey, from the time they quitted their boat at Wady Hafeh to the period of their return to it at Korosko (April 27th), occupied a period of ninety-nine days,* and was accompanied by much fatigue, and danger, and suffer-

* "These hundred days," he writes in his journal, "have been full of inci-

ing; and, alas! only two of the three Europeans returned to this place. Dr. Bromfield's young and amiable friend and fellow-traveller, Mr. Lakes, took the eruptive fever of the country, called Jeddereh or Jiddereh, very analogous in its symptoms to the small-pox, and died after a few days' illness at Berber. The same letter that conveyed this sad intelligence from Dr. Bromfield, contained also the particulars of the death of Mr. Melly, which had occurred only two months previously, in traversing the Desert between Berber and Abou Hamed: this gentleman died surrounded by his family, consisting of Mrs. Melly, a daughter, and two sons. These ladies are said to be the first English females that ever made their appearance in Khartoum and on the White Nile. These events must have cast a gloom over the minds of the surviving companions. Yet, and with debilitated constitutions owing to the fatigue and bad food, Dr. Bromfield and Mr. Pengilly visited a great number of interesting places on their return voyage; and in the midst of the ancient remains of Dendereh, he writes in his journal: "I had another object in again stopping at Dendereh, namely, to procure a cluster of the fruit of the Doum Palm, for the Botanical Museum at Kew Gardens, agreeably to a request from Sir William Hooker that I would if possible do so. In this I have fully succeeded; and, whilst the villagers were engaged in cutting me off a properly-sized cluster from the forests of the Palm which adorn the approach from the river to the Temple, I pushed on for the latter, returning with rather increased than diminished admiration of this beautiful and elaborately adorned structure."

The travellers returned to Cairo on the 4th of June, after an absence of seven months. Between that period and the 22nd of July, Dr. Bromfield accomplished the following trips, and saw the following lions. To Suez and back, 184 miles, on a donkey. To the mounds of Memphis, the quarries of Toorah, and the pyramids of Saccarah, visiting *en route* the chicken-hatching ovens of Ghizeh! To

dent;—of pleasure, mingled with much pain and inconvenience, and sometimes with great discomfort from the dirt, dust, heat, cold, bad lodging, and worse diet, which it was our lot to put up with at various times. Still these would have been light and transient evils, worthy only of being laughed at, and forgotten as soon as passed,—and indeed they mostly served as subjects for merriment to us at the moment of their occurrence;—but it pleased the Almighty to throw a deep gloom over the latter part of our Ethiopian journey by removing one of our little party of three by the hand of death."

Heliopolis; and, lastly, to the petrified forest. He afterwards made a voyage to Damietta, and the Lake Menzeleh, where he searched, but in vain, for the Papyrus, still reported, though on questionable authority, to flourish in that spot as of yore.

August 10th, Dr. Bromfield's letter is dated from Jaffa, the ancient Joppa, in Syria. August 15th, from Jerusalem. A postscript to that letter announces his future plans:—" I propose going to the Dead Sea and Jordan, of which I had a fine view, on the 15th, from the summit of the Mount of Olives; to Hebron and Bethlehem; then to Damascus (through Samaria and Galilee), where I intend introducing myself to General Guyon, now a pasha at Damascus, whose mother and brother I knew so well; thence to Bairout, over the Lebanon, taking the ruins of Baalbec in my way. Dismissing Saad (his servant) at Bairout, I shall take the first steamer for Smyrna and Constantinople. I shall try my utmost to leave Constantinople for Southampton on the 29th of September, so as to be home the middle of October."

The last letter he appears to have written home was from "Bairout, Syria, September 22nd" (received at Ryde, October 18th). In that he says, " I am momentarily expecting Dr. Bialloblotsky, to arrange for our journey to Damascus and Baalbec, which will conclude my Syrian travels; after which every successive day's journey will bring me nearer to dear Old England. I have had quite as much as I wish of Eastern travel; enough to furnish many pleasing reminiscences of past events and distant scenery. But I am not sorry my long pilgrimage is drawing to a close; and that, with God's blessing, I shall soon return to enjoy the smiles and comforts of ' home, sweet home.' "—Man indeed proposes, but God disposes. Only seventeen days after this letter, so full of thankfulness for the past and of hope for the future, was written, and nine days before it was received in England, our excellent friend breathed his last at Damascus. It was on the 9th of October, four days after his arrival there from Baalbec, that he expired. He was seized with typhus of the most malignant description. Everything that could possibly be done for him, both by medicine and attendance, was effected, there being two European doctors of great celebrity resident at Damascus. No servant, however, could be procured to wait upon him, owing to superstition and fear; but Mr. G. Moore, an English traveller, and Mr. James Barnett of the American Mission, appear to have been unremitting in their kindness and attentions. He became sensible three hours before his death, and requested that the Bible might be read to him, which was done by Mr. Moore. He then relapsed into insensibility, and died tranquilly two hours afterwards. He was buried in the Christian ground outside the city; and the greatest respect was shown by the consuls either attending in person or sending their "cavasses."

Two cases of his collections had reached Southampton long previous to the arrival of the sad intelligence just narrated; and these he had wished should remain in the custom-house till his return. As it was known they chiefly contained plants, they were, by the order of his sister, sent to the Royal Gardens at Kew; and, believing such to have been the wish of her brother, that lady has desired that this well-preserved and rather extensive collection of Egyptian and Nubian specimens should be shared by Mr. Bentham, Sir W. J. Hooker, Mr. Borrer, and Dr. Bell Salter. The seeds, at least the more hardy kinds, to be divided between the Royal Gardens and Mr. Lawrence of St. John's Wood, Isle of Wight, who was always the recipient of living plants and seeds from Dr. Bromfield's travels. The fine cluster of the Doum Palm is deposited in the Museum of the Royal Gardens. The rest of this portion of his collections, consisting only of a few geological specimens, fragments of remarkable ruins, a few curious insects, and some other objects of natural history, are retained by the bereaved family, as mementos of one so truly and so deservedly beloved.

Mr. Rucker's *Orchideous Plants*.

Botanists and horticulturists are alike interested in the conservation of the rarest and finest collection of Orchideous plants that has ever been amassed by any one individual. The palm for such an assortment has been universally awarded to S. Rucker, Esq., of West Hill, Wandsworth, where the collection is in all its integrity. Circumstances have happened, as the public is but too well aware, which require that these splendid specimens should find another possessor; but we must be allowed to express a most earnest hope that they will not be dispersed, but that some one among the noblemen and private gentlemen of fortune, of whom many have been distinguished by their love of science and of horticultural pursuits in this country, will be inclined to purchase it, and preserve it entire.

ALPHABETICAL INDEX

TO THE CONTENTS OF THE THIRD VOLUME OF

HOOKER'S JOURNAL OF BOTANY.

Algeria, Plants of, 317.
Amazon, Mr. Spruce's Botanical Mission on the, 84, 139, 239, 270, 335.
American Plants of Lindheimer and Fendler, 190.
Arnebia, new species of, 180.
Arnott, Dr. G. N. W.: Note on *Platynema*, 57.
Association Botanique Française d'Exploration, 93.

Babington, Rev. C.: Lichenes Arctici, 248.
Behr, Dr. H.: Character of the South Australian Flora; translated by R. Kippist, 129.
Bentham, George: *Florula Hongkongensis*, 255, 306, 326.
———— Second Report on Mr. Spruce's Collections of Dried Plants from North Brazil, 111, 161, 191.
Berkeley, Rev. M. J.: Decades of Fungi, 14, 39, 77, 167, 200.
Berkeley, Rev. M. J., and C. E. Broome, Esq.: On some facts tending to show the possibility of the Conversion of Asci into Spores in certain Fungi, 319.
Bicheno, J. E., Esq., Death of, 250.
Bourgeau, M., arrival of, in Paris, 31.
Bromfield, Dr. W. A., Death of, 373.
Brown Scale, or Coccus, of the Coffee-plant, 1.
Bulletin physico-mathématique de l'Académie Impériale de Saint Pétersbourg, 283.

Cedron, 59.
Cereus triangularis, 251.
Clarke, R. O.: Short notice of the African Plant *Diamba*, or Congo Tobacco, 9.
Colenso, Rev. W.: Letter relating to a second species of *Phormium*, 220.
Comptes rendus hebdomadaires des Séances de l'Académie des Sciences de Paris, 286.
Cryptogamic Plants collected by Prof. Jameson, Catalogue of, by William Mitten, 49, 351.

Dalzell, N. A.: Contributions to the Botany of Western India, 33, 89, 120, 134, 178, 206, 225, 279, 343.
De Berg: Physiognomy of Tropical Vegetation, 159.
De Vriese: *Angiopteris longifolia*, Grev. et Hook., and its synonyms, 323.
———— Dutch Botanical Archives, 126.
———— Descriptions et Figures des Plantes nouvelles et rares du Jardin Botanique de Leide, 251.
———— Letter to Robert Brown, Esq., on a new species of *Rafflesia*, 217.
———— Observations on the elevated temperature of the Male Inflorescence of Cycadeous plants, 186.
Diamba, or Congo Tobacco, Notice of, by R. O. Clarke, 9.
Dilpasand, account of, by Dr. J. E. Stocks, 74.

Flora Græca exsiccata, Prospectus of a, 253.
Florula Hongkongensis, by George Bentham, Esq., 255, 306, 326.
Fungi, Decades of, by the Rev. M. J. Berkeley, 14, 39, 77, 167, 200.

Gardner, George: Report on the Brown Scale or Coccus, of the Coffee-plant, 1.
———— Sale of the Herbarium and Books of, 188.
Gray, Asa: Characters of Gnaphalioid *Compositæ* of the division *Angiantheæ*, 97, 147.
———— Characters of a new genus of *Compositæ-Eupatoriaceæ*, 223.

Hooker, Sir W. J.: Catalogue of Mr. Geyer's Collection of Plants gathered in the Upper Missouri, &c., 287.
———— Description of a new species of *Ranunculus*, from the Rocky Mountains, 124.
———— Description of two species of *Bœhmeria*, 312.

INDEX.

Hooker, Dr. J. D., Letter from, on the Physical Geography of Sikkim-Himalaya, addressed to Baron Humboldt, 21.
Humboldt, Baron, extract of a Letter from, addressed to Sir W. J. Hooker, 21.

Icones Plantarum, by Sir W. J. Hooker, 128.
India, Contributions to the Botany of, by N. A. Dalzell, 33, 89, 120, 134, 178, 206, 225, 279, 343.

Junghun, F.: Plantæ Junghunianæ, 127.

Kunze, Professor, Death of, 190.

Ledebour, Professor, Death of, 283.
Lehmann, C.: Novarum et minus cognitarum stirpium Pugillus nonus generis *Potentillarum*, 350.
Lichenes Arctici: collected by Mr. Seemann; by the Rev. C. Babington, 248.
Link, Dr., Death of, 60.
Linnean Society, 63.
Lucæ, Dr., Sale of Herbarium and Drugs of, 153.

Martius, Dr. C. F. P. von: Sketch of the Royal Herbarium at Munich; translated by Dr. Wallich, 65, 102.
Microscopes of Professor Link, 318.
Mitten, William: Catalogue of Cryptogamic Plants collected by Prof. Jameson, 49, 351.
Museum Botanicum Lugduno-Batavum, 96.

Nees von Esenbeck, Dr., 95.
New Zealand, Collection of Plants from, 31.

Orchideous Plants, Mr. Rucker's, 382.

Palmyra Palm of Ceylon, by Wm. Ferguson, 63.
Papyrus of Sicily, 189.

Plant, N.: Natural History Journey in America, 125, 283.
Platynema, Note on, by G. N. Walker LL.D., 57.
Portugal, Welwitzsch's Plants of, 190.

Reichenbach, Professor, 91.
Requien, M., Death of, 250.
Royal Herbarium at Munich, Sketch of Dr. C. F. P. von Martius; translated Dr. Wallich, 65, 102.
Russia, Germany, &c., Plants of, 82.

Salicornia, 96.
Sanders, John: A practical treatise on Culture of the Vine, 252.
Schouw, Prof. J. F.: Origin of the Vegetable Creation, 11.
Seemann, Berthold: Sketch of the Vege of the Isthmus of Panamà, 283, 264, 8
Sikkim-Himalaya, Rhododendrons of, 154.
———— Physical Geography Letter from Dr. J. D. Hooker to Humboldt, 21.
Sinclair, Dr. Andrew: Vegetation, & Auckland, New Zealand, 212.
Spruce, R., Letter from, addressed to Bentham, Esq., 289.

Van Diemen's Land, Papers and of the Royal Society of, 348.
Vegetable Creation, Origin of the by Prof. Schouw, 11.
Victoria regia, by Sir W. J. Hooker, 12?
———— Letter on the successful vation of, in Philadelphia, 346.

Wight, Dr.: *Orchideæ* of the Neilgh 252.
Willkomm, Dr. Moritz: Botanic Gardens Madrid and Valencia, 181.

Lightning Source UK Ltd.
Milton Keynes UK
UKHW020135250119
336151UK00010B/594/P